Publication Bias in Meta-Analysis

Publication Bias in Meta-Analysis

Prevention, Assessment and Adjustments

Edited by

HANNAH R. ROTHSTEIN
Zicklin School of Business, Baruch College, USA

ALEXANDER J. SUTTON
University of Leicester, UK

MICHAEL BORENSTEIN
Biostat Inc, USA

John Wiley & Sons, Ltd

Other Wiley Editorial Offices

John Wiley & Sons Inc., 111 River Street, Hoboken, NJ 07030, USA

Jossey-Bass, 989 Market Street, San Francisco, CA 94103-1741, USA

Wiley-VCH Verlag GmbH, Boschstr. 12, D-69469 Weinheim, Germany

John Wiley & Sons Australia Ltd, 42 McDougall Street, Milton, Queensland 4064, Australia

John Wiley & Sons (Asia) Pte Ltd, 2 Clementi Loop #02-01, Jin Xing Distripark, Singapore 129809

John Wiley & Sons Canada Ltd, 22 Worcester Road, Etobicoke, Ontario, Canada M9W 1L1

Wiley also publishes its books in a variety of electronic formats. Some content that appears in print
may not be available in electronic books.

Library of Congress Cataloging in Publication Data

Rothstein, Hannah.
   Publication bias in meta-analysis : prevention, assessment and adjustments / Hannah Rothstein,
   Alex Sutton, Michael Borenstein.
      p. cm.
   Includes bibliographical references and index.
   ISBN 0-470-87014-1 (cloth)
   1. Meta-analysis.   2. Meta-analysis—Miscellanea.   3. Medicine—Research—Evaluation.
   I. Sutton, A.J. II. Borenstein, Michael. III. Title.
   R853.M48R68 2005
   610′.72—dc22

                                              2005021394

British Library Cataloguing in Publication Data

A catalogue record for this book is available from the British Library

ISBN-13 978-0-470-87014-3 (HB)
ISBN-10 0-470-87014-1 (HB)

Typeset in 10/12pt Times by Integra Software Services Pvt. Ltd, Pondicherry, India
Printed and bound in Great Britain by TJ International, Padstow, Cornwall
This book is printed on acid-free paper responsibly manufactured from sustainable forestry
in which at least two trees are planted for each one used for paper production.

This book is dedicated to the memory of Richard Tweedie, who played a key role in planning this volume but, sadly, did not live to see it completed. This book is part of his legacy.

Contents

Preface

Our goal in compiling this volume is to present a concise, yet detailed and readable account of publication bias – its origins, the problems it creates and the methods developed to address it. We believe that this book is the first to be wholly devoted to the topic, and hope that its appearance will increase awareness of the problem among researchers conducting primary studies and among editors as well as among systematic reviewers and meta-analysts. With luck, this will eventually lead to a reduction in the prevalence of publication bias through activities such as improved reporting of primary studies, widespread implementation of registries for trials and studies at their inception, changes in editorial practices, and the establishment of repositories for raw data. Additionally, we hope that it will encourage those carrying out systematic reviews and meta-analyses to conduct more thorough literature searches, and to routinely undertake assessments of publication bias as part of their data analyses. Hence, although this book is primarily written for those involved with evidence synthesis and decision-making, we believe that portions of it will, and should be, of interest to anyone involved with scientific research.

The authors who have contributed chapters to this book, as well as the reviewers of each of the chapters, are some of the world's leading authorities on publication bias issues. In addition, they (and we, the editors) come from a variety of scientific disciplines including medicine, public health, education, psychology and statistics. We feel this is a great strength of the book, because different approaches to the prevention and assessment of, and potential adjustments for, publication biases have developed in different research areas, and researchers do not generally communicate across disciplines. Although specific chapters often, naturally, have an emphasis on the authors' own discipline(s), we have worked hard to ensure that the book is as generic and relevant across disciplines as possible, in terms of both the issues it deals with and the language used to describe and discuss these issues. We are hopeful that the book will also find an interested audience in disciplines beyond those of the contributing authors (for example, we know that publication bias is starting to be addressed in areas of 'pure science' such as biology and in other social sciences such as economics). Indeed, we take great pride that the creation of the book has enabled us to tie together a literature which straddles numerous disciplines. The meetings which took place during the book's development, as well as the writing of the chapters, initiated a process of cross-pollination of ideas between leading thinkers in the several fields represented here. This appears to be continuing, and we hope it will bear fruit in the future development of methods and applications presented here. We are deeply indebted to all of the authors for their enthusiasm, creativity and diligence in producing chapters that push forward the boundaries of what is known, and what can be done about, publication bias in its various forms. We are also immensely grateful to the reviewers of each chapter, who (under tight time constraints) provided the authors and us with the thoughtful, detailed and constructive comments on early drafts that facilitated the production of the final versions contained here. The book simply would not have been as good without them. We apologize for stripping away some of their anonymity, but feel obligated to thank them by name. The chapter reviewers were: Keith Abrams, Doug Altman, Iain Chalmers, Harris Cooper, Julian Higgins, Jerard Kehoe, Satish Iyengar, David Jones, Trevor Sheldon, Anne Whitehead and David Wilson.

We have included several features designed to assist a cross-disciplinary, international audience in making its way through a cross-disciplinary, internationally authored book. First, we took note of the fact that authors sometimes used different terms for the same concept, and the same term for different concepts. (For example, where meta-analysts from the United States talk about coding of studies, those from the United Kingdom discuss data extraction. While those in health care reserve the term 'effect size' for standardized mean differences, social scientists use this term to refer to many indices of the size of a relationship.) Additionally, we recognized that some of the concepts and terms introduced by authors from a particular discipline would be unfamiliar to readers from other disciplines. We have therefore provided a glossary, which defines the key terms in each chapter. Second, although all authors clearly define all algebraic notation used, we have standardized the notation across chapters in order to facilitate understanding. A third feature which is intended to facilitate the book's coherence, as well as its applicability to a wide audience, is the use of three common data sets from the disciplines of medicine, education and psychology, particularly (but not exclusively) in Part B of the book. A description of the data sets can be found in Appendix A. Finally, a website has been developed as a companion for this book (www.Meta-Analysis.com/publication-bias) where readers can access the example data sets electronically, and further relevant information will be made available. The editors plan to update the website with additional examples and with other relevant material over the coming months and years.

We appreciate the support and enthusiasm shown by the editorial staff at John Wiley & Sons who managed this project from its inception to its publication. In particular, we would like to thank Rob Calver, Kathryn Sharples, Wendy Hunter and Jane Shepherd and Simon Lightfoot. Thanks also go to Reena Godi at Integra, and to Richard Leigh for his expert and fastidious copyediting.

We sincerely hope you find this volume both interesting and useful.

Hannah Rothstein
Alex Sutton
Michael Borenstein

Acknowledgements

The creation of this book was supported in large part by the generous funding provided by the National Institute on Aging (NIA), one of the National Institutes for Health (NIH) in the United States.

Funding was provided under the following Small Business Innovation Research (SBIR) grants:

Publication bias in meta analysis (AG20052)
Combining data types in meta analysis (AG021360)

We would especially like to take this opportunity to thank Dr Sidney M. Stahl, programme official at NIA, for arranging the funding, as well as for his commitment to and support for this work.

We would also like to thank Steven Tarlow, Shirley Rudolph and Vivian Vargas at Biostat for their valuable help in completing the manuscript.

Notes on Contributors

Betsy Jane Becker has just joined the faculty of the College of Education at Florida State University, where she is a professor in the program in measurement and statistics. For the previous 21 years she was in the measurement and quantitative methods program at Michigan State University. Dr. Becker has published widely in the area of meta-analysis. She serves as co-convener of the Methods Training Group for the Campbell Collaboration and also is a member of the Technical Advisory Group for the What Works Clearinghouse, an effort to produce research syntheses of studies of educational interventions, supported by a contract from the US Department of Education. Dr. Becker is a member of the National Assessment of Educational Progress (NAEP) Design and Analysis Committee, and is associate editor of the journal *Psychological Methods*. In the past she has also served on the editorial board of the *Journal of Educational and Behavioral Statistics* and the Applications and Cases Section of the *Journal of the American Statistical Association*.

Jesse Berlin has been extensively involved in methodological work related to meta-analysis and publication bias since the late 1980s, and has written some of the pioneering articles in this field. Dr. Berlin is a co-convener of the Cochrane Collaboration's Prospective Meta-analysis Methods Group. Dr. Berlin received the Thomas C. Chalmers, MD award at the 1997 Cochrane Colloquium for the best presentation addressing a methodological issue in research synthesis and meta-analysis. Dr. Berlin currently serves on the editorial board for *Clinical Trials*, the journal of the Society for Clinical Trials.

Michael Borenstein served as Director of Biostatistics at Hillside Hospital, Long Island Jewish Medical Center from 1980–2002, and as Associate Professor at Albert Einstein College of Medicine from 1992–2002. He has served on various review groups and advisory panels for the National Institutes of Health and as a member of the NIMH Data Safety Monitoring Board, and is an active member of the statistical advisory groups of the Cochrane and Campbell Collaborations. Since the mid-1990s, Dr Borenstein has lectured widely on meta-analysis. He is the PI on several NIH grants to develop software for meta-analysis and is the developer, with Larry Hedges, Julian Higgins, Hannah Rothstein and others, of *Comprehensive Meta Analysis*, a best-selling computer program for meta-analysis.

Sarah Burdett is a researcher in the Meta-analysis Group at the British MRC Clinical Trials Unit and has almost 10 years' experience working on international collaborative projects. The group has published major IPD meta-analyses on ovarian, lung, bladder, cervical, and oesophageal cancer and in glioma and soft tissue sarcoma. She is involved in the Cochrane Collaboration, and is a member of several Collaborative Review Groups including the IPD Meta-analysis Methods Group.

Ashley Busing is a doctoral candidate in Industrial and Organizational Psychology at the Department of Psychology, Weissman School of Arts and Sciences, Baruch College, City University of New York. Her research interests are in the areas of personality and goal-setting.

Mike Clarke has worked at the Clinical Trial Service Unit (CTSU) in Oxford since 1989 and is responsible for the collaborative overviews of cancer treatments coordinated by the CTSU. The largest of these investigates the treatment of operable breast cancer and has provided reliable estimates of the effects of treatments such as tamoxifen and chemotherapy. In April 1999 Professor Clarke became associate director (research) at the UK Cochrane Centre, with a special responsibility for the Centre's work in identifying reports of randomized trials and conducting research into the control of bias in research synthesis. Since October 2002 he has been director of the Centre, which is responsible for providing training and support to Cochrane entities and members in the UK, Ireland and several other countries. He is involved in many other aspects of the work of the Cochrane Collaboration and was co-chair of the international Steering Group until October 2004. In addition to his experience in systematic reviews, Professor Clarke has considerable experience in the conduct of large-scale, multicentre randomized trials and is actively involved with trials in pre-eclampsia, subarachnoid haemorrhage, breast cancer and poisoning – each of which is the largest ever randomized trial in these conditions. He became Professor of Clinical Epidemiology at the University of Oxford in October 2004.

Kay Dickersin is the director of the United States Cochrane Center. Dr. Dickersin's major research interests are related to randomized clinical trials, trials registers, meta-analysis, publication bias, women's health and the development and utilization of methods for the evaluation of medical care and its effectiveness. She is co-author of 'Publication bias in editorial decision making' which was published in the *Journal of the American Medical Association* in 2002, as well as numerous other articles on publication bias.

Sue Duval earned her doctorate from the University of Colorado Health Sciences Center on the topics of meta-analysis and publication bias. Along with her co-author, Richard Tweedie (deceased), she developed the trim and fill method for publication bias. This work was published in *Biometrics* and the *Journal of the American Statistical Association* in 2000 and continues to receive much attention both from researchers and applied meta-analysts.

Matthias Egger recently co-edited the second edition of *Systematic Reviews in Health Care: Meta-analysis in Context*, which was published by BMJ Books in 2001, and has co-authored an influential series on meta-analysis in the *British Medical Journal*. Dr. Egger has been working on methodological aspects of systematic reviews and meta-analysis for several years. His current research interests include publication bias, language bias and related reporting biases, the impact of the methodological quality of individual trials, and meta-analysis of observational epidemiological studies. His group developed a regression approach to quantifying and statistically testing for asymmetry in funnel plots. He is a convener of the Reporting Bias Methods Group of the Cochrane Collaboration.

Davina Ghersi is co-convener of the Prospective Meta-analysis Methods Group of the international Cochrane Collaboration. The role of the group is to provide guidance to members of the Collaboration on methodological issues related to the conduct of prospective meta-analyses. Dr. Ghersi is also director of the Systematic Review and Health Care Assessment Team, and research fellow at the NHMRC Clinical Trials Centre (CTC), University of Sydney.

Scott Halpern received his MD and PhD in epidemiology, and his Master of Bioethics from the University of Pennsylvania. His research focuses on several methodological and ethical issues regarding the design and conduct of randomized clinical trials. He is currently investigating whether the sponsorship of epidemiological studies of drug safety is associated with their statistical power to detect adverse events.

Larry Hedges is the Stella M. Rowley Distinguished Service Professor of Sociology, Psychology, and in the Harris Graduate School of Public Policy Studies at the University of Chicago, USA. He is a member of the National Academy of Education, a Fellow of the American Statistical Association and of the American Psychological Association. Dr. Hedges has been a leader in the development and application of non-parametric weight function models for publication bias. After the basic method was explicated in a 1992 paper by Hedges, he and Jack Vevea applied these methods to substantial data sets in papers from 1993 to 2000. They extended the methods to models including covariates and evaluated their small-sample properties in substantial simulation studies. In addition, Dr. Hedges, together with Ingram Olkin, co-authored one of the first textbooks on meta-analysis, *Statistical Methods for Meta-analysis*, and edited *The Handbook of Research Synthesis* together with Harris Cooper.

Sally Hopewell is a research scientist at the UK Cochrane Centre and is an active member of the Cochrane Collaboration. She has been at the Centre since 1999 and has recently been awarded a DPhil at the University of Oxford, and has been carrying out research into the effects of including grey literature and other unpublished literature in systematic reviews. Previously she worked as a paediatric nurse before undertaking an MSc in health care studies in 1997. Dr Hopewell is part of the methodology research team, at the UK Cochrane Centre, conducting empirical and descriptive research to improve the quality of Cochrane reviews and other evaluations of health care. Her research interests include the control of bias susceptibility in systematic reviews and she has published several methodology reviews on this subject. She is also a member of the Cochrane Methodology Review Group and is co-editor of the annual Cochrane Methods Groups Newsletter.

John Ioannidis is chair of the Department of Hygiene and Epidemiology at the University of Ioannina School of Medicine in Greece and is adjunct professor of Medicine at Tufts-New England Medical Center in Boston. His research interests include evidence-based medicine, clinical and molecular epidemiology and methodological issues in medical research.

Sue Mallett is a medical statistician working at the Centre of Statistics in Medicine in Oxford. She has studied biochemistry and immunology at Oxford University, followed by a Diploma in Statistics from the Open University. Sue has worked in a variety of areas including drug resistance in HIV and more recently a research project on the characteristics of a typical Cochrane review and the inclusion of grey and unpublished literature at the UK Cochrane Centre. She is currently conducting methodology research on systematic reviews of the accuracy of diagnostic tests in cancer, with Doug Altman and Jon Deeks. Her current projects also include working on a systematic review of virtual colonoscopy and analysis of clinical data on the accuracy of diagnostic tests in cervical cancer. Sue is a contributing author to the Cochrane Diagnostic Reviewers' Handbook.

Therese (Terri) Pigott's research interests center on statistical methods for meta-analysis and statistical analyses with missing data. With Larry Hedges, she has recently published an article on power analysis in meta-analysis in *Psychological Methods*, and was the author of the chapter on missing data in the *Handbook of Research Synthesis*. She is currently on the editorial board of the *Journal of Educational Psychology, Psychological Methods*, and *Psychological Bulletin*.

Hannah Rothstein is co-chair of the Methods Group of the Campbell Collaboration, and a member of the Collaboration's Steering Group. She is also a member of the Cochrane Collaboration's reporting bias methods group. Dr. Rothstein has been first author of four published meta-analyses of employment selection methods and has written many articles on methodological issues in meta-analysis. She has authored a chapter on meta-analysis that appeared in *Measuring and Analyzing Behavior in Organizations*, and has completed a 25-year retrospective on the contributions of meta-analysis to the field of industrial and organizational psychology that appeared in *Validity Generalization: A Critical Review*. With Michael Borenstein, and others, she is the author of computer software for meta-analysis and power analysis.

Jonathan Sterne's first degree was in Mathematics from the University of Oxford, and he obtained a MSc and PhD in Statistics from University College London. He has worked in the Department of Social Medicine at the University of Bristol, UK, since 1999. His research interests include meta-analysis and systematic reviews, statistical methods for epidemiology and health services research, and causal models. Particular current research interests include the epidemiology of bias in systematic reviews in medicine, systematic reviews of studies of diagnostic test accuracy, and systematic reviews of observational studies with results expressed as estimates of dose-response. He has written a number of meta-analysis software routines. With Betty Kirkwood, he is co-author of 'Essential Medical Statistics'. He is a convener of the Reporting Bias Methods Group of the Cochrane Collaboration.

Lesley Stewart is head of the Meta-analysis Group at the British MRC Clinical Trials Unit in London. She has been responsible for designing and running individual patient data (IPD) meta-analyses and associated methodological research for more than 15 years. The group's research portfolio is mostly in the cancer field, having published major IPD meta-analyses in ovarian, lung, bladder, cervical, and oesophageal cancer and in glioma and soft tissue sarcoma, but recently has expanded into other healthcare areas including pre-eclampsia and prion disease. The methodological research done by the group has included a number of studies of potential bias in systematic reviews exploring how obtaining IPD can circumvent biases related to relying on information as presented in publications. Dr. Stewart, is a member of several Cochrane Collaborative Review Groups and is co-convener of the Cochrane IPD Meta-analysis Methods Group.

Alex Sutton has published extensively on meta-analysis methodology generally, and on publication bias specifically in recent years, including a major systematic review on the topic of the methodology that has been developed for meta-analysis. He currently has an active interest in the area of partially reported study information, which is currently under-researched. Dr. Sutton is co-author of a textbook on meta-analysis (*Methods for Meta Analysis in Medical Research*), which was published by Wiley in 2000.

Jayne Tierney is a researcher in the Meta-analysis Group at the British MRC Clinical Trials Unit and has more that 10 years' experience working on international collaborative IPD meta-analyses and associated methodological projects. The group has published major IPD meta-analyses on ovarian, lung, bladder, cervical and oesophageal cancer and in glioma and soft tissue sarcoma. Dr. Tierney is involved in the Cochrane Collaboration, and is a member of several Collaborative Review Groups and is co-convener of the IPD Meta-analysis Methods group.

Thomas Trikalinos is assistant professor of Medicine at Tufts-New England Medical Center in Boston and a research associate at the Department of Hygiene and Epidemiology at the University of Ioannina School of Medicine in Greece. His research interests include empirical methodological evaluations and clinical and molecular epidemiology.

Jack Vevea received his PhD in measurement, evaluation, and statistical analysis from the University of Chicago in 1996. He is currently on the psychology faculty at the University of California at Santa Cruz. Dr. Vevea specializes in cognitive and quantitative psychology. His primary research interest is the development of new statistical methods in meta-analysis, with special emphasis on mixed-effects models and publication bias. He is also active in applied meta-analytic research. In the domain of cognitive psychology, he studies biases in memory for sensory stimuli.

CHAPTER 1

Publication Bias in Meta-Analysis

Hannah R. Rothstein
Department of Management, Zicklin School of Business, Baruch College, New York, USA

Alexander J. Sutton
Department of Health Sciences, University of Leicester, UK

Michael Borenstein
Biostat, Inc., USA

PUBLICATION BIAS AS A THREAT TO VALIDITY

Publication bias is the term for what occurs whenever the research that appears in the published literature is systematically unrepresentative of the population of completed studies. Simply put, when the research that is readily available differs in its results from the results of *all* the research that has been done in an area, readers and reviewers of that research are in danger of drawing the wrong conclusion about what that body of research shows. In some cases this can have dramatic consequences, as when an ineffective or dangerous treatment is falsely viewed as safe and effective. This can be illustrated by two events that received much media attention as this book was going to press in late 2004. These are the debate surrounding Merck's recall of Vioxx, a popular arthritis drug (Merck maintained that it recalled Vioxx as soon as the data indicated the high prevalence of cardiovascular events among those who took Vioxx for more than 18 months, while media reports said that Merck hid adverse event evidence for years), and the use of selective serotonin reuptake inhibitor (SSRI) anti-depressants among adolescents (Elliott Spitzer, attorney general of New York State, filed a 2004 lawsuit against GlaxoSmithKline,

Publication Bias in Meta-Analysis – Prevention, Assessment and Adjustments Edited by H.R. Rothstein, A.J. Sutton
and M. Borenstein   © 2005 John Wiley & Sons, Ltd

charging that they had concealed data about the lack of efficacy and about the increased likelihood of suicide associated with the use of Paxil for childhood and adolescent suicide). In most cases, the causes of publication bias will not be as clear, nor the consequences as serious as in these examples. Nevertheless these examples highlight why the topic is critically important.

Publication bias is a potential threat in all areas of research, including qualitative research, primary quantitative studies, narrative reviews, and quantitative reviews, that is, meta-analysis. Although publication bias has likely been around for as long as research has been conducted and reported, it has come to prominence in recent years largely with the introduction and widespread adoption of the use of systematic review and meta-analytic methods to summarize research. In part, this is because, as methods of reviewing have become more scientific and quantitative, the process of reviewing (and synthesizing) research has been increasingly seen as paralleling the process of primary research. Parallels to the threats to the validity of primary research have been uncovered at every step of the systematic review process (Cooper, 1998; Shadish et al., 2002). Furthermore, as methods of reviewing have become more systematic and quantitative, it has been possible to empirically demonstrate the existence of publication bias and to quantify its impact. Thus, a problem that was viewed hazily through the looking glass of traditional reviews came into sharp focus under the lens of meta-analysis.

In meta-analysis, publication bias is a particularly thorny issue because meta-analysis has been put forward as providing a more accurate appraisal of a research literature than is provided by traditional narrative reviews (Egger et al., 2000), but if the sample of studies retrieved for review is biased, then the validity of the results of a meta-analytic review, no matter how systematic and thorough in other respects, is threatened. This is not a hypothetical issue: evidence that publication bias has had an impact on meta-analyses has been firmly established by several lines of research (see Chapters 2 and 10 of this volume).

Since systematic reviews are promoted as providing a more objective appraisal of the evidence than traditional narrative reviews, and since systematic review and meta-analysis are now generally accepted in many disciplines as the preferred methodology for summarizing a literature, threats to their validity must be taken very seriously. Publication bias must be taken especially seriously, as it presents perhaps the greatest threat to the validity of this method. On the other hand, the vulnerability of systematic review and meta-analysis to publication bias is not an argument against their use, because such biases exist in the literature irrespective of whether systematic review or other methodology is used to summarize research findings. In fact, we suggest that the attention given to objectivity, transparency and reproducibility of findings in systematic reviews and meta-analyses has led to the first serious attempt to confront the problems that have always existed because of publication biases, and to ameliorate them. As demonstrated by this volume, there are now several tools available with which meta-analysts can assess the potential magnitude of bias caused by selective publication. When the potential for severe bias exists in a given analysis, this can now be identified, and appropriate cautionary statements about the meta-analytic results can be made. When potential bias can effectively be ruled out, or shown not to threaten the results and conclusions of a meta-analysis, the validity and robustness of these results and conclusions are strengthened.

Publication bias was originally defined as the publication or non-publication of studies depending on the direction and statistical significance of the results, and the first systematic investigations of publication bias focused on this aspect of the problem. However, as readers will appreciate as they work through the book, there are numerous potential information suppression mechanisms that go well beyond the simple definition given above, including: language bias (selective inclusion of studies published in English); availability bias (selective inclusion of studies that are easily accessible to the researcher); cost bias (selective inclusion of studies that are available free or at low cost); familiarity bias (selective inclusion of studies only from one's own discipline, and outcome bias (selective reporting by the author of a primary study of some outcomes but not others, depending on the direction and statistical significance of the results). All of these biases lead to the same consequence, namely that the literature located by a systematic reviewer will be unrepresentative of the population of completed studies; hence all present the same threat to a review's validity. For this reason, it has been suggested that a single, broadly encompassing term, dissemination bias (Song *et al.*, 2000), be used to refer to the problem. We agree with this sentiment, but the widespread and established use of the term publication bias has made us hesitant to tamper with, and potentially confuse, the current terminology. Readers should bear in mind that when they read 'publication bias' the broader but more cumbersome 'publication bias and associated dissemination biases' is implied.

ORGANIZATION OF THE BOOK

The book is split into three parts, and there are three appendices. Part A contains a set of chapters which together provide a non-technical introduction to publication bias and describe how it can be minimized in future research. Part B presents each of the currently available methods for assessing or adjusting for publication bias in a meta-analytic context; these chapters also illustrate each method using the data sets described in Appendix A. The chapters in Part C discuss several advanced and emerging issues that have not yet received much attention elsewhere in the literature. Finally, Appendix B is an annotated bibliography that provides illuminating further reading on publication bias; it is presented in chronological order to allow the reader to see how the field has developed over time. While we have attempted to compile and edit the book so that the chapters are integrated (with as much detailed cross-referencing of chapters as possible), we believe that each chapter can stand on its own. A more detailed outline of the contents of each section of the book is given below.

Outline of Part A

In Chapter 2, Kay Dickersin begins with a thoughtful review of the causes and origins of publication bias, after which she presents a comprehensive overview of the history and findings of publication bias research. Her chapter concludes with some suggestions for minimizing publication bias in the future, which are elaborated upon in Chapter 3 by Jesse Berlin and Davina Ghersi. These authors propose two strategies which, if widely adopted, would go a long way towards alleviating publication bias in trials-based research areas. The first is prospective registration of clinical trials, which would create an 'unbiased sampling frame for subsequent meta-analyses'. As Berlin and Ghersi point out, however, to avoid publication bias, this would need to be coupled with open access to the results of these trials. The second is prospective meta-analysis, whereby multiple groups of investigators conducting ongoing trials agree, prior to knowing the results of their studies, to combine their findings when the trials are complete. In a variant of this strategy, the meta-analysis is *designed* prospectively to standardize the instruments used to measure specific outcomes of interest across studies. In Chapter 4, Sally Hopewell, Mike Clarke and Sue Mallett describe how to minimize publication bias by attempting to locate and retrieve grey and unpublished literature. They also point out the problems associated with the retrieval and inclusion of this literature, namely that it is time-consuming and difficult, and that its methodological quality can be hard to assess. They conclude by suggesting criteria for weighing the potential benefits and costs of grey literature searches.

Outline of Part B

Despite the excellent suggestions made in Chapters 3 and 4, it is safe to say that publication bias will remain a problem in many disciplines for the foreseeable future. The second part of this book presents several statistical methods that have been developed to identify, quantify and assess the impact of publication bias on meta-analyses. Essentially three kinds of techniques have been developed to help analysts deal with publication bias in meta-analysis. One set of techniques is designed to detect publication bias. This set of techniques includes graphical diagnostics such as the funnel plot and explicit statistical tests for the statistical significance of publication bias. In Chapter 5, Jonathan Sterne, Betsy J. Becker and Matthias Egger define funnel plots and demonstrate how they can be used as a graphical tool to for the investigation of publication bias. Specific issues discussed in this chapter are the effects of choice of axes for these plots, and the need to consider explanations in addition to publication bias for funnel plot asymmetry. In Chapter 6, Jonathan Sterne and Matthias Egger describe and illustrate two statistical tests for funnel plot asymmetry, the Begg and Mazumdar (1994) non-parametric test based on the rank correlation between intervention effect estimates and their sampling variances, and the Egger *et al.* (1997) regression method, which tests for a linear association between the intervention effect and its standard error. Sterne and Egger also provide information about the power of these tests, and caution against their use when a meta-analysis contains only a small number of studies.

The second set of techniques is designed to assess the sensitivity of conclusions of an analysis to the possible presence of publication bias. Chapter 7, by Betsy J. Becker, describes and illustrates the first statistical method developed for the assessment of publication bias, file-drawer analysis, originally described by Robert

Rosenthal (1979). This method has been commonly referred to as the failsafe N, a term coined by Harris Cooper (1979). Becker also reviews other approaches to file-drawer analysis, including one that was intended to overcome some of the limitations of the original. Her conclusion is that all of these methods should be abandoned in favour of the more sophisticated methods described in the other chapters in this part of the book.

The third set of techniques is designed to adjust estimates for the possible effects of publication bias under some explicit model of publication selection. In Chapter 8, Sue Duval describes trim and fill, a method that she and the late Richard Tweedie developed to estimate and adjust for the number of missing studies (due to publication bias) in a meta-analysis. The trim and fill method is illustrated using a detailed worked example, in addition to its application to the three common data sets used throughout Part B. Chapter 9, by Larry Hedges and Jack Vevea, explicates the general selection model approach to the assessment of publication bias, and demonstrates how it is used to correct for bias. The authors show how their method, as well as that of John Copas, can be used to detect and correct for bias in the three common data sets used throughout the book. As the reader will see, trim and fill is relatively simple to implement and involves relatively little computation, while the Hedges–Vevea and Copas methods involve considerable computation.

In Chapter 10, Alex Sutton summarizes the results of empirical investigations that have been carried out to assess the extent of publication bias present in various scientific literatures, using the methods described earlier in Part B. He concludes on the basis of these results that publication bias assessment should become a routine part of every meta-analysis. In Chapter 11, Michael Borenstein discusses a number of computer programs that can be used to address publication bias, and shows how these would be used to apply the statistical procedures discussed throughout this volume. This chapter should be of great practical value for any researcher who wishes to investigate whether publication bias is likely to be a problem in his or her meta-analysis.

While Part B of the book is necessarily more technical then the other parts, we believe that it is generally accessible to the non-statistically minded reader. Particularly technical sections of these chapters, which can be skipped by those who are not interested in statistical fine points without loss of continuity, are identified with an asterisk.

Outline of Part C

The final part of the book describes advanced and emerging issues related to publication bias. Chapter 12, by Alex Sutton and Terri Pigott, provides a taxonomy of types of missing data. Sutton and Pigott describe and evaluate the application of standard missing-data imputation techniques to meta-analysis, and outline the need for the development of new methods in this area. Chapter 13, by Tom Trikalinos and John Ioannidis, considers how treatment effects in medicine evolve over time and the impact that selective publication may have on such evolution. In Chapter 14, Lesley Stewart, Jayne Tierney and Sarah Burdett discuss the advantages of obtaining individual participant data, rather than relying on published summary results, in combating publication bias. They suggest that high-quality individual participant data meta-analyses may offer a 'gold standard' for research synthesis, at least in the domain of randomized controlled trials. One of the difficulties in trying to diagnose whether publication bias is present in a meta-analysis data set is that the influence of other factors may mimic the appearance of publication bias. In Chapter 15, John Ioannidis considers how to attack the difficult task of disentangling such factors from true publication bias. In Chapter 16, Scott Halpern and Jesse Berlin reflect on data suppression that may occur for other reasons than those traditionally considered to cause publication bias. These factors include the financial, political, ideological and professional competing interests of investigators, research sponsors, journal editors and other parties. Notable events in the pharmaceutical industry, which we mentioned at the beginning of this chapter, and that received much attention as this book was being completed, suggest that these issues merit serious attention from research synthesists.

OUR MODEST PROPOSAL

We hope that reading this book will convince our audience that is imperative for every meta-analysis to include an analysis of publication bias and that this should be reported as a standard part of the results. We suggest that such reports focus on the practical impact of publication bias. To discuss the practical impact of publication bias it helps to consider three levels of impact, based on the concordance between (1) the results that *are* reported and (2) our best guess (informed by the results of our publication bias analyses) of what the results might look like if all relevant studies had been included. The impact of bias could be called 'minimal' when the two versions of the analysis would yield essentially similar estimates of the effect size. The impact could be considered 'modest' when the effect size would change substantially but the key finding (that the treatment is or is not effective) would remain in force. The impact could be labelled 'severe' when the basic conclusion of the analysis (for example, that the treatment is clinically useful, or that it is not) is called into question. The surveys on this topic, as reviewed by Sutton, in Chapter 10, suggest that publication bias exists in most published meta-analyses, but that the conclusions are nevertheless valid in most cases. In the meta-analyses surveyed so far, the impact of bias is minimal in approximately 50 %, modest in about 45 %, and severe in only 5 % of the analyses surveyed. It also appears to be the case that the amount of bias varies substantially between fields of research. For example, we suspect that publication bias may be more likely in the social sciences, which are characterized by many small and isolated studies, than in medicine, where studies are more likely to be larger, better funded, and better publicized. The prevalence of bias will also likely vary with the experience and resources of the researchers conducting the meta-analysis. The bias cited in the surveys reviewed by Sutton is based primarily on meta-analyses from the Cochrane database, whose researchers are trained to do extensive searches, and which typically include some 30 % more studies than meta-analyses on the same topic that appear in journals. Therefore, the bias cited in these surveys is probably less than the bias one would expect in other fields.

In cases where publication bias analyses suggest that severe bias may exist, this can serve as a warning to researchers and practitioners to regard the initial results cautiously, and to avoid potentially serious mistakes such as recommending an intervention or policy that could be useless or even harmful. Based on the existing state of knowledge in the field, we are hopeful that, in the majority of cases, the publication bias analysis will show that bias probably had little impact. This is also critically important information, as it allows us to have confidence that the meta-analysis is valid.

Finally, we note that it is important to address bias not only to ensure the integrity of the individual meta-analysis, but also to ensure the integrity of the field. When a meta-analysis is published that ignores the potential for bias and is later found to be incorrect, the perception is fostered among editors and researchers that meta-analyses cannot be trusted. By encouraging the prevention and assessment of, and adjustments for, publication bias, we hope to further the use and usefulness of meta-analysis.

REFERENCES

Begg, C.B. and Mazumdar, M. (1994). Operating characteristics of a rank correlation test for publication bias. *Biometrics*, **50**, 1088–1101.

Cooper, H.M. (1979). Statistically combining independent studies: A meta-analysis of sex differences in conformity research. *Journal of Personality and Social Psychology*, **37**, 131–146.

Cooper, H.M. (1998). *Synthesizing Research: A Guide for Literature Reviews*, 3rd edition. Thousand Oaks, CA: Sage.

Egger, M., Davey Smith, G. and Altman, D. G. (2000). *Systematic Reviews in Health Care: Meta-analysis in Context*. London: BMJ Books.

Egger, M., Davey Smith, G., Schneider, M. and Minder, C. (1997). Bias in meta-analysis detected by a simple, graphical test. *British Medical Journal*, **315**, 629–634.

Rosenthal, R. (1979). The 'file drawer problem' and tolerance for null results. *Psychological Bulletin*, **86**, 638–641.

Shadish, W.R., Cook, T.D. & Campbell, D.T. (2002). *Experimental and Quasi-experimental Designs for Generalized Causal Inference*. Boston: Houghton-Mifflin.

Song, F., Easterwood, A., Gilbody, S., Duley, L. and Sutton, A.J. (2000) Publication and other selection biases in systematic reviews. *Health Technology Assessment*, **4**(10).

PART A

Publication Bias in Context

CHAPTER 2

Publication Bias: Recognizing the Problem, Understanding Its Origins and Scope, and Preventing Harm

Kay Dickersin

Department of Epidemiology, Bloomberg School of Public Health, Johns Hopkins University, USA

A negative result may be dull but often it is no less important than the positive; and in view of that importance it must, surely, be established by adequate publication of the evidence.

(Bradford Hill, 1959)

THE BOSTON
Medical and Surgical Journal.

———◆———

THURSDAY, AUGUST 19, 1909

D. C. HEATH & COMPANY,
120 BOYLSTON STREET, BOSTON, MASS.

THE REPORTING OF UNSUCCESSFUL CASES.

IT is somewhat proverbial that medical literature abounds in publications of successful results, particularly from the therapeutic standpoint. It is natural that men should be eager to present to the world their successes rather than

Publication Bias in Meta-Analysis – Prevention, Assessment and Adjustments Edited by H.R. Rothstein, A.J. Sutton and M. Borenstein © 2005 John Wiley & Sons, Ltd

their failures, and furthermore, that a certain over-enthusiasm should see results where none actually exist. It is not to be questioned, therefore, that many papers are continually appearing written with an entirely conscientious motive, but which present an exaggerated or wholly distorted statement of the actual facts. This is notably true in the exploitation of new remedies and of new forms of treatment of recognized diseases. In the hands of one man certain drugs or combinations of drugs or other therapeutic measures apparently attain results which are absolutely incapable of verification at the hands of other equally conscientious investigators. Doctors, unfortunately, are no more completely endowed with the judicial faculty than persons in other walks of life, and it is certainly not strange that distortions of facts, put out with apparent scientific accuracy, are everywhere in evidence. The situation would be the more unfortunate were it not for the fact that a certain healthy skepticism serves continually as an offset to over-enthusiasm. The result is that many discussions occur and controversies are raised which have small permanent significance and which yield nothing of value in the search for ascertained truth.

In view of this very general and, in certain respects it must be admitted, praiseworthy enthusiasm to ventilate new theories and exploit new methods, it would be well if leaders in the profession gave more attention to the publication of their mistakes and failures. Certainly there is no more valuable lesson to the practitioner than a complete realization of his errors. In fact, by this means alone is further knowledge possible. In spite, however, of the manifest truth of this statement, it is relatively rare that surgeons or physicians publish detailed reports of their mistakes and of the lessons which these mistakes inculcate. We too commonly see reports of "so many successful cases," with a certain inevitable emphasis on the word "successful." Such papers naturally have their value and also their manifest dangers. The case of certain surgical procedures is, for example, greatly exaggerated in the minds of less trained practitioners through the report of a large series of successful operations. A certain false security is thereby encouraged, to the end that operations of significance are no doubt performed by really incompetent persons. There is unquestionably a false emphasis in all such publications, tending to increase the reputation of the writer, but not to render the public more secure. We have no proper balance to this very natural tendency to publish our successes except through the more frequent publication of the errors and failures which likewise mark the path of every successful practitioner. Such papers, written by men of experience and standing, would do much toward overcoming the tendency to over-security and would certainly serve an educational purpose which the ordinary publication so often fails to attain. No greater benefit could be secured for the medical public than a series of papers illustrating the dangers and pitfalls of medical practice, written by men to whom the actual experience has come.

———◆———

(Editors, 1909)

KEY POINTS

- There is now strong evidence that publication bias exists in the social and biomedical sciences, for both observational and experimental studies.
- Investigators appear to be its main source, although bias at the editorial level has not been well studied.
- Examination of the dissemination process as a whole has shed light on the points at which research findings get lost to knowledge, but more research is needed to fully understand the contributing factors.
- Registers of all research undertaken are required to mitigate the potentially harmful consequences of publication bias.

BACKGROUND

It is curious that publication is not inextricably bound to the performance of biomedical research. While studies do have distinct phases, the natural divide between collecting and analyzing data and publication of the findings has evolved into a worrisome schism. Perhaps this is because publication, and not the research itself, has been given such a prominent position in the academic reward system. Or perhaps it is because funding for research often fails to acknowledge the time and effort needed for writing up and disseminating results. Regardless of its origins, the prevailing divide has led to a surprisingly large fraction of research findings never being published.

There are numerous reasons to be concerned about the failure to report research findings. One is that it is unethical to use precious research resources without delivering the promised return on the investment – carrying out the investigation and sharing the findings. Another reason, particular to the case of studies with human participants, is that it is unethical to ask the participants to contribute to research without providing the yield they were almost certainly promised would result: a contribution to human knowledge. These reasons apply to privately as well as publicly funded research.

Scientific issues are also a concern, most notably the likelihood that failure to publish is a non-random event, that is, that there is a selection bias favoring publication of certain types of research findings over others. This bias can be related to characteristics of the research itself, the investigators, or those responsible for the decision to publish. It has potentially grave implications if the literature generally and systematic reviews specifically do not serve as an accurate reflection of the totality of knowledge on a topic.

'Publication bias', used here to refer to a tendency toward preparation, submission and publication of research findings based on the nature and direction of the research results, has been the area of greatest concern to date. While this definition is broadly understood, it actually encompasses a number of different overlapping, and even conflicting, concepts: the existence of a difference between two or more exposures or outcomes being compared, the direction of that difference, the size of the difference, and the statistical significance of the difference. The term can refer to results analyzed using either quantitative or qualitative methods.

The 'importance' of a study's findings is inherent in the definition of publication bias, but importance itself is not easily defined. Indeed, because to some extent results only exist once they are disseminated, importance has been equated roughly with newsworthiness, although the definition of 'newsworthy' has not been adequately addressed either. While newsworthy findings include those that support the efficacy of a new or experimental intervention ('positive results'), they also include unexpected findings that a new or existing intervention is no better than the standard intervention, no intervention, or placebo ('null results'). Also potentially newsworthy is a finding that one intervention is better than another, and that the observed difference between the two interventions, however small, is statistically significant (also referred to as 'positive results').

The term 'negative results' has been broadly used to refer both to the direction of findings (that is, results favoring the comparison intervention or null results favoring neither intervention), and to the lack of statistical significance of the findings, regardless of whether they are positive, negative, or null. Even negative results could be newsworthy, but the term is not usually applied when this is the case.

Sometimes, reference is made to a 'positive' or 'negative' study (Angell, 1989), meaning one with positive or negative results. This pejorative terminology has been discouraged but is pervasive, and reflects the way in which the direction or statistical significance or study findings can color our sense of the importance of a study, and even perhaps its methodology.

Since investigators have conceived and defined positive results in so many different ways, this chapter will not necessarily differentiate among them when describing the available evidence. This is an area in need of attention in the future, however, since positive results may mean something entirely different to a drug company considering publication of study findings than to a journal editor considering a manuscript for acceptance.

EARLY EVIDENCE OF PUBLICATION BIAS

There are many examples of centuries-old cautions against publication bias by prominent scientists (Ferriar, 1792; Hall, 1965; Tröhler, 2000, 2003a, 2003b). By the eighteenth century, individuals were referring to failures to report results (see Box 2.1). It is only recently, however, that research has been undertaken to estimate the size of the problem and its source. There is now a fairly large literature confirming that publication bias exists.

Box 2.1   Voices from the Past: Concern about Publication Bias[a]

'But the worst inconvenience of all is yet to be mentioned, and that is, that whilst this vanity of thinking men obliged to write either systems or nothing is in request, many excellent notions are, by sober and honest men, suppressed.' (Robert Boyle, 1661, quoted in Hall, 1965)

'When we attempt to introduce any new and important deviations from the common mode of practice into general use, and particularly in a point of such consequence, as the directing almost a total change in mode of performing and after-treating one of the principal operations in surgery, the public have a right to be fully acquainted with the author's reasons and motives for such attempt; and such trials should likewise previously have been made, as are sufficient to demonstrate, that the doctrine recommended will bear the test of general experience.' (Alanson, 1782)

'It would have been an easy task to have given select cases, whose successful treatment would have spoken strongly in favour of the medicine, and perhaps been flattering to my own reputation. But Truth and Science would condemn the procedure. I have therefore mentioned every case in which I have prescribed the Foxglove, proper or improper, successful or otherwise. Such a conduct will lay me open to censure of those who are disposed to censure, but it will meet the approbation of others, who are best qualified to be judges. ... After all, in spite of opinion, prejudice or error, TIME will fix the real value upon this discovery, and determine whether I have imposed upon myself and others, or contributed to the benefit of science and mankind.' (Withering, 1785)

'We have no proper balance to this very natural tendency to publish our success except through a more frequent publication of the errors and failures which likewise mark the path of every successful practitioner.' (Editors, 1909)

'By academic freedom I understand the right to search for truth and to publish and teach what one holds to be true. This right implies also a duty: one must not conceal any part of what one has recognized to be true. It is evident that any restriction on academic freedom acts in such a way as to hamper the dissemination of knowledge among the people and thereby impedes national judgment and action.' (Einstein, 1930)

[On 'cargo cult science'] 'It's a kind of scientific integrity, a principle of science that corresponds to a kind of utter honesty – a kind of leaning over backwards. The idea is to try to give all of the information to help others judge the value of your contribution, not just the information that leads to your judgment in one particular direction or another ... If you've made up your mind to test a theory, or you want to explain some idea, you should always

---

Box 2.1   (Continued)

decide to publish it whichever way it comes out. If we only publish results of a certain kind, we can make the argument look good. We must publish both kinds of results'. (Feynman, 1985)

'Few people outside science (and not nearly enough people inside) recognize the severe effects of biased reporting. The problem is particularly acute, almost perverse, when scientists construct experiments to test for an expected effect. Confirmations are joyfully reported; negative results are usually begrudgingly admitted. But null results – the failure to find any effect in any direction – are usually viewed as an experiment gone awry.' (Gould, 1987)

'Failure to publish an adequate account of a well-designed clinical trial is a form of scientific misconduct which can lead to those caring for patients to make inappropriate treatment decisions.' (Chalmers, 1990)

[a] See the James Lind Library (http://www.jameslindlibrary.org) for historical material on reporting biases, including full text of early quotations.

Early cross-sectional studies by Sterling (1959) and Smart (1964) showed that a sizable majority of published articles in psychology and education, respectively, had statistically significant findings. The results of these studies have been confirmed by many others, including those examining journals in the social and behavioral sciences (Glass *et al.*, 1981; Smith, 1980a, 1980b; White, 1982) and those examining medical journals (Ernst and Pittler, 1997; Moscati *et al.*, 1994; Sterling *et al.*, 1995, Vickers *et al.*, 1998; Yaphe *et al.*, 2001). A few studies have not shown this imbalance, however (Dimick *et al.*, 2001; Djulbegovic *et al.*, 2000).

Studies showing that positive results dominate the literature do not demonstrate that a selection bias exists. It is theoretically possible that by the time most research is undertaken, enough preliminary evidence has accumulated that only studies affirming the hypothesis they test are undertaken. This scenario is unlikely, however.

Additional early research has suggested publication bias, but again has provided little solid evidence. For example, Chalmers *et al.* (1965) found widely disparate data on fatality rates for serum hepatitis – from 0.3 % to 63 % – in the 23 publications he identified that provided this information. Because the smaller studies were associated with higher fatality rates, the authors suggested that this reflected a tendency for unusual findings to be reported.

Surveys of peer reviewers and investigator-authors, mainly conducted in the 1970s and 1980s, have indicated that studies with results rejecting the null hypothesis are more likely to be submitted or recommended for publication (Coursol and Wagner, 1986; Dickersin *et al.*, 1987; Greenwald, 1975; Shadish *et al.*, 1989; Sommer, 1987). These findings, although potentially unreliable because they are based solely on survey responses, have been consistent across a number of fields, including the social sciences and medicine: Greenwald (1975) surveyed reviewers and authors associated with the *Journal of Personality and Social Psychology*; Coursol and Wagner (1986) surveyed members of the American Psychological Association; Sommer (1987) surveyed members of the Society for Menstrual Cycle Research; Hetherington *et al.* (1989) surveyed 42 000 obstetricians and pediatricians from 18 countries; and Dickersin *et al.* (1987) surveyed authors of published clinical trials. In the last of these studies, investigators reported that the major reasons why they had not published their completed trials ($n = 204$) were negative results (28 %), lack of interest (12 %), a manuscript intended or under way (12 %), and sample size problems (11 %).

Experimental studies are the ideal way to examine publication bias, for example, by submission of similar manuscripts with positive and negative findings to journals or referees to examine whether there are differences in recommendations to publish. Two such studies have been performed, one involving 75 reviewers used by a single psychology journal (Mahoney, 1977) and one involving 33 social work journals (Epstein, 1990). Both showed that positive results were associated with a higher rate of manuscript acceptance.

IS PUBLICATION BIAS IMPORTANT?

Publication bias was perceived as a somewhat arcane issue, mainly of academic interest, until recently, when systematic reviews and meta-analysis assumed increased importance for policy setting, particularly in the area of intervention studies. As a result, reporting biases have received more serious attention. Clinicians who depend on systematic reviews and meta-analyses based solely on published data for deciding on a patient's treatment course could be causing harm if studies that could influence the review results are omitted. Simes (1986) published a groundbreaking study demonstrating not only the existence of publication bias, but also its potential influence on the treatment of a life-threatening disease. Simes compared the pooled results from published trials with clinical trials registered with the International Cancer Research Data Bank. He found a statistically significant survival advantage for combination over single-agent chemotherapy for advanced ovarian cancer when he combined results of published trials, but not when he combined results of registered (published plus unpublished) trials.

Other striking examples of publication bias related to actual or potential patient harm come from cardiovascular disease (Chalmers, 2001) and thyroid disease (Rennie, 1997).

Sterne *et al.* (2002) identified 122 meta-analyses published in *The Cochrane Library* that were based on 'comprehensive' literature searches. Thirty-nine of these included unpublished trials and thus allowed the authors to examine the contributions of unpublished and published trials to the overall findings of the meta-analyses. Overall, published trials estimated a greater beneficial effect of the intervention compared to unpublished trials (ratio of odds ratios = 1.12 [95 % confidence interval (CI) 0.99 to 1.26]), and this association was strengthened after controlling for factors associated with trial quality.

Systematic reviews summarize all available data meeting prespecified inclusion criteria, including the 'grey' literature (e.g., conference proceedings, dissertations and technical reports). These sources lie somewhere between being published and unpublished because they are relatively inaccessible and may contain only limited information. Indeed, 56 % of Cochrane reviews contain grey literature studies (Mallett *et al.*, 2002). Inclusion of the grey literature in meta-analyses reduces effect sizes (McAuley *et al.*, 2000), providing yet more evidence that publication bias may influence decision-making.

IMPROVED UNDERSTANDING OF THE PROCESS OF RESEARCH DISSEMINATION AND PUBLICATION

Understanding the research-to-publication continuum

In response to concerns generated by Simes' (1986) study, new research was undertaken to study publication bias and to estimate the size of the problem. Special conferences dedicated to peer review (Rennie, 2002) and the formation of Cochrane methods groups (see http://www.cochrane.org for more information) have been instrumental in stimulating this area of investigation, starting in the 1990s. Focus has evolved from documenting publication bias to achieving a better understanding of each of the steps in the research dissemination and publication process. That is, what happens to a research project once it has been initiated? What proportion of results is at least reported at conferences and in the form of abstracts? And what are the factors that determine whether the results will be published in full? Are there characteristics of a study that are associated with publication, such as its design (experimental vs. observational), its funding source (industry vs. government vs. none or other), its size, its country of origin, or its findings? The rest of this chapter will focus on one part of the evolving and increasingly rich literature related to reporting of study findings: publication associated with positive results.

Cohort studies of all initiated studies

At this writing, the research process, from study initiation to dissemination of results, has been examined in seven cohort studies of biomedical research, six reported in full these studies identified and followed research projects approved by biomedical research ethics review boards (Easterbrook *et al.*, 1991; Dickersin *et al.*, 1992; Stern and Simes, 1997; Wormald *et al.* 1997) and studies funded by the National Institutes of Health (NIH) (Dickersin and Min, 1993; Ioannidis, 1998). Because detailed results are not available for one of the studies (Wormald *et al.*, 1997), the discussion will focus on six of them, two presented in a single publication (Dickersin *et al.*, 1992). The methods used, including definitions of positive and negative outcomes, were fairly consistent across the studies (see Table 2.1).

These six investigations found that between 59 % and 100 % of studies had been published. The highest proportion of published studies was for the two cohorts of clinical trials funded by the NIH. All of the 66 completed HIV/AIDS trials supported from 1986–1996 through the NIH AIDS Clinical Trials Group and followed up to

Table 2.1 Definitions of main study outcomes in six cohort studies of publication bias.

	Easterbrook et al. (1991)	Dickersin et al. (1992)	Dickersin and Min (1993)	Stern and Simes (1997)	Ioannidis (1998)
Cohort(s) of studies followed	Approved by Central Oxford Research Ethics Committee in 1984–1987	Approved by (1) Johns Hopkins Medical School research ethics review board (RERB) in 1980 and (2) Johns Hopkins School of Hygiene and Public Health RERB in 1980	Phase I and II clinical trials funded by NIH in 1979 (National Cancer Institute trials excluded)	Approved by Royal Prince Alfred Hospital Ethics Committee in 1979–1988	Multicenter randomized efficacy trials conducted by two HIV/AIDS trials groups and funded by NIH in 1986–1996
How main study outcome determined	Interview with or questionnaire completed by principal investigator (PI)	Interview with PI	Interview with PI	Questionnaire completed by PI	Abstraction of archived trial reports at NIH
How main outcome defined	Main study findings	Results for primary outcome	Results for primary outcome	Results for main research questions posed at study outset	Main endpoint or any major efficacy endpoint
Outcome definition – Quantitative studies	Statistically significant ($p < 0.05$)	Statistically significant ($p < 0.05$)	Statistically significant ($p < 0.05$)	Statistically significant ($p < 0.05$)	Statistically significant ($p < 0.05$) in favor of experimental therapy over control or standard therapy; main endpoint or any endpoint
	Not significant but a trend ($p \geq 0.05$)	Not significant but a trend ($p \geq 0.05$)	Not significant but a trend ($p \geq 0.05$)	Not significant but a trend ($0.10 \geq p \geq 0.05$)	Non-significant findings or favoring the control arm
	Null (no difference)	No trend or difference	No trend or difference	Non-significant or null ($p \geq 0.10$)	–

Table 2.1 (Continued).

	Easterbrook *et al.* (1991)	Dickersin *et al.* (1992)	Dickersin and Min (1993)	Stern and Simes (1997)	Ioannidis (1998)
Outcome definition – Qualitative studies	Observations were. . . Striking Definite but not striking Null findings	Perceived importance: Great Moderate Little	Perceived importance: Great Moderate Little	Findings were. . . Striking Important and definite Unimportant and negative	Not applicable – – –
Clinical or scientific importance	1 (not important) to 10 (extremely important)	–	–	–	–
Analysis of quantitative studies and qualitative studies/no statistical testing done	Together (?) (no methods identified)	Together ('significant' = [significant + great importance]; 'non–significant' = [trend but not significant + moderate importance] + [no difference + little importance])	Together ('significant' = [significant + great importance]; 'non–signficant' = [trend but not significant + moderate importance] + [no difference + little importance])	Separately	Not applicable

1996 were eventually published (Ioannidis, 1998; Ioannidis, personal communication, 2003), and 93 % of NIH trials funded in 1979 had been published at the time of
interview (Dickersin and Min, 1993). There was a greater range in the proportion of
studies published for the cohorts identified through institutional research ethics review
board approvals. At Johns Hopkins Medical School, for example, 80.4 % of studies
were published in full length or abstract form (69.4 % if only full publications were
counted), and at Johns Hopkins School of Public Health 65.3 % were published (61.2 %
if only full publications were counted). One might expect that the proportion published would be even lower in other settings, given the high level of funded research
activity associated with the research ethics review boards at the institutions for these
cohorts. Indeed, Pich *et al.* (2003) found that only 21 % of closed clinical trials that
had been approved by a major hospital ethics committee in Spain had been published
three years later in biomedical journals that were included in Science Citation Index.

The main factor associated with failure to publish in all of these cohort studies
was negative or null findings (Dickersin, 1997). Investigators reported that the
major reason for not publishing was that they never wrote up and submitted a
manuscript because 'the results were not interesting' or the results were negative.
Other reasons included co-investigator or other operational problems, additional
analyses needed, and 'publication was not an aim'. Only a small proportion of
studies were not published because of rejection by journals. Other factors were
less clearly associated with failure to publish, for example, having industry funding
(Easterbrook *et al.*, 1991), no 'external' funding (Dickersin *et al.*, 1992; Stern and
Simes, 1997), and being randomized (Stern and Simes, 1997).

A cohort study using a similar design replicated these findings generally in
the field of psychology. Cooper *et al.* (1997) followed the publication outcome of
159 studies approved by the research ethics review board in 1986–1988 for the
Department of Psychology at the University of Missouri. Of the 117 studies that
carried out data analysis and significance testing, 62 % had statistically significant
results. While 50 % of studies with statistically significant results were submitted
for presentation at conferences, only 7 % of statistically non-significant results were
submitted. Furthermore, 74 % of studies with significant results were submitted for
publication in books or journals, compared to 4 % of non-significant results. The
reasons for failure to produce or submit a written summary were similar to those
given for biomedical research, including 'publication was not an aim', 'results not
interesting', and statistically non-significant results.

Despite the consistent findings that only a small fraction of studies are not
published because they are turned down by journals, investigators have persisted in
naming bias at the editorial level as the main reason why negative or null results
are not published. A recent study examining editorial review at the *Journal of the
American Medical Association* (*JAMA*) found that if an editorial bias exists, it is
small (Olson *et al.*, 2002). There is no reason to think that editorial practices at a
major journal such as *JAMA* are similar to those at smaller and specialty journals,
however, and this issue should be further investigated. A 1996 survey of 36 editors
of English-language dental journals, for example, found that 'the significance and
importance of the research work' was valued most highly, above factors such as
the validity of the experimental and statistical method (Radford *et al.*, 1999).

From presentation at a meeting to full publication

While it is common practice for research results to be presented at conferences before final analyses and full-length manuscripts are prepared, the typical sequence in the dissemination process is not well understood. The existence of publication bias at the time of acceptance for presentation, the proportion of study findings disseminated as conference abstracts, and the likelihood that being presented predicts publication are all areas of concern.

Dudley (1978) reported on follow-up of 51 surgical studies initially presented at the Surgical Research Society in Great Britain in 1972 and published as abstracts, to learn about subsequent full publication (Dudley, 1978). It was the first investigation of its type, and almost 80 others have been published since then, in many fields of biomedicine (Scherer *et al.*, 2005). Scherer's Cochrane review of these 79 'abstract follow-up studies' found that only about 53 % of all studies initially reported as abstracts, and 63 % of those reporting controlled trials, reach full publication.

Whereas earlier investigations of full publication of abstracts focused on simply estimating rates of full publication, recent studies have attempted to assess the impact of additional factors. Scherer's review showed that full publication is more likely for studies accepted for meeting presentation compared to those rejected RR = 1.78; 95 % CI = 1.50 to 2.12; randomized study design (RR = 1.24; 95 % CI = 1.14 to 1.36); and basic research compared to clinical research (RR= 0.79; 95 % CI = 0.70 to 0.89). 'Positive' results defined as any 'significant' result showed an association with full publication (RR = 1.30; CI 1.14 to 1.47), as did 'positive' results defined as a result favoring the experimental treatment (RR= 1.17; CI 1.02 to 1.35), and 'positive' results emanating from randomized or controlled clinical trials (RR= 1.18, CI 1.07 to 1.30).

Timmer and her colleagues (Timmer *et al.*, 2002) found that abstracts reporting statistically significant results were more likely than all others be reported in full in higher impact journals, even though they found no evidence of publication bias *per se*.

In a related area, various studies over the years have shown that dissertations and theses are more likely to be published in full if results are positive (Smart, 1964; Vogel and Windeler, 2000; Zimpel and Windeler, 2000), and that, on average, dissertations that remain unpublished have lower effect sizes than the published literature (Smith, 1980b).

Time to publication

The faster study results are published, the faster doctors and consumers can benefit from the findings, sometimes by avoiding harm. Examination of the association between study results and time taken to publish research findings has yielded mixed results, perhaps because of the many components involved in 'time to publication'. There is the amount of time taken by the authors before submission for publication, and this can be counted from the time funding was received, the study was approved by the research ethics review board, the first participant was enrolled, the last participant was enrolled, data collection was completed, analysis was completed, or other starting points. There is also the amount of time taken by the editors and peer reviewers, in conjunction with the authors if revision is required, between a manuscript's submission and publication.

Different studies have also used different denominators – either all studies initiated or all studies published. Analysis of time to publication for all studies initiated includes studies never published, which most research has shown include disproportionate numbers of studies with negative results.

A number of studies have reported that negative findings take longer to reach publication, regardless of whether the time interval measured was between publication and funding start date (Misakian and Bero, 1998), research ethics review board approval (Stern and Simes, 1997), enrollment of the first patient, or completion of follow-up (Ioannidis, 1998). Even in a group of initiated HIV/AIDS studies which were all eventually published, those with negative findings took longer (Ioannidis, 1998).

No difference in time to publication was seen in a study of controlled trials accepted by *JAMA*, a study that measured time between submission and publication and included only published trials (Dickersin *et al.*, 2002). Tierney *et al.* (2000) also found no evidence of publication bias nor difference in time to publication among 38 individual patient data meta-analyses in the field of cancer, 30 published and eight unpublished. Finally, mixed results have been obtained by four studies in different fields examining whether abstracts reporting positive or statistically significant results are published in full more quickly than those reporting negative findings (Eloubeidi *et al.*, 2001; Evers, 2000; Petticrew *et al.*, 1999; Timmer *et al.*, 2002).

There are several factors contributing to time to publication. One is that investigators may take longer to write up and submit manuscripts for publication when the results are not interpreted as interesting. In this case, one might expect to see a pattern such that positive findings on a topic appear first in the literature, often from small studies. Studies with negative and null results would appear later, either because they are now more interesting since they refute existing knowledge or because they were undertaken in response to results from the initial reports. There are examples of this pattern from the HIV/AIDS (Ioannidis, 1998), genetics (Ioannidis *et al.*, 2003), and ecology and evolution literature (Jennions and Møller, 2002). Some investigators may delay or not bother to write up their findings at all if they feel their data would not affect overall knowledge, for example if considerable evidence indicating a positive association has already been published.

Trials with statistically significant results may be completed more quickly, for example they may be stopped early. Ioannidis (1998) has shown that larger trials of HIV/AIDS may take longer to complete, but they are published sooner. The length of time it takes to perform a study may also be related to the availability and source of funding; indeed, external funding has been shown to be associated with publication in several studies (Dickersin, 1997).

OTHER FACTORS ASSOCIATED WITH FAILURE TO PUBLISH AND DIRECTION OF RESULTS

Because systematic reviews depend on identification of all relevant literature, efforts have been made to develop effective and efficient ways to identify unpublished and grey literature studies. The various approaches tried have required a lot of work and none has been very successful (Hetherington *et al.*, 1989; McManus *et al.*, 1998). This had led investigators to try to identify those most likely to be harboring research findings in their file drawers, so as to focus searching efforts in that direction. Research has mainly focused on bias related to studies funded by industry and studies published in languages other than English.

Substantial research exists showing that published reports from studies funded by industry are more likely to show positive results, compared to reports from studies funded by other sources. Yaphe *et al.* (2001) examined randomized controlled trials of drugs published in five major general medical journals between October 1, 1992 and October 1, 1994, and found that of the 314 trials, 77 % reported results likely to promote use of the drug and 67 % reported industry funding. Eighty-seven percent of industry-supported studies and 65 % of non-industry supported studies reported a positive outcome. Others have reported similar results (Bero and Rennie 1996; Cho and Bero 1996; Davidson, 1986; Djulbegovic *et al.*, 2000; Kjaergard and Als-Nielsen, 2002; Rochon *et al.*, 1994).

There are several possible explanations for these differences. Industry may selectively publish findings supporting a product's efficacy. It is also possible that industry is more likely to undertake studies with a high likelihood of a positive outcome, for example, by selecting a comparison population likely to yield results favoring the product (Mann and Djulbegovic, 2004). Neither of these actions would be ethical. There is evidence from two cohort studies described earlier that industry funding is inversely associated with publication (Easterbrook *et al.*, 1991; Dickersin *et al.*, 1992). The results presented may also vary by dissemination medium: articles published in journals were more likely to present results suggesting adverse outcomes from environmental tobacco smoke than articles in non-industry-supported symposium proceedings or tobacco industry-supported proceedings – 80 % vs. 59 % vs. 28 %, respectively (Bero *et al.*, 1994).

It is also possible that journals are more likely to publish studies with negative results if they are funded by government sources. No evidence of selective publication by source of funding was seen, however, in a study of controlled trials submitted to *JAMA* between 1996 and 1999 (Olson *et al.*, 2002).

There is reason to be concerned about the possibility of selective publication of industry-supported studies. Investigations on this topic are challenging, however, because it is difficult to identify an inception cohort of initiated industry trials (Manheimer and Anderson, 2002), and even if it were possible to do so, a sizable proportion of trials have funding from more than one source. In addition, methodological studies have typically had small sample sizes, limiting both subgroup analyses of publication by funding source, and statistical power.

There are conflicting results about whether English-language journals publish positive results in different proportions than non-English-language journals. Egger *et al.* (1997) demonstrated that, for pairs of reports of clinical trials by the same author, one published in English the other in German, only statistically significant results predicted publication in an English-language journal. A subsequent study, however, found that studies published in languages other than English were of lower overall methodological quality and were more likely to report statistically significant results (Jüni *et al.*, 2002). Publishing practices may vary by country and health field (Grégoire *et al*, 1995; Ottenbacher and Difabio, 1985; Vickers *et al.*, 1998), and at this time one should probably not assume that publication practices are language-related.

The possible influence of the quality of studies and reports of studies on the direction and strength of results is of long-standing research interest (Emerson, 1990; Bero and Rennie, 1996; Balk *et al.*, 2002; Kunz and Oxman, 1998), but is too complex to be addressed here. While most would agree that studies of poor quality should not be published, a valid definition of 'sufficient' quality has not been agreed, even in the case of intervention studies where randomized controlled trials are the accepted gold standard. In the case of observational studies, it is generally accepted that 'fishing expeditions' are likely to lead to false positive associations (Swaen *et al.*, 2001). The interaction between study quality and positive findings, and the relationship to publication, an issue of concern in this chapter, has not been explored sufficiently, probably because of the difficulty of defining quality and because large sample sizes would be required.

There is evidence that the strength and direction of study findings influence either the author's choice of journal or the likelihood of acceptance. Easterbrook *et al.* (1991) found that biomedical journals publishing statistically significant results compared to those publishing null results had, on average, a higher impact factor (1.62 vs. 0.9). In another study of 38 individual patient data meta-analyses, those with statistically significant or more impressive results tended to be published in journals with higher impact factors compared to those with null or unimpressive findings (Tierney and Stewart, 2000). Publication practices related to NIH-funded HIV/AIDS clinical trials may help to explain these findings. Trials published in general medical journals were more likely than trials in specialty journals to report statistically significant results for at least one primary endpoint (28/37 vs. 14/34); perhaps general journals have higher impact than specialty journals, on average (Ioannidis *et al.*, 1997).

BIASED REPORTING OF OUTCOMES

There has long been concern that study outcomes may be selectively reported, depending on the nature and direction of results. At this writing, few studies investigating this problem have been published, however (Hutton and Williamson, 2000; Tannock, 1996). A pilot project that compared the analysis plan submitted to the research ethics review board to the subsequent study report found that seven out of eight studies did not follow their original plan, and indicated possible selective reporting of outcomes (Hahn *et al.*, 2002). Study plans can legitimately change between protocol submission and implementation, so investigations of selective reporting of outcomes should be interpreted with caution. Ioannidis reported that even studies reporting a negative finding for a primary outcome typically reported their results in a positive frame, and included positive results for subgroups and surrogate outcomes (Ioannidis *et al.*, 1997).

Selective reporting of suspected or confirmed adverse treatment effects is an area with potential for patient harm and one where one might well imagine a publication bias on the part of industry. For example, in a study of adverse drug events submitted to Scandinavian drug licensing authorities, reports for published studies were less likely than unpublished studies to record adverse events – for example, 56 vs. 77 % respectively for Finnish trials involving psychotropic drugs (Hemminki, 1980). A recent study by MacLean *et al.* (2003) found that while, indeed, industry fails to publish a majority of its findings on adverse events, there is no indication of a publication bias. Their study of new drug applications for non-steroidal anti-inflammatory drugs (NSAIDs) found that only one out of 37 studies summarized in Food and Drug Administration reviews was published. However, they observed no meaningful difference in the summary risk ratios for dyspepsia (an adverse event) among unpublished versus published studies, nor were there differences in sample size, sex of participants, indications for drug use, or methodological quality.

SELECTIVE CITATION OF POSITIVE RESULTS

Although not strictly part of the research process, citation of published findings and the relationship to the nature and direction of the outcome has concerned researchers for decades (Callaham *et al.*, 2002; Christiansen-Szalanski and Beach, 1984; Gotzsche, 1987; Hutchison & Lloyd, 1995; Kjaergard and Gluud 2002; Ravnskov, 1992, 1995). In particular, citation of a study may make it more easily identified for systematic reviews and meta-analysis, particularly when the study is unpublished or published in the grey literature. Studies have found mixed results, with some suggesting selective citation of positive findings (Gotzsche, 1987; Kjaergard and Gluud, 2002; Ravnskov, 1992, 1995) and others not (Callaham *et al.*, 2002; Christensen-Szalanski and Beach, 1984; Hutchison & Lloyd, 1995). Positive results appear to be published more often in duplicate, which can lead to overestimates of a treatment effect (Timmer *et al.*, 2002).

SOLUTIONS TO PUBLICATION BIAS

Clearly, publication bias has great potential for promoting false conclusions and for patient harm, and many have called for publication of the results of all research undertaken (see, for example, Chalmers, 1990; Horton, 1997). There have been special sections in journals for publishing negative findings (for example, *JAMA* had a section called Negative Results in the mid-1960s and *Cancer Epidemiology Biomarkers and Prevention*, in 2003, had a section call Null Results in Brief), and there is now a *Journal of Negative Results in Biomedicine*. Research ethics review boards, department chairs, funding agencies, and others have been challenged to hold investigators accountable (Antes and Chalmers, 2003; Savulescu *et al.*, 1996).

For over 25 years, repeated calls have been made to register clinical trials at inception and assign unique trial identification numbers; among other things, a register would provide basic information about trials that remain unpublished (*Ad hoc* Working Party of the International Collaborative Group on Clinical Trials Registries, 1993; Antes and Chalmers, 2003; Chalmers, 1977; Dickersin, 1988; Dickersin and Rennie 2003; Hetherington *et al.*, 1989; Institute of Medicine, 2000, 2002; Simes, 1986). Although they will not themselves solve the problem of publication bias or biased reporting of outcomes, research registers are the best potential solution proposed so far. Despite some progress, including newly developed registers (Anderson *et al.*, 2001; Chalmers, 2000; McCray, 2002; Tonks, 2002), most believe that publication bias is a real threat. Indeed, the basic scientists have recently joined the list of those concerned enough to start taking action (Knight, 2003).

Registers of clinical trials and observational studies are necessary to circumvent the inadequacies of the published literature, which can be hard to search (Dickersin *et al.*, 1994), minimally informative (for example, conference abstracts), selective in presentation of results, or missing information entirely for a study. It is increasingly evident that directives and legislation will be required to ensure establishment of comprehensive research registers, similar to that used in specific disease areas (Health Omnibus Programs Extensions Act of 1988 (Public Law 100-607); McCray, 2001). The fact that a legal approach has been taken and will likely be taken again shows recognition of the seriousness of publication bias as a potential impediment to the advancement of knowledge.

CONCLUSIONS

The road from study start to dissemination of results provides many opportunities for slowdowns and even complete stops. Data collection may not be started or completed, data may not be analyzed, results may simply be presented or written up in an abstract, an internal report, or thesis, or they may be published in full, once or multiple times. Aside from wasting precious resources, failing to disseminate study results is unethical, especially if human volunteers were involved. The human consequences of failure to report study findings may be great if there are reporting biases, especially biases based on study findings.

There is now good evidence for publication bias and recognition that those performing systematic reviews should be aware of the problem and consider methods for addressing the issue. If studies were registered at inception, reviewers and others would at least know when relevant studies had been done, and could pursue the responsible investigators to obtain study results.

ACKNOWLEDGEMENTS

Thanks to Sir Iain Chalmers for reviewing the manuscript and many helpful suggestions. The James Lind Library (http://www.jameslindlibrary.org), which he edits, is the source of many of the older quotations referring to reporting biases. Thanks to Doug Altman for the quotation from Sir Austin Bradford Hill, Mike Clarke for calling my attention to 'Null Results in Brief', and Harry Marks for locating historical material.

REFERENCES

Ad Hoc Working Party of the International Collaborative Group on Clinical Trials Registries. International Collaborative Group on Clinical Trial Registries (1993). Position paper and consensus recommendations on clinical trial registries. *Clinical Trials Metaanalysis*, **28**, 255–266.

Alanson, E. (1782). *Practical observations on amputation, and the after-treatment*, 2nd edition. London: Joseph Johnson.

Anderson, D., Costa, I. and Dickersin, K. (2001). Building Trials Central, an online register of clinical trials registers. *Controlled Clinical Trials*, **22**(Suppl. 2), 40S.

Angell, M. (1989). Negative studies. *New England Journal of Medicine*, **321**, 454–466.

Antes, G. and Chalmers, I. (2003). Under-reporting of clinical trials is unethical. *Lancet*, **361**, 978–979.

Balk, E.M., Bonis, P.A., Moskowitz, H., Schmidt, C.H., Ioannidis, J.P.A., Wang, C. and Lau, J. (2002). Correlation of quality measures with estimates of treatment effect in meta-analyses of randomized controlled trials. *Journal of the American Medical Association*, **287**, 2973–2982.

Bero, L.A. and Rennie, D. (1996). Influences on the quality of published drug studies. *International Journal of Technology Assessment in Health Care*, **12**, 209–237.

Bero, L.A., Glantz, S.A. and Rennie, D. (1994). Publication bias and public health policy on environmental tobacco smoke. *Journal of the American Medical Association*, **272**, 133–136.

Bradford Hill, A. (1959). Discussion of a paper by D. J. Finney. *Journal of the Royal Statistical Society, Series A*, **119**, 19–20.

Callaham, M., Wears, R.L. and Weber, E. (2002). Journal prestige, publication bias, and other characteristics associated with citation of published studies in peer-reviewed journals. *Journal of the American Medical Association*, **287**, 2847–2850.

Chalmers I. (1990). Underreporting research is scientific misconduct. *Journal of the American Medical Association*, **263**, 1405–1408.

Chalmers, I. (2000). Current Controlled Trials: An opportunity to help improve the quality of clinical research. *Current Controlled Trials*, **1**, 3–8.

Chalmers, I. (2001). Using systematic reviews and registers of ongoing trials for scientific and ethical trial design, monitoring and reporting. In: M. Egger, D. Smith and D.G. Altman (eds), *Systematic Reviews in Health Care: Meta-analysis in Context*, 2nd edition. London: BMJ Books.

Chalmers, T.C. (1977). Randomize the first patient! *New England Journal of Medicine*, **296**, 107.

Chalmers, T.C., Koff, R.S. and Grady, G.F. (1965). A note on fatality in serum hepatitis. *Gastroenterology*, **49**, 22–26.

Cho, M.K. and Bero, L.A. (1996). The quality of drug studies published in symposium proceedings. *Annals of Internal Medicine*, **124**, 485–489.

Christensen-Szalanski, J.J.J. and Beach, L.R. (1984). The citation bias: Fad and fashion in the judgment and decision literature. *American Journal of Psychology*, **39**, 75–78.

Cooper, H., DeNeve, K. and Charlton, K. (1997). Finding the missing science: The fate of studies submitted for review by a human subjects committee. *Psychological Methods*, **2**, 447–452.

Coursol, A. and Wagner, E.E. (1986). Effect of positive findings on submission and acceptance rates: A note on meta-analysis bias. *Professional Psychology Research & Practice*, **17**, 136–137.

Davidson, R.A. (1986). Source of funding and outcome of clinical trials. *Journal of General Internal Medicine*, **1**, 155–158.

Dickersin, K. (1988). Report from the Panel on the Case for Registers of Clinical Trials at the Eighth Annual Meeting of the Society for Clinical Trials. *Controlled Clinical Trials*, **9**, 76–81.

Dickersin, K. (1997). How important is publication bias? A synthesis of available data. *AIDS Education and Prevention*, **9**(1 Suppl), 15–21.

Dickersin, K. and Min, Y.I. (1993). NIH clinical trials and publication bias. *Online Journal of Current Clinical Trials*, Doc. No. 50.

Dickersin, K. and Rennie, D. (2003). Registering clinical trials. *Journal of the American Medical Association*, **290**, 516–523.

Dickersin, K., Chan, S., Chalmers, T.C., Sacks, H.S. and Smith, H. Jr. (1987). Publication bias and clinical trials. *Controlled Clinical Trials*, **8**, 343–353.

Dickersin, K., Min, Y.I. and Meinert, C.L. (1992). Factors influencing publication of research results. Follow-up of applications submitted to two institutional review boards. *Journal of the American Medical Association*, **267**, 374–378.

Dickersin, K., Scherer, R. and Lefebvre, C. (1994). Identifying relevant studies for systematic reviews. *British Medical Journal*, **309**, 1286–1291.

Dickersin, K., Olson, C.M., Rennie, D., Cook, D., Flanagin, A., Zhu, Q., Reiling, J. and Pace, B. (2002). Association between time interval to publication and statistical significance. *Journal of the American Medical Association*, **287**, 2829–2831.

Dimick, J.B., Diener-West, M. and Lipsett, P.A. (2001). Negative results of randomized clinical trials published in the surgical literature: Equivalency or error? *Archives of Surgery*, **136**, 796–800.

Djulbegovic, B., Lacevic, M., Cantor, A., Fields, K.K., Bennett, C.L., Adams, J.R., Kuderer, N.M. and Lyman, G.H. (2000). The uncertainty principle and industry-sponsored research. *Lancet*, **356**, 635–638

Dudley, H.A.F. (1978). Surgical research: Master or servant. *American Journal of Surgery*, **135**, 458–460.

Easterbrook, P.J., Berlin, J.A., Gopalan, R. and Matthews, D.R. (1991). Publication bias in clinical research. *Lancet*, **337**, 867–872.

Editors (1909). The reporting of unsuccessful cases (editorial). *The Boston Medical and Surgical Journal*, **161**, 263–264.

Egger, M., Zellweger-Zahner, T., Schneider, M., Junker, C., Lengeler, C. and Antes, G. (1997). Language bias in randomized controlled trials published in English and German. *Lancet*, **350**, 326–329.

Einstein, A. (1930). What I believe. *Forum and Century*, **84**, 183–194.

Eloubeidi, M.A., Wade, S.B. and Provenzale, D. (2001). Factors associated with acceptance and full publication of GI endoscopic research originally published in abstract form. *Gastrointestinal Endoscopy*, **53**, 275–282.

Emerson, J.D., Burdick, E., Hoaglin, D.C., Mosteller, F. and Chalmers, T.C. (1990). An empirical study of the possible relation of treatment differences to quality scores in controlled randomized clinical trials. *Controlled Clinical Trials*, **11**, 339–352.

Epstein, W.M. (1990). Confirmational response bias among social work journals. *Science, Technology, and Human Values*, **15**, 9–37.

Ernst, E. and Pittler, M.H. (1997). Alternative therapy bias [letter]. *Nature*, **385**, 480.

Evers, J.L. (2000). Publication bias in reproductive research. *Human Reproduction*, **15**, 2063–2066.

Ferriar, J. (1792). *Medical Histories and Reflexions* (Vol. 1). London: Cadell and Davies.

Feynman, R.P. (1985). *Surely You're Joking Mr. Feynman*. New York: Norton.

Glass, G.V., McGaw, B. and Smith, M.L. (1981). *Meta-analysis in Social Research*. Beverly Hills, CA: Sage Publications.

Gøtzsche, P.C. (1987). Reference bias in reports of drug trials. *British Medical Journal (Clinical Research Edition)*, **195**, 654–656.

Gould, S.J. (1987).*Urchin in the Storm. Essays about Books and Ideas*, New York: Norton.

Greenwald, A.G. (1975). Consequences of prejudice against the null hypothesis. *Psychological Bulletin*, **82**, 1–20.

Grégoire, G., Derderian, F. and Le Lorier, J. (1995). Selecting the language of the publications included in a meta-analysis: Is there a Tower of Babel bias? *Journal of Clinical Epidemiology*, **48**, 159–163.

Hahn, S., Williamson, P.R. and Hutton, J.L. (2002). Investigation of within-study selective reporting in clinical research: Follow-up of applications submitted to a local research ethics committee. *Journal of Evaluation in Clinical Practice*, **8**, 353–359.

Hall, M.B. (1965). In defense of experimental essays. In: *Robert Boyle on Natural Philosophy* (pp. 119–131). Bloomington: Indiana University Press.

Hemminki, E. (1980). Study of information submitted by drug companies to licensing authorities. *British Medical Journal*, **280**, 833–836.

Hetherington, J., Dickersin, K., Chalmers, I. and Meinert, C.L. (1989). Retrospective and prospective identification of unpublished controlled trials: Lessons from a survey of obstetricians and pediatricians. *Pediatrics*, **84**, 374–380.

Hopewell, S., Clarke, M., Stewart, L. and Tierney, J. (2003). Time to publication for results of clinical trials. *The Cochrane Database of Methodology Reviews* 2001, Issue 3. Art No.: MR000011. DOI 10.1002/14651858.MR000011. John Wiley & Sons, Ltd.

Horton, R. (1997). Medical editor trial amnesty. *Lancet*, **350**, 756.

Hutchison, B.G. and Lloyd, S. (1995). Comprehensiveness and bias in reporting clinical trials: Study of reviews of pneumococcal vaccine effectiveness. *Canadian Family Physician*, **41**, 1356–1360.

Hutton, J.L. and Williamson, P.R. (2000). Bias in meta-analysis due to outcome variable selection within studies. *Applied Statistics*, **49**, 359–370.

Institute of Medicine Committee on Assessing the System for Protecting Human Research Participants. Responsible Research (2002). *A Systems Approach to Protecting Research Participants*. Washington, DC: National Academy Press.

Institute of Medicine Committee on Routine Patient Care Costs in Clinical Trials for Medicare Beneficiaries (2000). *Extending Medicare Reimbursement in Clinical Trials*. Washington, DC: National Academy Press.

Ioannidis, J.P. (1998). Effect of the statistical significance of results on the time to completion and publication of randomized efficacy trials. *Journal of the Americal Medical Association*, **279**, 281–286.

Ioannidis, J.P.A., Cappelleri, J.C., Sacks, H.S. and Lau, J. (1997). The relationship between study design, results, and reporting of randomized clinical trials of HIV infection. *Controlled Clinical Trials*, **18**, 431–444.

Ioannidis, J.P.A., Trikalinos, T.A., Ntzani, E.E. and Contopoulos-Ioannidis, D.G. (2003). Genetic associations in large versus small studies: An empiric assessment. *Lancet*, **361**, 567–571.

Jennions, M.D. & Møller, A.P. (2002). Relationships fade with time: A meta-analysis of temporal trends in publication in ecology and evolution. *Proceedings of the Royal Society of London, Series B: Biological Sciences*, **269**, 43–48.

Jüni, P., Holstein, F., Sterne, J., Bartlett, C. and Egger, M. (2002). Direction and impact of language bias in meta-analyses of controlled trials: An empirical study. *International Journal of Epidemiology*, **31**, 115–123.

Kjaergard, L.L. and Gluud, C. (2002). Citation bias of hepato-biliary randomized clinical trials. *Journal of Clinical Epidemiology*, **55**, 407–410.

Kjaergard, L.L. and Als-Nielsen, B. (2002). Association between competing interests and authors' conclusions: Epidemiological study of randomized clinical trials published in the *British Medical Journal*. *British Medical Journal*, **325**, 249–252.

Knight, J. (2003). Null and void. *Nature*, **422**, 554–555.

Kunz, R. and Oxman, A.D. (1998). The unpredictability paradox: review of empirical comparisons of randomised and non-randomised clinical trials. *British Medical Journal*, **317**, 1185–1190.

MacLean, C.H., Morton, S.C., Ofman, J.J., Roth, E.A. and Shekelle, P.G. for the Southern California Evidence-based Practice Center (2003). How useful are unpublished data from the Food and Drug Administration in meta-analysis? *Journal of Clininical Epidemiology*, **56**, 44–51.

Mahoney, M.J. (1977). Publication prejudices: An experimental study of confirmatory bias in the peer review system. *Cognitive Therapy & Research*, **1**, 161–175.

Mallett, S., Hopewell, S. and Clarke, M. (2002). The use of the grey literature in the first 1000 Cochrane reviews. In 4th *Symposium on Systematic Reviews: Abstract no. 5. Pushing the Boundaries*. Oxford, UK.

Manheimer, E. and Anderson, D. (2002). Survey of public information about ongoing clinical trials funded by industry: Evaluation of completeness and accessibility. *British Medical Journal*, **325**, 528–531.

Mann, H. and Djulbegovic, B. (2004) *Biases due to differences in the treatments selected for comparison (comparator bias)*. James Lind Library, Retrieved on January 14, 2004 from www.jameslindlibrary.org.

McAuley, L., Pham, B., Tugwell, P. and Moher, D. (2000). Does the inclusion of grey literature influence estimates of intervention effectiveness reported in meta-analysis? *Lancet*, **356**, 1228–1231.

McCray, A.T. (2001). Better access to information about clinical trials. *Annals of Internal Medicine*, **133**, 609–614.

McManus, R.J., Wilson, S., Delaney, B.C., Fitzmaurice, D.A., Hyde, C.J., Tobias, R.S., Jowett, S. and Hobbs, F.R. (1998). Review of the usefulness of contacting other experts when conducting a literature search for systematic reviews. *British Medical Journal*, **317**, 1562–1563.

Misakian, A.L. and Bero, L.A. (1998). Publication bias and research on passive smoking. Comparison of published and unpublished studies. *Journal of the American Medical Association*, **280**, 250–253.

Moscati, R., Jehle, D., Ellis, D., Fiorello, A. and Landi, M. (1994). Positive-outcome bias: Comparison of emergency medicine and general medicine literatures. *Academic Emergency Medicine*, **1**, 267–271.

Olson, C.M., Rennie, D., Cook, D., Dickersin, K., Flanagin, A., Hogan, J., Zhu, Q., Reiling, J. and Pace, B. (2002). Publication bias in editorial decision making. *Journal of the American Medical Association*, **287**, 2825–2828.

Ottenbacher, K. and Difabio, R. P. (1985). Efficacy of spinal manipulation/mobilization therapy: A meta-analysis. *Spine*, **10**, 833–837.

Petticrew, M., Gilbody, S. and Song, F. (1999). Lost information? The fate of papers presented at the 40th Society for Social Medicine Conference. *Journal of Epidemiology and Community Health*, **53**, 442–443.

Pich, J., Carné, X., Arniaz, J.A., Gómez, B., Trilla, A. and Rodés, J. (2003). Role of a research ethics committee in follow-up and publication of results. *Lancet*, **361**, 1015–1016.

Radford, D.R., Smillie, L., Wilson, R.F. and Grace, A.M. (1999). The criteria used by editors of scientific dental journals in the assessment of manuscripts submitted for publication. *British Dental Journal*, **187**, 376–379.

Ravnskov, U. (1992). Cholesterol lowering trials in coronary heart disease: Frequency of citation and outcome. *British Medical Journal*, **305**, 15–19.

Ravnskov, U. (1995). Quotation bias in reviews of the diet heart idea. *Journal of Clinical Epidemiology*, **48**, 713–719.

Rennie, D. (1997). Thyroid storm. *Journal of the American Medical Association*, **277**, 1238–1243.

Rennie, D. (2002). Fourth International Congress on peer review in biomedical publication. *Journal of the American Medical Association*, **287**, 2759–2760.

Rochon, P.A., Gurwitz, J.H., Simms, R.W., Fortin, P.R., Felson, D.T., Minaker, K.L. and Chalmers, T.C. (1994). A study of manufacturer supported trials of non-steroidal anti-inflammatory drugs in the treatment of arthritis. *Archives of Internal Medicine*, **154**, 157–163.

Savulescu, J., Chalmers, I. and Blunt, J. (1996). Are research ethics committees behaving unethically? Some suggestions for improving performance and accountability. *British Medical Journal*, **313**, 1390–1393.

Scherer, R.W. and Langenberg, P. The Cochrane Library (2005). *Full publication of results initially presented in abstracts. The Cochrane Database of Methodology Reviews* 2005, Issue 2. Art No.: MR000005. DOI: 10.1002/14651858.MR000005.pub2. John Wiley & Sons, Ltd.

Shadish, W.R., Doherty, M. and Montgomery, L.M. (1989). How many studies are in the file drawer? An estimate from the family/marital psychotherapy literature. *Clinical Psychology Review*, **9**, 589–603.

Simes, R.J. (1986). Publication bias: The case for an international registry of clinical trials. *Journal of Clinical Oncology*, **4**, 1529–1541.

Smart, R.G. (1964). The importance of negative results in psychological research. *Canadian Psychologist*, **5**, 225–232.

Smith, M.L. (1980a). Publication bias and meta-analysis. *Evaluating Education*, **4**, 22–24.

Smith, M.L. (1980b). Sex bias in counseling and psychotherapy. *Psychological Bulletin*, **87**, 392–407.

Sommer, B. (1987). The file drawer effect and publication rates in menstrual cycle research. *Psychology of Women Quarterly*, **11**, 233–242.

Sterling, T.D. (1959). Publication decisions and their possible effects on inferences drawn from tests of significance – or vice versa. *Journal of the American Statistical Association*, **54**, 30–34.

Sterling, T.D., Rosenbaum, W.L. and Weinkam, J.J. (1995). Publication decisions revisited: The effect of the outcome of statistical tests on the decision to publish. *American Statistician*, **49**, 108–112.

Stern, J.M. and Simes, R.J. (1997). Publication bias: Evidence of delayed publication in a cohort study of clinical research projects. *British Medical Journal*, **315**, 640–645.

Sterne, J.A.C., Jüni, P., Schulz, K.F., Altman, D.G., Bartlett, C. and Egger, M. (2002). Statistical methods for assessing the influence of study characteristics on treatment effects in 'meta-epidemiological' research. *Statistics in Medicine*, **21**, 1513–1524.

Swaen, G.G., Teggeler, O. and van Amelsvoort, L.G. (2001). False positive outcomes and design characteristics in occupational cancer epidemiology studies. *International Journal of Epidemiology*, **30**, 948–954.

Tannock, I.F. (1996). False-positive results in clinical trials: Multiple significance tests and the problems of unreported comparisons. *Journal of the National Cancer Institute*, **88**, 206–207.

Tierney, J.F., Clarke, M. and Stewart, L.A. (2000). Is there bias in the publication of individual patient data meta-analyses? *International Journal of Technology Assessment in Health Care*, **16**, 657–667.

Timmer, A., Hilsden, R.J., Cole, J., Hailey, D. and Sutherland, L.R. (2002). Publication bias in gastroenterological research – a retrospective cohort study based on abstracts submitted to a scientific meeting. *BMC Medical Research Methodology*, **2**, 7.

Tonks, A. (2002). A clinical trials register for Europe. *British Medical Journal*, **325**, 1314–1315.

Tröhler, U. (2000). *'To improve the evidence of medicine'. The 18th century British origins of a critical approach*. Edinburgh: Royal College of Physicians of Edinburgh.

Tröhler, U. (2003a) *Withering's 1795 appeal for caution when reporting on a new medicine*. James Lind Library. Retrieved on April 7, 2003, from www.jameslindlibrary.org.

Tröhler, U. (2003b) *John Clark 1780 & 1792: Learning from properly kept records*. James Lind Library. Retrieved on April 7, 2003, from www.jameslindlibrary.org.

Vickers, A., Goyal, N., Harland, R. and Rees, R. (1998). Do certain countries produce only positive results? A systematic review of controlled trials. *Controlled Clinical Trials*, **19**, 159–166.

Vogel, U. and Windeler, J. (2000). E in flu ßfaktoren auf die publikation shaiifigkeit klinischer forschung sergebnisse am Beispiel medizinischer Dissertationen. [Factors modifying the frequency of publications of clinical research results exemplified by medical dissertations]. *Deutsche Medizinische Wochenschrift*, **125**, 110–113.

White, K.R. (1982). The relationship between socioeconomic status and academic achievement. *Psychological Bulletin*, **91**, 461–481.

Withering, W. (1785). *An account of the foxglove and some of its medical uses: With practical remarks on dropsy and other diseases*. London: J. and J. Robinson.

Wormald, R., Bloom, J., Evans, J. and Oldfield, K. (1997). Publication bias in eye trials. Paper presented at the Fifth Annual Cochrane Colloquium, Amsterdam.

Yaphe, J., Edman, R., Knishkowy, B. and Herman, J. (2001). The association between funding by commercial interests and study outcome in randomized controlled drug trials. *Family Practice*, **18**, 565–568.

Zimpel, T. and Windeler, J. (2000). Veröffentlichungen von Dissertationen zu unkonventionellen medizinischen Therapie- und Diagnoseverfahren – ein Beitrag zum 'publication bias'. *Forschung zu Komplementärmedizin und Klassischer Naturheilkunde*, **7**, 71–74.

CHAPTER 3

Preventing Publication Bias: Registries and Prospective Meta-Analysis

Jesse A. Berlin
Center for Clinical Epidemiology and Biostatistics,
Department of Biostatistics and Epidemiology,
University of Pennsylvania School of Medicine, USA

Davina Ghersi
NHMRC Clinical Trials Centre,
University of Sydney, Australia

KEY POINTS

- The potential for bias arising from knowledge of study results prior to the conduct of a systematic review may influence the definition of the review question, the criteria for study selection, the treatment and patient groups evaluated, and the outcomes to be measured. This potential for bias argues in favor of prospective identification of trials for inclusion in meta-analyses.
- Prospective registration of trials, prior to their inception, or at least prior to the availability of their results, is one strategy that has been proposed to avoid publication bias. This registration creates an unbiased sampling frame for subsequent meta-analyses. However, registration alone does not guarantee access to data.
- A number of registries of clinical trials exist, most of which are accessible through the internet. However, studies of registries have found them to

Publication Bias in Meta-Analysis – Prevention, Assessment and Adjustments Edited by H.R. Rothstein, A.J. Sutton and M. Borenstein © 2005 John Wiley & Sons, Ltd

be incomplete. The existence of multiple registries and the absence of any unified approach to identifying trials across them make for a potentially bewildering situation for researchers attempting to locate trials in a given area, or for patients seeking access to information about ongoing clinical trials. In addition, registration alone does not guarantee access to data from trials.

- A prospective meta-analysis is a meta-analysis of studies, generally randomized control trials, identified, evaluated, and determined to be eligible for the meta-analysis before the results of any of those trials become known. It is possible, and some would argue advantageous, to plan a prospective meta-analysis so that important elements of study design complement each other across studies.

INTRODUCTION: GENERAL CONCERNS ABOUT SYSTEMATIC REVIEWS

A properly conducted systematic review defines the question to be addressed in advance of the identification of potentially eligible trials. However, systematic reviews are by nature retrospective, as the trials included are usually identified after the results have been reported. Knowledge of results may influence the definition of the review question, the criteria for study selection, the treatments and patient groups evaluated, the outcomes to be measured, and the decision to publish the results of a trial (i.e., the potential for publication bias). While such influences do not necessarily lead to over- or underestimation of intervention effects, the potential for bias arising from these sources makes appealing the idea of prospective identification of trials for inclusion in meta-analyses.

This chapter addresses strategies for avoiding publication bias and related biases through the prospective identification of studies. Although the general principles we discuss apply across a broad range of academic disciplines, most of the methodological work and all of the existing examples with which we are familiar are related to medicine. Thus, medical research will provide the framework for the bulk of this chapter. We do describe some preliminary efforts currently under way in the social sciences and education later in this chapter.

In this chapter, we discuss three levels of prospective identification. As a first step, trial registration is simply the prospective identification of studies in a database accessible to concerned parties. Patients and physicians, seeking information on potential experimental therapies, are potential users of such registries, in addition to researchers interested in identifying trials for inclusion in systematic reviews. At the next strategic level, investigators conducting ongoing trials may, prior to knowing the results of those studies, agree to combine their findings when the trials are complete. This process, known as prospective meta-analysis (PMA), overcomes the issue with registries that mere registration does not guarantee access to the data that might be necessary for a systematic review. Finally, we present the concept of the prospective design of meta-analyses. This involves planning a series of studies at their inception, and can overcome some of the data analytic issues that can arise even in PMAs. We next present each of these approaches in detail.

PREVENTION OF PUBLICATION BIAS THROUGH REGISTRIES

What are registries?

Prospective registration of trials, prior to their inception, or at least prior to the availability of their results, is one strategy that has been proposed to avoid publication bias. This registration creates an unbiased sampling frame for subsequent meta-analyses. However, registration alone does not guarantee access to data.

The case for registries: why register trials?

Registration of trials accomplishes a number of goals. Registries facilitate communication among researchers, and between researchers and consumers, including both physicians and patients who are seeking information on experimental therapies that might be available for particular health conditions. Collaborative research may be stimulated by registries. For example, if a physician is potentially interested in participating in a clinical trial by referring patients as potential research participants. This could have the effect of stimulating recruitment to ongoing trials. By making other researchers aware of ongoing trials, registration can help prevent unnecessary duplication. In fact, a consequence of publication bias, in general, might be the unnecessary duplication of studies finding no effect of an intervention. Thus, the unbiased sampling frame provided by registries could lead to more efficient expenditures of research funding. In the context of this book, registries also enable meta-analysts to identify studies in a particular area of science for inclusion in systematic reviews.

While a number of authors have argued, in principle, in favor of registries (Chalmers *et al.*, 1992), some have provided specific evidence of the benefits of trial registration. Simes (1986, 1987), for example, argues that by restricting attention to registered studies, one can be reasonably assured of avoiding publication bias. The goal of one of his examples was to summarize the evidence from randomized trials comparing two chemotherapy strategies in the treatment of advanced ovarian cancer. He used literature searches to identify 20 randomized trials. Using an international trial registry available at the time, he found an additional six trials. Twelve trials were published but not registered; eight were both published and registered, and, as just mentioned, six were registered but not published. Three of the four statistically significant trials were published but not registered. The unpublished studies were all non-significant. The standard meta-analysis, using only published trials, showed a statistically significant advantage of one therapeutic approach over the other ($p = 0.02$, with a 16% increase in the median survival in the better therapy). When the analysis was restricted to the registered studies, the improvement was only 6% and was no longer statistically significant ($p = 0.24$).

The case for registries has been made by many other authors (Dickersin, 1988, 1992; Horton, 1999; Meinhert, 1988; Moher & Berlin, 1997; Rennie, 1999; Sykes, 1998). Iain Chalmers (1990; see also Savulescu *et al.*, 1996) has proposed registration through institutional review boards or research ethics committees and has called the failure to report research a form of scientific misconduct. Under his proposed scheme, registration would occur at the time the study receives ethical approval.

One other advantage of study registries is that they give us the ability to conduct research on the research process. For example, they provide a framework that allows us to track changes over time in the way that trials have been conducted.

Existing registries

A number of registries of clinical trials exist, most of which are accessible through the internet and seem, at least to our eyes, oriented toward helping patients and physicians find ongoing clinical trials in which to participate. Nevertheless, having a registry available does provide the kind of unbiased sampling frame that would be useful for systematic reviews of medical studies.

A large number of existing registries can be found through links provided by TrialsCentral (http://www.trialscentral.org). One can search for trials by disease and by location at the state level in the USA. One registry to which this site provides links is http://clinicaltrials.gov, which indexes all federally sponsored, and some privately funded, clinical trials being conducted around the USA. That site is maintained by the US National Institutes of Health. Another registry of registries (also linked to TrialsCentral site) is the Current Controlled Trials metaRegister and Links Register (http://www.controlled-trials.com). Based in the UK, this is a searchable international database of clinical trials in all areas of health care, many of which are being conducted outside the USA.

In fact, what the interested researcher or consumer seems to be facing now is not a lack of registries, but a bewildering excess of registries, making it an ongoing challenge for physicians and patients to locate trials in a particular therapeutic area (Shepperd, *et al.*, 1999). TrialsCentral, for example, provides access to over 200 websites of registries (Anderson *et al.*, 2001). The Current Controlled Trials metaRegister links many of the other registries, but because of lack of adequate funding, has begun charging a registration fee, which has apparently caused some groups to decline further participation (K. Dickersin, personal communication, 2002). Perhaps paradoxically, the increased number of registries does not, as we note below in a discussion of problems with registries, imply that comprehensive registration is in place. The issue seems to be lack of coordination of the registration process. This could, in part, be solved by a unique, international identifier, but in any case some way of centralizing these registries would seem to be an important next step, to provide a single source through which to locate ongoing trials.

It is worth noting that clinical trials seem to be more amenable to registration than studies that do not involve interventions of some kind. It may also be true that publication bias may be worse for non-randomized studies. (Easterbrook *et al.*, 1991) We believe the following example illustrates one of the key reasons for this disparity, although there might be many reasons, as well. The Nurses' Health

Study, which is still active and still generating scientific publications, began as a study of over 122 000 nurses, who were to be followed over time to examine associations between various 'exposures' and the risk of various diseases (Belanger *et al.*, 1978). Since the study began, the exposures have included numerous aspects of diet, use of oral contraceptives, use of hormone replacement therapy, and many others. The diseases have included heart disease and many types of cancer. While we think there is inherent value in registering such a study (e.g., as a cohort study of nurses), it would only have been possible to register it as addressing the hypotheses specified at the inception of the study. Thus, newer hypotheses, generated after the study's inception, no longer fit neatly into the concept of prospective registration as it applies to studies evaluating a particular intervention. We suspect that similar considerations affect non-intervention studies in the social sciences and education as well.

The Campbell Collaboration (C2) is an international group of scientists whose mission (http://www.campbellcollaboration.org) is to generate, maintain, and promote access to systematic evidence on the effects of interventions implemented in the USA and abroad (Davies and Boruch, 2001). In particular, its focus is to provide evidence on 'what works' in crime and justice, social welfare, education, and other social and behavioral sectors, based primarily on systematic reviews of research findings generated mostly by experiments, and to a lesser extent based on high-quality quasi-experiments. A related activity is the What Works Clearinghouse, which was established in August 2002 by the US Department of Education. This clearinghouse has similar goals to those of C2, with a narrower focus on systematic reviews of studies on the effects of US education policies, programs, and practices (see http://w-w-c.org). The What Works Clearinghouse is developing four web-based registries, including one on education interventions, all intended to provide systematic reviews of programs, products and practices aimed at improving performance of students. A linked register will provide information on people and organizations that do scientific evaluation studies. A database, known as the Sociological, Psychological, Educational, and Criminological Trials Register (C2-SPECTR) is a resource for both the Campbell Collaboration and the What Works Clearinghouse. Studies are included in C2-SPECTR if they can be classified as a randomized controlled trial; a cluster, place, or group randomized controlled trial; or a possible randomized controlled trial. Turner *et al.* (2003) have provided a full description of the approach to 'populating' the C2-SPECTR, with comments on the importance, and increasing conduct of, randomized trials in the social sciences.

Problems with registries

Does the existence of so many registries mean that all clinical trials are being registered? According to at least one recent paper, the answer is a definitive 'no'. Manheimer and Anderson (2002) conducted a study to see whether information about trials of therapies for prostate and colon cancer was available using the online registries of trials in the USA. The authors used two types of data sources to locate trials. These were industry-based sources about drugs in development (which they termed 'pipeline sources') and publicly available sources of information about ongoing drug trials (which they termed 'online trials registers'). Based on the pipeline sources, they found 12 drugs under development for prostate cancer and 20 drugs under development for colon cancer. Few of the drugs for either cancer were listed in all pipeline sources, and about half were listed in only one source. None of the online trials registers listed trials for all of the 12 prostate cancer drugs or all of the 20 colon cancer drugs that had been identified by the pipeline sources. One site mentioned above, clinicaltrials.gov, was the most comprehensive of the online registers, but still listed only seven of the prostate cancer drugs and 10 of the colon cancer drugs.

Three of the 12 prostate cancer drugs and eight of the 20 colon cancer drugs were not found in any of the online registers. The authors confirmed that trials in the USA had been, or were being, conducted for two of these 'missing' prostate cancer drugs and five of the colon cancer drugs. For the rest, a drug company contact either said explicitly that no eligible trial had yet been conducted (one prostate cancer drug and one colon cancer drug) or the contact could not locate information about any appropriate trials (two colon cancer drugs).

One could argue, though, that progress is being made. For example, in the USA, the Food and Drug Administration Modernization Act of 1997, Section 113 (Public Law 105–115), requires prospective registration of all clinical trials of efficacy conducted in the USA for serious or life-threatening conditions (Manheimer, 2002). In the UK, all trials funded by the National Health Service must be registered (Department of Health, 2002; Manheimer, 2002). It was the requirement by the Food and Drug Administration that prompted the creation of clinicaltrials.gov in the USA (Manheimer, 2002; McCray, 2000). Manheimer and Anderson (2002) note, though, that even that register did not include listings for nearly half of the drugs investigated in their study.

It is important to note, at this point, that a registry could still provide an unbiased sampling frame for meta-analyses whether or not it is comprehensive. Ideally, of course, most meta-analysts would prefer a complete enumeration of studies over an incomplete list. However, strictly from the perspective of preventing publication bias in systematic reviews, or, more particularly, preventing the selective availability of study results based on the direction and magnitude of those results, we would argue that completeness is desirable but not required. Because registration occurs prior to knowledge of the results of a study, a registry cannot, by definition, be affected by selection of studies into the registry based on their results. Publication bias is thereby avoided, regardless of how comprehensive the registry may be. Again, we are not arguing against striving for completeness, but simply stating that bias can still be avoided even with an incomplete registry. From the perspective of a consumer interested in identifying all ongoing trials, completeness of the registration process would still be essential.

Another potential problem is that registration *per se* does not guarantee access to trials. Another approach is for investigators not only to register trials, but also to commit, before knowing the results of their trials, to providing data to be included in a systematic review. This concept is addressed in the remainder of this chapter.

WHAT IS A PROSPECTIVE META-ANALYSIS?

A prospective meta-analysis is a meta-analysis of studies, generally randomized control trials, identified, evaluated, and determined to be eligible for the meta-analysis before the results of any of those trials become known. PMA can help to overcome some of the recognized problems of retrospective meta-analyses by enabling hypotheses to be specified *a priori*, ignorant of the results of individual trials; prospective application of selection criteria for trials; and prospectively defined analysis plans, before the results of trials are known. This avoids potentially biased, data-dependent emphasis on particular subgroups or particular endpoints.

In addition to the inherent lack of bias provided by PMA, this approach offers other advantages. For example, most PMAs will collect and analyze individual patient data. As pointed out in Chapter 14, the use of patient-level, as opposed to group-level, data permits an analysis based on all randomized patients for the relevant outcomes. This approach is generally considered to be the appropriate analytic strategy for randomized trials, but is not always the analysis reported in published trials. Use of patient-level data also permits the exploration of subgroups of patients for whom treatment may be more effective. Such analyses should be motivated by conceptual, theoretical, and/or biological considerations, and specified *a priori*, but are at least feasible using patient-level data but not, generally, using published data (Stewart and Clarke, 1995; Stewart and Parmar, 1993; Berlin *et al.*, 2002). The risk of 'fishing expeditions' in the absence of large numbers and prespecification of analyses is high, and the temptation to publish subgroup findings that show 'striking' effects, either favoring intervention or favoring control, is great. Again, a thorough discussion of the advantages and disadvantages of using patient-level data is presented in Chapter 14.

Standardization of data collection across studies is also facilitated with PMA. For example, the same instruments might be used to measure a particular outcome, such as academic performance, in a PMA of randomized studies of an educational intervention. When a number of scales or endpoints are available to measure the same attribute, participants in a PMA might all agree to use one or two common measures, in addition to other instruments that might be unique to each trial. So, for example, all participants might agree to use scores on a particular standardized test of mathematical performance, while each study might use a variety of additional measures, such as class grades, or one of an array of other standardized tests.

The question arises, in discussions of PMA, as to what makes it different from a multicenter trial. In truth, PMA and multicenter trials share many common features. Both amount to multiple 'studies' of the same intervention or class of interventions. In general, the centers in multicenter studies would follow a common protocol with common entry and exclusion criteria. A central steering committee would make all decisions about the conduct of the study, often with the advice of the investigators at each of the individual sites, but in the end with relatively little autonomy granted explicitly to the individual centers. What distinguishes a PMA from a multicenter trial is that there is no requirement in a PMA for the protocols to be identical across studies. Variety in the design of the studies may be viewed by some as a desirable feature of PMA. The Frailty and Injuries: Cooperative Studies of Intervention

Techniques (FICSIT) is an example of a preplanned meta-analysis. FICSIT included eight clinical sites in separate trials designed to test the effectiveness of several exercise-based interventions in a frail elderly population (Schechtman and Ory, 2001). Patient-level data were available from these studies, which were chosen in advance of their results being available. Importantly, the eight FICSIT sites defined their own interventions using site-specific endpoints and evaluations and differing entry criteria (except for the fact that all participants were elderly). FICSIT also includes a coordinating center, collecting data from all the sites. The extensive database contains data common to all, or at least to several, FICSIT sites. In fact, one might consider the deliberate introduction of *systematic* variability in design, according to a 'meta-experimental design', as a possible approach to PMA (Berlin and Colditz, 1999). We expand on this concept later in this chapter.

The Prospective Pravastatin Pooling Project (PPP) provides a prominent example of a PMA (Simes, 1995). Pravastatin is a member of a class of drugs, known collectively as 'statins', that are used to lower cholesterol levels, with the intention of preventing heart disease. This PMA is comprised of three large-scale trials of pravastatin versus placebo: LIPID, CARE, and WOSCOPS. The objectives of the PMA were: to examine the effects of pravastatin on total mortality and cause-specific mortality; and to examine the effects of pravastatin on total coronary events within specific subpopulations, for example, primary versus secondary prevention, elderly, men versus women, current smokers, etc.

We next present a general discussion of the conduct of a PMA, during which we will return to the PPP example for illustrative purposes.

CONDUCTING A PMA

Developing a protocol for a PMA is the same conceptually as doing so for a single trial. One might view a PMA as a kind of 'study of studies', with 'studies' playing the role of participants. As in any protocol, the first important step is to define the hypotheses. The hypotheses for the PPP study are given above. Again, as with any protocol, a major next step is to establish eligibility criteria for studies. For example, we might require that the enrolled studies use random assignment of participants to interventions. We might specify certain attributes of the participating populations that need to be present, what treatment comparisons need to be made, and what endpoints need to be measured, or any number of other features of study design.

Development of a statistical analysis plan is a next important step. This would generally include sample size or power calculations, a plan for possible interim analyses, and the prespecification of subgroup analyses (which are best if biologically motivated).

One would generally try, in a PMA, to identify *all* ongoing trials, so as to avoid bias that might be introduced by excluding particular studies, although remember that the studies are being enrolled prior to knowledge of their results. To certify a study as a PMA, one might also require a statement confirming that, at the time of submission for registration as part of the PMA, trial results were not known outside the trial's own data monitoring committee. From a practical perspective, it is also helpful to have an explicit (and signed) agreement by each of the trial groups to collaborate. Generally, the idea is to encourage substantive contributions by the individual investigators and to get 'buy-in' to the concept of the PMA and the details of the protocol.

Another practical set of issues revolves around publication policy. For example, we consider it essential to have a policy regarding authorship (e.g., specifying that publication will be in the group name, but also include a list of individual authors). A policy regarding manuscript preparation is also essential. One might specify, for example, that drafts of papers be circulated to all trialists for comment, prior to submission for publication. There might be a writing committee, like those that are often formed within cooperative study groups. A unique issue that arises in the context of the PMA (which would generally not arise for a multicenter study) is whether or not individual studies should publish on their own. Given the demands of most academic careers, our assumption would be that most investigators would definitely want to publish their own studies individually, in addition to contributing to the PMA. We assume that such publication would surely need to happen before the PMA is published, so as to avoid issues related to duplicate publication of the same data. In a similar spirit, though, any PMA publication(s) should clearly indicate the sources of the included data and refer to prior publications of the same data.

PMA can make it possible to have adequate statistical power to study questions that could not be addressed in individual studies, for example, studying uncommon clinical outcomes, safety endpoints, and planned subgroup analyses. The ability to study subgroups, and to specify subgroup analyses in advance, with plans to publish those analyses, is a key step in avoiding publication bias. The selective publication of 'interesting' subgroup findings is avoided by both the increased power and the prespecification of particular subgroup analyses. We next return to our example, to illustrate the various aspects of a PMA.

The design features of the PPP trials are summarized in Table 3.1. Note that the inclusion and exclusion criteria, as well as the endpoints studied, varied across the three trials. For example, LIPID and CARE both included patients with prior heart attacks (prior MI), whereas WOSCOPS was limited to people with no history of a heart attack. LIPID and CARE both included women, whereas WOSCOPS was limited to men. Thus, certain questions, such as effectiveness in women or in patients with prior MI, can only be answered in LIPID and CARE, whereas the question of effectiveness in men can be addressed by all three studies.

Individually these trials do not have adequate power to analyze total mortality, non-cardiovascular mortality, or to perform separate analyses within subgroups of patients. Combined, the three studies include over 19 000 patients, so that the PPP collectively does have the power to examine effects of treatment on total mortality, coronary mortality, cancer incidence, and total coronary events in important subgroups.

To perform power calculations for PPP, the investigators calculated a projected number of events based on the original assumptions of each protocol, the estimated event rates for different categories of endpoint, and the assumption that each trial continues until complete. As an example, for cardiovascular mortality, the PPP

Table 3.1 Design features of the trials included in the PPP.

Feature	LIPID	CARE	WOSCOPS
Sample size	9014	4159	6595
Primary endpoint	CVM[a]	CVM+AMI[b]	CVM+AMI
Prior MI[c] (%)	64	100	0
Women (%)	17	14	none
Age (yr)	31–75	21–75	45–64
Total cholesterol	155–271	<240	N/A
LDL cholesterol	N/A	115–174	155–232

[a] CVM = cardiovascular mortality.
[b] AMI = acute myocardial infarction.
[c] MI = myocardial infarction.

assumed an average reduction in total cholesterol of 18 % and a 1% reduction in coronary artery disease mortality for each 1 % reduction in cholesterol achieved. The power to detect the effect of interest under these assumptions is given in Table 3.2.

Under a similar set of assumptions, one can calculate the power of the PPP to detect effects on total mortality. These calculations are presented in Table 3.3. For these calculations, it was also assumed that pravastatin would have no effect, either beneficial or harmful, on non-cardiovascular mortality. We note here that, despite the increased number of projected events, as compared with cardiovascular mortality alone, the power is reduced relative to the more restricted endpoint, because the benefit is 'washed out' by the lack of effect on non-cardiovascular endpoints. (Note that the reduction in overall mortality is only about 12 %, compared with an 18 % reduction in cardiovascular mortality). Other power calculations are available for a variety of other planned analyses, including non-cardiovascular events (to be examined because of suggestions from prior studies of possible increases in cancer and trauma) and total cardiovascular events (fatal and non-fatal) in subgroups.

There are other examples of potential applications of PMA. A hypothetical example would be a planned analysis of trials of non-steroidal anti-inflammatory

Table 3.2 Statistical power for cardiovascular mortality in the PPP.

Group	No. of patients	Projected events	Power to detect (% reduction)
All patients	19 768	1 100	0.92 (18 %)
Prior MI	9 911	650	0.73 (18 %)

Table 3.3 Statistical power for total mortality in the PPP.

Group	No. of patients	Projected events	Power to detect (% reduction)
All studies	19 768	1 600	0.78 (12 %)
LIPID + CARE	13 173	1 400	0.77 (13 %)

drugs. This class of drugs includes many over-the-counter pain medications, such as ibuprofen and naproxen. Studies of these drugs are most often powered to detect benefits in terms of pain relief, not differences with respect to potential adverse effects. One might consider prospectively planning to combine data from a series of trials designed to examine pain relief, in order to study effects of these drugs on adverse event rates, such as gastrointestinal bleeding. For example, detecting a difference in means between one of these drugs and placebo, on a visual analog scale created to measure pain, of 0.25 standard deviations with 90 % power requires 338 patients per group (already quite a large study for this class of drugs). However, detecting an increase in risk of gastrointestinal bleeding from 4 % to 6 % with 90 % power requires 2500 per group.

The authors of this chapter were involved in the planning of a yet-to-begin prospective meta-analysis of zinc in the adjuvant treatment of diarrhea in infants. Investigators met in Newcastle, Australia, to plan five studies with nearly identical protocols (identical except for the populations being enrolled). The study was being planned in four countries, under the guidance of Dr. Michael Dibley. The main comparison of interest is zinc versus placebo.

Why is a PMA required for this question? Beyond the overall benefits of zinc, there is strong interest in examining benefits in subgroups defined by the underlying cause of diarrhea. While one can power a study, at each one of the sites, to examine the continuous outcome, duration of diarrhea, there is clinical interest in a dichotomous version of the endpoint, specifically, duration longer than 7 days. A third reason for a PMA is an interesting twist in the overall design of the PMA. A third intervention group of interest is zinc combined with copper. The reason for the interest in the third group is that zinc is known to deplete copper in the blood. There are problems involved in this comparison since: copper in the blood is difficult to measure accurately; and nobody is sure of the clinical consequences of depleted copper. Reduced copper levels are thought to be harmful, but may, in fact, be the mechanism by which zinc confers a benefit to the immune system. Individual studies cannot be powered for this third arm given the limited resources available to these particular sites, but a small third treatment arm could be added at each site, providing power for this third group when combined across all the sites.

PMA raises some ethical concerns, among which is whether it is appropriate to continue randomization in 'later' studies after, one might argue, the overall benefit of a therapy has already been proven. When results are not known in the subgroups of clinical interest, or for less common endpoints, one has to be willing to proceed with the study to obtain further information, in the face of existing evidence of the efficacy of an intervention.

DESIGNING A PROSPECTIVE META-ANALYSIS *DE NOVO*: THE 'META-EXPERIMENT'

The PPP involved several ongoing trials and obtained agreement from the individual investigators to provide data when their respective studies were completed. The proposed zinc study is an example of planning a PMA *de novo*, that is, with none of the component studies under way at the time of the design of the PMA. We noted earlier the confounding of certain aspects of study design and study populations in the PPP PMA. If possible, it seems best to try to avoid such confounding of important features of design with 'study' or with each other. For example, the Physicians' Health Study studied the benefits of aspirin in middle-aged male physicians (Steering Committee, 1989). The women's counterpart to that study is using a lower dose of aspirin (Buring and Hennekens, 1992). There might be many reasons why such a reduced dose is more appropriate clinically than the higher dose used in the men's study. Nevertheless, when the women's study is done, if it obtains different results from the all-male physicians' study, we will have no way of knowing whether the effect of aspirin differed because the dose was lower, because the study population was women and not men, or for some other reason related to the many unmeasured nuances that distinguish one study from another. From the statistical perspective, it seems preferable to stratify within study, for example, to conduct studies that enroll both men and women. This could lead to larger sample sizes for early studies (e.g., in conditions that occur much more commonly in men than in women) compared to studies done in more homogeneous high-risk groups.

DESIGNING MULTIPLE PHASE II STUDIES

Phase II studies in medicine are generally small, often (but not always) uncontrolled studies used to show proof of principle about the efficacy of a new drug. Phase I studies are used to establish a safe dose. In Phase II, studies are designed to demonstrate benefits of therapy, often on intermediate endpoints. For example, with so-called 'cytotoxic' cancer therapies (i.e., therapies that are used in the attempt to kill cancer cells), Phase I studies would be used to determine a maximum tolerable dose, on the theory that 'more is better' when it comes to killing cancer cells. Phase II studies would then be used to examine endpoints such as 'response rate', or shrinking of the tumor. Response rate is assumed to be predictive of subsequent survival. In some trials, especially trials of newer drugs, so-called 'cytostatic' drugs that are intended to prohibit further growth of tumors but not necessarily to shrink them, Phase II studies tend to use the progression-free survival interval as an endpoint.

One application of the design of meta-experiments might be to these Phase II studies. As a hypothetical example, we might ask whether the response rate for patients treated with a particular drug differ in patients with a certain genotype. A single Phase II study to address response rate differences would be atypically and probably prohibitively large for the usual Phase II study. We might ask, however, if we could design a series of Phase II studies, each examining a new (but related) question and measuring this genotype. Clearly, we would want to do this in a sensible manner, designing studies that are similar enough to be combined. For example, we might design two studies, one using the drug of interest plus another drug given every three weeks, and another study using the drug of interest plus the same other drug but given every week. We would then have two studies, using the same drugs (albeit using different protocols for the administration of one of the drugs), that could be combined to address the question of whether genotype influences response to the drug of interest.

CONCLUSIONS

We have argued that PMA provides a way of ensuring freedom from publication and other biases (provided, of course, that the component studies are well designed). PMA provides statistical power to examine important questions about uncommon events (including safety) and subgroups. The ability to study subgroups is not necessarily unique to PMA, as any meta-analysis using individual patient data could examine subgroups. However, the *a priori* definition of subgroups in a PMA offers the advantage of avoiding the bias involved in the *post hoc* definition of subgroups often employed. Because new questions may be asked at the level of the meta-analysis, PMA may involve collecting data that would otherwise not have been collected (e.g., genotypes, in our Phase II studies). We have, at the same time, suggested that PMA raises ethical concerns about continued randomization in the face of existing evidence of efficacy. We also propose that designed 'meta-experiments' may provide additional structure to PMA that may permit unconfounded comparisons of interest to be made. Controlling the variability among studies while still allowing autonomy to individual investigators becomes a new challenge potentially imposed by PMA.

For those interested, there is a PMA register maintained by a PMA methods group. Access is through the Cochrane Library (if one is registered through a Cochrane Review Group). The PMA website can be found at http://www.cochrane.de/cochrane/pma.htm.

REFERENCES

Anderson, D., Costa, I. and Dickersin, K. (2001). Building Trials Central, an online register of clinical trials registers. *Controlled Clinical Trials*, **22**, 40–41S.

Belanger, C., Hennekens, C.H., Rosner, B. and Speizer, F.E. (1978). The nurses' health study. *American Journal of Nursing*, **78**, 1039–1040.

Berlin, J. and Colditz, G. (1999). The role of meta-analysis in the regulatory process for foods, drugs, and devices. *Journal of the American Medical Association*, **281**, 830–834.

Berlin, J., Santanna, J., Schmid, C., Szczech, M. and Feldman, H. Anti-lymphocyte Antibody Induction Therapy Study Group (2002). Individual patient versus group-level data meta-regressions for the investigation of treatment effect modifiers: Ecological bias rears its ugly head. *Statistics in Medicine*, **21**, 371–387.

Buring, J.E. and Hennekens, C.H. (1992). The Women's Health Study Research Group. The Women's Health Study: Summary of the study design. *Myocardial Ischemia*, **4**, 27–29.

Chalmers, I. (1990). Underreporting research is scientific misconduct. *Journal of the American Medical Association*, **263**, 1405–1408.

Chalmers, I., Dickersin, K. and Chalmers, T.C. (1992). Getting to grips with Archie Cochrane's agenda. *British Medical Journal*, **305**, 786–788.

Davies, P. and Boruch, R. (2001) The Campbell Collaboration does for public policy what Cochrane does for health. *British Medical Journal*, **323**, 294–295.

Department of Health (2002) Research governance: Science. http://www.dh.gov.uk/Policy AndGuidance/ResearchAndDevelopment/ResearchAndDevelopmentAZ/Research Governance/ResearchGovernanceArticle/fs/en?CONTENT_ID = 4002130&chk = pebh9u (accessed 16 May 2005).

Dickersin, K. (1988). Report from the panel on the Case for Registers of Clinical Trials at the Eighth Annual Meeting of the Society for Clinical Trials. *Controlled Clinical Trials*, **9**, 76–81.

Dickersin, K. (1992). Why register clinical trials? – Revisited. *Controlled Clinical Trials*, **13**, 170–177.

Easterbrook, P.J., Berlin, J.A., Gopalan, R. and Matthews, D.R. (1991). Publication bias in clinical research. *Lancet*, **337**, 867–872.

Horton, R.S.R. (1999). Time to register randomized trials. The case is now unanswerable. *British Medical Journal*, **319**, 865–866.

Manheimer, E. and Anderson, D. (2002). Survey of public information about ongoing clinical trials funded by industry: Evaluation of completeness and accessibility. *British Medical Journal*, **325**(7363), 528–531.

McCray, A. (2000). Better access to information about clinical trials. *Annals of Internal Medicine*, **133**, 609–614.

Meinert, C.L. (1988). Toward prospective registration of clinical trials. *Controlled Clinical Trials*, **9**, 1–5.

Moher, D. and Berlin, J. (1997). Improving the reporting of randomized controlled trials. In A. Maynard and Chalmers (eds), *Non-random Reflections on Health Services Research* (pp. 250–271). London: BMJ Publishing Group.

Rennie, D. (1999). Fair conduct and fair reporting of clinical trials. *Journal of the American Medical Association*, **282**, 1766–1768.

Savulescu, J., Chalmers, I. and Blunt, J. (1996). Are research ethics committees behaving unethically? Some suggestions for improving performance and accountability. *British Medical Journal*, **313**, 1390–1393.

Schechtman, K. and Ory, M. (2001). The effects of exercise on the quality of life of frail older adults: A preplanned meta-analysis of the FICSIT trials. *Annals of Behavioral Medicine*, **23**, 186–197.

Shepperd, S., Charnock, D. and Gann, B. (1999). Helping patients access high quality health information. *British Medical Journal*, **319**, 764–766.

Simes, R.J. (1986). Publication bias: The case for an international registry of clinical trials. *Journal of Clinical Oncology*, **4**, 1529–1541.

Simes, R.J. (1987). Confronting publication bias: A cohort design for meta-analysis. *Statistics in Medicine*, **6**, 11–29.

Simes, R.J. (1995). Prospective meta-analysis of cholesterol-lowering studies: The Prospective Pravastatin Pooling (PPP) Project and the Cholesterol Treatment Trialists (CTT) Collaboration. *American Journal of Cardiology*, **76**, 122C–126C.

Steering Committee of the Physicians' Health Study Group (1989). Final report on the aspirin component of the ongoing Physicians' Health Study. *New England Journal of Medicine*, **321**, 129–135.

Stewart, L.A. & Clarke, M.J. (1995). Practical methodology of meta-analyses (overviews) using updated individual patient data. *Statistics in Medicine*, **14**, 2057–2079.

Stewart, L.A. and Parmar, M.K. (1993). Meta-analysis of the literature or of individual patient data: Is there a difference? *Lancet*, **341**, 418–422.

Sykes, R. (1998). Being a modern pharmaceutical company involves making information available on clinical trial programs. *British Medical Journal*, **317**, 1172.

Turner, H., Boruch, R., Petrosino, A., de Moya, D., Lavenberg J. and Rothstein, H.R. (2003). Populating an international register of randomized trials. *Annals of the American Academy of Political and Social Sciences*, **589**, 203–225.

CHAPTER 4

Grey Literature and Systematic Reviews

Sally Hopewell, Mike Clarke and Sue Mallett

UK Cochrane Centre, Oxford, UK

KEY POINTS

- The validity of a systematic review is highly dependent on the results of the underlying data. The aim of a good literature search is to generate as comprehensive a list as possible of studies that might be suitable for answering the questions posed in the systematic review.
- The search for, and inclusion of, grey literature in a systematic review is an important way to help overcome some of the problems of publication bias.
- Grey literature is defined as 'that which is produced on all levels of government, academia, business and industry in print and electronic formats, but which is not controlled by commercial publishers'.
- Evidence in health and social sciences research shows that there is a systematic difference between the results of published studies and those found in the grey literature.
- Identifying relevant studies in the grey literature and including them in a systematic review can be particularly time-consuming and difficult, and their methodological quality can sometimes be difficult to assess.
- The optimal extent of the search is very much dependent on the subject matter and the resources available for the search. It is important to document in the review what has been searched so that the reader can assess the validity of its conclusions based on the search that has been conducted.

Publication Bias in Meta-Analysis – Prevention, Assessment and Adjustments   Edited by H.R. Rothstein, A.J. Sutton and M. Borenstein   © 2005 John Wiley & Sons, Ltd

No librarian who takes his job seriously can today deny that careful attention has also to be paid to the 'little literature' and the numerous publications not available in normal bookshops, if one hopes to avoid seriously damaging science by neglecting these.

(Minde-Pouet, 1920, cited by Schmidmaier, 1986)

INTRODUCTION

Health-care providers, researchers and policy-makers are inundated with vast and unmanageable amounts of information (Mulrow, 1994). This information is constantly changing and expanding and, in many areas, it would be impossible for individuals to read, critically evaluate and synthesize current knowledge, let alone keep this synthesis up to date (Egger *et al.*, 2001). The rationale for systematic reviews is firmly grounded in the need to identify and refine these large amounts of otherwise unmanageable information into a usable format and to do this in ways to minimize bias (Mulrow, 1994). Systematic reviews, therefore, need to employ explicit methods to identify, select and critically appraise the relevant research and to collect and analyse data from studies that are included in the review (Clarke and Oxman, 2003).

The validity of the results of a systematic review is highly dependent on the results of the underlying data and as such requires the identification of as unbiased and complete a set of relevant studies as possible. The identification of relevant studies for inclusion in a systematic review is therefore an important part of the systematic review process. This chapter starts by describing why a comprehensive search is important to help minimize bias and describes different ways in which bias can be introduced into a systematic review during the search process and how this can best be avoided. It also describes the searching done for three example reviews that provide a common thread throughout this book. The second part of the chapter focuses on how the inclusion of studies found in the grey literature can help overcome some of the problems of publication bias. The current use of grey literature in systematic reviews is described and its effect on the overall results of a systematic review is examined. We also explore some of the problems of searching for studies found in the grey literature and the difficulties, faced by those conducting systematic reviews, of including these studies in a review once they have been identified. Finally, we will make some recommendations on how a comprehensive search, which aims to minimize publication bias, should best be conducted. The chapter draws heavily on work in health care, but many of the points made and lessons learned will be relevant in social care. Where there are important differences, these are highlighted.

THE IMPORTANCE OF MINIMISING BIAS THROUGH A COMPREHENSIVE SEARCH

The identification of relevant studies through an unbiased search is a crucial part of the systematic review process. The aim of a good search in a systematic review is to generate as comprehensive a list as possible of studies, both published and unpublished, which might be suitable for answering the questions posed in the

Table 4.1   Sources of bias affecting the search process.

Type of bias	Definition
Publication bias	Studies with statistically significant results are more likely to be published than those with statistically non-significant or null results.
Time-lag bias	Studies with statistically significant results are more likely to be stopped earlier than originally planned and published quicker.
Language bias	Studies with statistically significant results are more likely to be published in English.
Duplication bias	Studies with statistically significant results are more likely to be published more than once.
Citation bias	Studies with statistically significant results are more likely to be cited by others.

review. However, there are a number of different ways in which bias can be introduced into a systematic review during the search process. As has been discussed by Dickersin (Chapter 2, this volume), there is now considerable evidence to show that failure to publish is strongly linked to the significance of the results of the research (Dickersin, 1997). Studies with statistically significant results are more likely to be published than those with non-statistically significant or null results (Scherer and Langenberg, 2003). Studies with statistically significant results are also more likely to be stopped earlier than originally planned and published quicker (Hopewell *et al.*, 2003a), more likely to be published in English (Egger *et al.*, 1997), more likely to be published more than once and more likely to be cited by others (Gotzsche, 1987); see Table 4.1. Much of this evidence comes from health-care research and specifically the conduct and reporting of randomized controlled trials. Evidence in the area of education and the social sciences is somewhat limited and is still very much in its infancy.

Those carrying out systematic reviews need to ensure they conduct as comprehensive a search as possible to help avoid introducing bias into their review. It is important that the authors of the review do not restrict their searches to those studies published in the English language, as evidence suggests that trials published in languages other than English may have different results than those published in English (Egger *et al.*, 1997; Jüni *et al.*, 2002). It is also important not to rely solely on the references cited by key relevant studies, as the more a study is cited the easier it will be to identify it and the more likely it is to be included in a systematic review (Gotzsche, 1987).

ATTEMPTING TO AVOID BIAS: HOW A SEARCH MIGHT PROCEED

Searching the *Cochrane Central Register of Controlled Trials* and C2-SPECTR

A search for relevant studies usually begins with the searching of electronic bibliographic databases. The Cochrane Collaboration, an international organization that helps people make informed decisions about health-care interventions, has done a considerable amount of work to identify reports of randomized trials and thereby minimize the risk of introducing bias into a systematic review. The *Cochrane Central Register of Controlled Trials* (CENTRAL), published as part of the output of the Cochrane Collaboration, is now the most comprehensive source of records relating to controlled trials of health-care interventions (Dickersin *et al.*, 2002). As of January 2004, this register contains over 400 000 citations of reports of randomized trials and other studies potentially relevant for inclusion in systematic reviews of health-care interventions.

CENTRAL includes reports of randomized trials that are not indexed in electronic databases such as MEDLINE or EMBASE, as well as those that are; it includes citations published in languages other than English, citations published only in conference proceedings and citations from other sources which are difficult to locate (Dickersin *et al.*, 2002). Over 2200 journals as well as hundreds of conference proceedings have been, or are being, handsearched (i.e., read from cover to cover) within The Cochrane Collaboration to identify reports of controlled trials to be included in CENTRAL (http://www.cochrane.org/Cochrane/hsearch.htm). In addition, a highly sensitive search strategy has been developed to identify reports of controlled trials in MEDLINE so that these records can also be included in CENTRAL (Dickersin *et al.*, 1994). A similar project is currently under way to identify reports of randomized trials published in EMBASE and to include these in CENTRAL. As of December 2003, nearly 70 000 relevant reports have been identified in EMBASE and submitted to CENTRAL (*The Cochrane Library*, 2004).

The Campbell Collaboration is developing a similar register of reports of randomized trials of social, psychological, educational and criminological interventions. As of January 2004 the register, which is called C2-SPECTR, contains approximately 12 000 reports of randomized or possibly randomized trials and serves as an important resource for identifying studies for inclusion in the Campbell Collaboration's systematic reviews. Methods similar to those used by The Cochrane Collaboration have been deployed to identify trials, including searching bibliographic databases such as the Educational Resources Information Center (ERIC), PsycINFO, Sociological Abstracts and Criminal Justice Abstracts and the handsearching of journals and conference proceedings (Petrosino *et al.*, 2000).

Searching electronic databases and the merits of handsearching

Searching electronic bibliographic databases such as MEDLINE, EMBASE, ERIC and PsycINFO is an important means of identifying studies for possible inclusion in a systematic review. However, it is important not to restrict a search solely to electronic databases as this also has the potential to introduce bias into a review. A recent systematic review of studies compared the results of handsearching with those of searching electronic databases such as MEDLINE and EMBASE to identify reports of randomized trials. Handsearching identified 92–100 % of the total number of reports of randomized trials found. The Cochrane Highly Sensitive Search Strategy, which is used to identify reports of randomized trials in MEDLINE (Dickersin *et al.*, 1994), identified 80 % of the total number of reports of trials found. Electronic searches categorized as 'complex' found 65 % and those categorized as 'simple' found 42 %. The retrieval by an electronic search was higher when the search was restricted to English-language journals: 62 % versus 39 % for journals published in other languages. A major reason for the failure of the electronic searches to identify all the reports of randomized trials was that some reports were published as abstracts and/or in journal supplements, which were not routinely indexed by the electronic databases. When the search was restricted to full reports, the retrieval for complex searches improved to 82 %. Another reason for failure of the electronic searches to identify all the reports of randomized trials was that some reports were published before 1991 when the National Library of Medicine's system for indexing trial reports was not as well developed as it is today (Hopewell *et al.*, 2003b). Turner *et al.* (2004) also have conducted research comparing the ability of handsearches versus searches of electronic databases to identify randomized controlled trials in 12 education and education-related journals. Their results indicate that electronic searches, on average, identified only one-third of the trials identified by handsearching.

Searching conference proceedings

Conference proceedings are also an important source for identifying studies for inclusion in systematic reviews. These abstracts are not usually included in electronic databases, and relevant studies are only identified by searching the proceedings of the conference. There is considerable evidence to show that failure to publish in full a study reported at a conference is strongly linked to the significance of its results. A recent systematic review of 79 separate assessments of this found that 44.5 % of research studies presented as abstracts at scientific conferences were published in full. This figure was higher (63.1 %), for those abstracts which only presented the results of randomized trials. Most studies, if they were published at all, were published in full within three years of presentation at the meeting (Scherer *et al.*, 2005). This means that for a large proportion of trials and their results, the only available source of information might be an abstract published in the proceedings of a conference. These findings are also consistent in areas of research other than health care. For example, a study of conference abstracts for research into the methods of health-care research found that approximately half of the abstracts had not been or were unlikely ever to be published in full (Hopewell and Clarke, 2001). The most common reason given by investigators as to why research submitted to scientific meetings is not subsequently published in full was lack of time. Other reasons given included that the researchers thought that a journal was unlikely to accept their study, or that the authors themselves perceived that the results were not important enough (Callaham *et al.*, 1998; Weber *et al.*, 1998; Donaldson & Cresswell, 1996).

Contact with researchers

Some studies will never be published in any form and, given the evidence that unpublished studies are likely to have different results than published studies, it is important to identify and include them in the review process to minimize bias

(Dickersin, 1997). Contacting researchers, subject specialists and, in the case of health care, the pharmaceutical industry, can be an important means of identifying ongoing or unpublished studies. This is, however, dependent on the goodwill of the researchers and the time they have available to respond to these requests. Evidence suggests that requests for additional information about a study are not always successful (http://www.mrc-bsu.cam.ac.uk/firstcontact/pubs.html). To overcome this problem, some pharmaceutical companies are making records of their ongoing and completed trials publicly available, either through the metaRegister of Controlled Trials (http://www.controlled-trials.com) or by providing registers of research on their own website. For example, an initiative by Glaxo Wellcome has led to the development of a register of Phase II, III and IV studies on newly registered medicines (http://ctr.glaxowellcome.co.uk). However, it is important to note that the knowledge of a possible study does not necessarily mean that the data required for a review will be made available (see Chapter 3).

Searching research registers

Research registers are also an important way of identifying ongoing research, some of which may never be formally published. There are many national registers such as the National Research Register in the UK, which contains ongoing health research of interest to, and funded by, the UK National Health Service (http://www.doh.gov.uk/research/nrr.htm). There are also numerous local and specialist registers each of which is specific to its own area of interest. Identifying ongoing studies is important so that whenever a review is updated, these studies can be assessed for possible inclusion. The identification of ongoing studies also helps the reviewer to assess the possibility of time-lag bias.

Searching the Internet

The Internet may be another important source of information about ongoing and completed research, particularly that which has not been formally published. However, searching the Internet can be a major undertaking. Many of the general search engines do not allow sophisticated searching and may identify thousands of web sites to check. Few studies have been conducted to establish whether searches on the World Wide Web are a useful means of identifying additional unpublished and ongoing trials. One such study is by Eysenbach *et al.* (2001) who adapted search strategies from seven systematic reviews in an attempt to find additional reports of trials. They reviewed 429 web pages in 21 hours and found information about 14 unpublished, ongoing or recently published trials. At least nine of these were considered relevant for inclusion in four of the seven systematic reviews.

Taking account of differences in searching between the health and social sciences

The potential for introducing publication bias during the search process, when conducting a systematic review, may be even greater in areas outside health care, such as the social sciences. However, the empirical evidence in this area is limited and is still very much in its infancy. The medical literature is predominantly published in peer-reviewed journals, some of which are accessible through large, sophisticated, bibliographic databases such as MEDLINE and EMBASE. In comparison, the social sciences literature is much more diverse. Grayson and Gomersall (2003) propose that one of the major differences between health and the social sciences is in the publication of research. Social science research is published in a much broader set of peer-reviewed journals; additionally, a larger proportion of social sciences research is exclusively published in books, governmental and organizational reports, and other sources that are not as widely available as are peer-reviewed journals. Some research is indexed in bibliographic databases such as Sociological Abstracts, the Social Sciences Citation Index and the Applied Social Sciences Index and Abstracts (ASSIA) database, but the coverage and quality of these databases vary greatly. There is a lack of specific indexing terms available in many of these databases to identify research studies, and none of the social science databases indexes by methodology.

Whatever is searched, it is important that the authors of the systematic review are explicit and document in the review exactly which sources have been searched and which strategies they have used so that the reader can assess the validity of the conclusions of the review based on the search that has been conducted. The studies described in Box 4.1 are those of the three example systematic reviews that provide a common thread throughout this book (see Appendix A for more details on these illustrative examples). In each of these three studies the search process was documented but was not extensive as the authors relied heavily on searching a single source such as MEDLINE, ERIC or a US-based study register. Each of these studies also relied heavily on the citations of key studies and those of other colleagues. It is also not clear when the searches were conducted and, therefore, how up to date the findings are. While these studies do include studies found in the published and unpublished literature, the relatively limited scope of their searches may mean that the conclusions of these reviews are biased.

Box 4.1   Three examples of searches carried out in systematic reviews

Passive smoking data set (Hackshaw *et al.*, 1997)
The aim of this study was to estimate the risk of lung cancer in lifelong non-smokers exposed to environmental tobacco smoke. Thirty-seven epidemiological studies of environmental tobacco smoke and lung cancer were identified from searching MEDLINE, from the citations in each study and from consultation with colleagues. The 37 studies included 27 studies published as journal articles, four books, two doctoral theses, three studies published in conference proceedings and one study which was an official report from a scientific organization.

Box 4.1 (Continued)

Interview validities data set (McDaniel et al., 1994)

This study was a systematic review investigating the validity of the employment interview. Studies were identified by searching a database of studies collected by the US Office of Personnel Management and by examining the references from five published literature reviews. One hundred and forty studies were identified: 60 were published journal articles, 23 were theses, 6 were published in conference proceedings, 30 were governmental or organizational reports, 3 were book chapters, 8 were personal communication and 10 were unpublished manuscripts. Two of the 60 journal articles were published in languages other than English.

Teacher expectancy data set (Raudenbush, 1984)

This study was a meta-analysis assessing the effects of teacher expectancy on pupil IQ. Eighteen relevant studies were identified by searching the reference lists of four published literature reviews and by searching ERIC. These 18 studies included six studies published as journal articles, eight theses, one conference abstract, two book chapters and one study which was referenced by a published journal article and a thesis.

INCLUDING STUDIES FOUND IN THE GREY LITERATURE

Many of the ways to help minimize bias in a systematic review, described above, are examples of searching the grey literature. The search for, and inclusion of, grey literature should therefore form an important part of the systematic review process. The definition of what constitutes grey literature varies, however, and the terminology can often be confusing (Van Loo, 1985; Alberani *et al.*, 1990; Cook *et al.*, 1993; McAuley *et al.*, 2000; Song *et al.*, 2000). Grey literature is generally assumed to include literature that has not been formally published, has limited distribution or is not available through conventional channels (Auger, 1998). Examples of grey literature include conference abstracts, research reports, book chapters, unpublished data, dissertations, policy documents and personal correspondence.

The emergence of the idea of grey literature is not new, and the concept of grey literature is beautifully illustrated in the quote by Minde-Pouet in 1920 with which this chapter began. Acceptance of the term 'grey literature' dates back to 1978. A seminar in York in the UK, was recognized as a milestone for grey literature as a primary information source and resulted in the creation of the System for Information on Grey Literature in Europe (SIGLE) database which is now managed by the European Association for Grey Literature Exploitation (EAGLE). In November 1997, the term grey literature was redefined at the Third International Conference on Grey Literature (Auger, 1998), the so known as the 'Luxembourg Convention' after the city in which the conference was held: grey literature is 'that which is produced on all levels of government, academia, business and industry in print and electronic formats, but which is not controlled by commercial publishers'.

Sources of grey literature

There are a number of electronic resources available which are relatively easily accessible and which may prove useful in searching for grey literature. Four key types of these electronic resources are: grey literature databases, which only include studies found in the grey literature; general bibliographic databases, which contain studies published in the grey and conventional literature; specific bibliographic databases, which contain studies published in the grey and conventional literature in a specific subject area; and ongoing trial registers that contain information about planned and ongoing studies. Examples of some key sources in each of these categories are given in Box 4.2 (see also Chapter 3 for more information).

Box 4.2   Where to begin to search for grey literature – some key electronic sources in the biomedical and social sciences

The Cochrane Central Register of Controlled Trials (CENTRAL) – includes references to trial reports in conference abstracts, handsearched journals, and other resources (http://www.cochrane.org).

The Campbell Collaboration Social, Psychological, Educational, and Criminological Trials Register (C2-SPECTR and C-PROT) – includes references to trials reported in abstracts, handsearched journals and ongoing trials (http://www.campbellcollaboration.org/).

Grey literature databases
BiomedCentral – contains meetings abstracts and protocols of ongoing trials (http://www.biomedcentral.com/).

British Library's Inside Web – includes references to papers presented at over 100 000 conference proceedings (http://www.bl.uk/services/current/inside.html).

Dissertation Abstracts Online – includes US dissertations since 1861 and British dissertations since 1988 (http://www.lib.umi.com/dissertations/).

Economic & Social Data Service, UK Data Archive – contains an extensive range of key economic and social data, both quantitative and qualitative (http://www.esds.ac.uk/).

Meeting Abstracts – includes meeting abstracts relevant to health care and health technology assessment (http://gateway.nlm.nih.gov/gw/Cmd).

Regard (Economic & Social Research Council UK) – includes descriptions of ESRC-funded research projects in the social sciences and links to researchers' websites and publications available online (http://www.regard.ac.uk/regard/about/).

Box 4.2 (Continued)

System for Information on Grey Literature in Europe (SIGLE) – includes research and technical reports, preprints, working papers, conference papers, dissertations and government reports (http://www.ovid.com/site/products/fieldguide/sigl/About_SIGLE.jsp).

Databases which contain grey literature

Biological Abstracts – indexes papers presented at over 1500 meetings, symposia and workshops worldwide (http://www.biosis.org/products_services/ba.html).

Science Citation Index – indexes meeting abstracts not covered in MEDLINE or EMBASE relevant to sciences and technology (http://www.isinet.com/isi/products/citation/sci).

Sociological Abstracts – covers social and behavioural sciences and includes dissertations, conference proceedings, book chapters from 1963 to date (http://www.csa.com/csa/factsheets/socioabs.shtml).

Social Sciences Citation Index – includes conference abstracts relevant to the social sciences from 1956 to date (http://www.isinet.com/isi/products/citation/sci).

Social Services Abstracts & InfoNet – includes dissertations in social work, social welfare, social policy and community development research from 1980 to date (http://www.csa.com/csa/factsheets/socserv.shtml).

Subject specific databases which contain grey literature

Caredata – includes government reports, research papers in social work and social care literature (http://www.elsc.org.uk/caredata/caredata.htm).

CINAHL – includes nursing theses, government reports, guidelines, newsletters, and other research relevant to nursing and allied health (http://www.cinahl.com/).

ERIC (Education Resources Information Center) – includes report literature, dissertations, conference proceedings, research analyses, translations of research reports relevant to education and sociology (http://www.askeric.org/Eric/).

PAIS (Public Affairs Information Service) – includes report literature, government publications, Internet resources and other grey literature sources from 1972 to date (http://www.pais.org/).

PsycINFO – includes dissertations, book chapters and academic/government reports relevant to psychology and sociology (http://www.apa.org/psycinfo/).

Ongoing trials registers

BiomedCentral - publishes trials together with their International Standard Randomized Controlled Trial Number (ISRCTN) and publishes protocols of trials (http://www.biomedcentral.com/clinicaltrials/).

ClincalTrials.gov – covers trials in the USA of life-threatening diseases (http://clinicaltrials.gov).

Current Controlled Trials metaRegister of Controlled Trials – contains records from multiple trial registers including the UK National Research Register, the Medical Editors' Trials Amnesty, the UK MRC Trials Register, and links to other ongoing trials registers (http://www.controlled-trials.com).

Database of Promoting Health Effectiveness Reviews (DoPHER) – produced by the EPPI-Centre and includes completed as well as unpublished randomized and non-randomized trials in health promotion (http://eppi.ioe.ac.uk/EPPIWeb/home.aspx).

The Lancet (Protocol Reviews) – contains protocols of randomized trials (http://www.thelancet.com).

TrialsCentral – over 200 US-based registers of trials (http://www.trialscentral.org).

For more extensive lists of potential sources, the following websites have very useful pointers:

UK National Health Service Centre for Reviews and Dissemination - finding studies for systematic reviews: a basic checklist for researchers (http://www.york.ac.uk/inst/crd/revs.htm).

Health Technology Assessment – databases and research registers (http://www.york.ac.uk/inst/crd/htadbase.htm).

School of Health and Related Research (ScHARR) UK – netting the evidence – evidence-based resources in health care (http://www.shef.ac.uk/~scharr/ir/netting/).

Trawling the net – free databases of interest to health-care professionals (http://www.shef.ac.uk/~scharr/ir/trawling.html).

Information Gateways and Virtual Libraries – BIOME – Internet resources in health and life sciences (http://biome.ac.uk/).

OMNI – Internet resources in health and medicine (http://omni.ac.uk/).

NeLH – National electronic Library for Health (http://www.nelh.nhs.uk/).

SOSIG - Social Sciences Information Gateway – Internet resources in the social sciences (http://www.sosig.ac.uk/).

Box 4.2 **(Continued)**

Evidence Network – Economic & Social Research Council's UK Centre for Evidence-Based Policy and Practice Research – database resources available in the social sciences (http://www.evidencenetwork.org/cgi-win/enet.exe/resourcesmain).

Electronic Library for Social Care – Social Care Institute for Excellence (http://www.elsc.org.uk/socialcareresources.htm).
Centre for Evidence-based Social Services – critically appraised research resources (http://www.be-evidence-based.com/).

Use of grey literature in systematic reviews

One of the criteria that distinguishes systematic reviews from traditional narrative reviews is the comprehensive search for all relevant studies in both the published and grey literature. Estimates of the use and type of grey literature in systematic reviews vary. Clarke and Clarke (2000) conducted a study of the references included in Cochrane protocols and reviews published in *The Cochrane Library* in 1999. Of the references to studies included in the reviews, 91.7 % were references to journal articles, 3.7 % were conference proceedings that had not been published in journals, 1.8 % were unpublished material (e.g., personal communications, in press documents and data on file), and 1.4 % were books and book chapters.

Mallett *et al.* (2000) looked at the sources of grey literature included in the first 1000 Cochrane systematic reviews published in Issue 1 (2001) of *The Cochrane Library*. Almost all of these reviews (n = 998) were analysed; 12 reviews were excluded, as the text of the review had been withdrawn from publication. All sources of information other than journal publications were counted as grey literature sources and were identified from the reference lists and details provided within the review. The grey information that was included ranged from details of study design, which helped to determine its eligibility for the review, to unpublished patient data included in the review analysis. The 988 reviews contained 9723 studies in the 'included studies' section of the review. Fifty-one per cent (*n* = 513) of Cochrane reviews included trials with information from the grey literature. Table 4.2 shows the number of trial citations for different sources of grey information. The majority (55 %) of the grey literature was from unpublished information supplementing published journal articles reporting trials. A random sample of 94 of these Cochrane reviews was examined in detail to determine the nature of this unpublished information, and it was found that 80 % of the reviews listed the trial authors as the source of the unpublished information, with the remaining 20 % of reviews not indicating the source of the unpublished information. The high proportion of unpublished information in Cochrane reviews contrasts with other studies where the major source of grey information was conference abstracts (Egger *et al.*, 2003; McAuley *et al.*, 2000). This is possibly because Cochrane reviewers are specifically encouraged to search for and record unpublished information.

Table 4.2 Citations for grey literature in 513 Cochrane reviews (a total of 6266 trials).

Grey literature source	Number of trial references[a]
Unpublished information	1259 (55 %)
Conference abstracts	805 (35 %)
Government reports	78 (4 %)
Company reports	66 (3%)
Theses/dissertations	63 (3 %)
Total grey citations (in 1446 trials)	2271 (100 %)

Source: Mallett *et al.* (2000).
[a] Trials can be referenced by more than one grey literature source; 4820/6266 trials were referenced only by published journal articles. Seventeen trials had missing citations.

Within reviews containing grey literature, studies containing some form of grey literature contributed a median of 28 % of the included studies (interquartile range 13% to 50%). In 40 reviews more than 95% of studies contained information from the grey literature and in 67 reviews more than 75 % of studies contained information from the grey literature.

Nine per cent of the trials (588/6266) included in the 513 Cochrane reviews using grey literature were referenced only by grey literature sources and not by journal articles. These trials were analysed to determine how many of them would be identified if different grey literature sources were searched. The trials were categorized by a primary reference assigned according to the following hierarchy going from the least to the most grey: conference abstracts, government reports, company reports, theses/dissertations, unpublished source. Figure 4.1 shows that 68 % of trials were referenced by conference abstracts, with primary references for 8 % being government reports, 7 % being company reports and 4 % being theses or dissertations. The remaining 19 % of trials were referenced from unpublished reference sources alone (e.g., personal communication with trialists, in press documents and data on file). A previous analysis of 102 trials referenced from grey literature sources in 38 meta-analyses found 62 % referenced by conference abstracts and 17 % from unpublished sources, with similar proportions to the Mallett *et al.* study coming from government and company reports and theses (McAuley *et al.*, 2000). A study of 153 trials in 60 meta-analyses by Egger *et al.* (2003) found 45 % from conference abstracts, 14 % from book chapters, 3 % from theses and 37 % from other grey literature sources including unpublished, company and government reports. Overall, the results are similar across the studies. A lower overall percentage of trials referenced only from grey literature sources was found in the Mallett *et al.* study: 9 % compared to 22 % in the study by McAuley *et al.* (2000) and 20 % in the study by Egger *et al.* (2003). This could be due to different inclusion criteria for meta-analyses within the three studies and also to The Cochrane Collaboration policy of encouraging reviewers to regularly update reviews, which might lead to grey literature references being replaced as published journal articles become available.

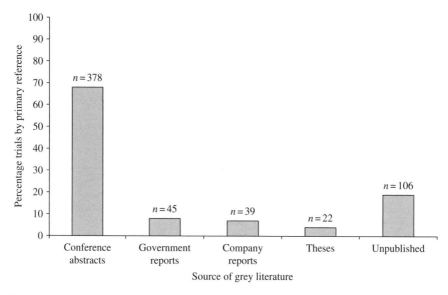

Figure 4.1   Source of primary ('least grey') reference for trials cited by grey literature alone (Mallett *et al.*, 2000).

Further evidence of the use of grey literature in systematic reviews is demonstrated in a study assessing the source of references in 75 reviews of interventions for neurological conditions. Seventy-nine per cent of references were from articles published in journals indexed in MEDLINE or EMBASE, 10 % were from articles published in journals not indexed by these databases and 11 % were from other sources such as books, pharmaceutical companies, data on file and unpublished studies (Taus *et al.*, 1999).

Estimates of the amount of grey literature included in systematic reviews vary depending on the area of health care. Alberani *et al.* (1990) examined the type of references cited in articles published in six Italian health science journals in 1987 and 1988. A total of 1398 articles were identified, and 67 % of the articles cited references found in the grey literature. Sixteen per cent of the total number of references identified were found in the grey literature. Of these, 69 % were references to reports, 16 % were references to conference proceedings, 11 % were references to theses and 5 % were references to other types of grey literature (these included unpublished data, personal communication and translations).

The use of grey literature in the social sciences and educational research may be even higher, although empirical evidence in this area is somewhat limited. In a review of the medical and psychological effects of unemployment, 72 % of studies were published through conventional channels and 28 % were found in the grey literature (Van Loo, 1985). In a systematic review by Petrosino *et al.* (2000) of randomized experiments in crime reduction, around one-third of the 150 trials identified were reported in governmental documents, dissertations and technical reports. Earlier, Boruch *et al.* (1978), in a review of early delinquency experiments, reported similar results in that about 25 % of the included studies were unpublished.

Impact of grey literature in the social sciences

Much of the early work on assessing the impact of grey literature and how it might affect the overall results of a systematic review was in the social sciences. In social sciences research, it has been known for some time that omitting unpublished studies from a systematic review can magnify the effect of the intervention. Glass *et al.* (1981) analysed 11 meta-analyses of the experimental literature published between 1976 and 1980. These included meta-analyses of psychotherapy, the effects of television on anti-social behaviour and sex bias in counselling. In each of the nine instances in which a comparison could be made, the average experimental effects from studies published in journals were larger than the corresponding effects estimated from theses and dissertations. The findings reported in journals were on average 33 % more favourable to the hypothesis of the investigators than the findings reported in theses or dissertations. In one meta-analysis of sex bias in counselling and psychotherapy, the magnitude and the direction of the effect were different in the published and unpublished studies. A further study assessing socio-economic status and academic achievement also found that the treatment effects reported in journals were greater than the corresponding effects reported in dissertations (White, 1982).

It is perhaps not surprising, given what is known about publication bias, that those studies that are most accessible are most likely to be included in systematic reviews, and are more likely to show larger treatment effects. A study of meta-analyses of psychological, educational and behavioural treatment research found that published studies were more likely to produce more positive results than unpublished studies. In the 92 meta-analyses studied, published studies yielded mean effect sizes that were on average 0.14 standard deviations larger than unpublished studies (Lipsey and Wilson, 1993). De Smidt and Gorey (1997) compared the results of a meta-analysis containing published studies on the evaluation of social work interventions, with those of a replicated meta-analysis containing only studies found in the grey literature. The grey literature studies were doctoral dissertations and master's theses, which reported empirical findings on social interventions. Published studies showed the interventions to be more effective than studies found in the grey literature; however, this difference was only marginal. The meta-analysis containing only published studies found that 78 % of clients in the experimental group did better than those in the comparison group, compared to 73 % for the meta-analysis containing only theses and dissertations.

Impact of grey literature in health sciences

There is now evidence in health research which shows that there is a systematic difference between the results of published studies and those found in the grey literature. An early example is a meta-analysis of preoperative parental nutrition for reducing complications from major surgery and fatalities (Detsky *et al.*, 1987). In this study, the results of randomized trials published only as abstracts found the treatment to be more effective than trials published in papers; this is the opposite result to what might have been expected.

The exclusion of grey literature from meta-analyses can also lead to an exaggeration of the effect of treatment. A recent systematic review compared the effect of the inclusion and exclusion of grey literature on the results of meta-analyses of randomized trials (Hopewell *et al.*, 2003c). Eight studies met the inclusion criteria for the review (Table 4.3). Four studies (Burdett *et al.*, 2003; Egger *et al.*, 2003; Fergusson *et al.*, 2000; McAuley *et al.*, 2000) contained multiple meta-analyses; one of these four studies was also an individual patient data meta-analysis (Burdett *et al.*, 2003). Four studies (Bhandari *et al.*, 2000; Horn and Limburg, 2002; Jeng *et al.*, 1995; Man-Son-Hing *et al.*, 1998) contained only one meta-analysis in which a sensitivity analysis had been done to assess the impact of grey literature. The two studies containing the greatest number of meta-analyses were the studies by Egger *et al.* (60 meta-analyses) and McAuley *et al.* (41 meta-analyses).

The number of trials included in the meta-analyses ranged from 7 to 783. The studies by Egger *et al.* (783 trials) and McAuley *et al.* (467 trials) contained the greatest number of trials. In most cases there were more published trials included in the meta-analyses than grey trials (median 46 (IQR 4–300) versus 5.5 (IQR 4–88)). The median number of participants included in the trials was also larger in the published trials than in the grey trials. The study by Egger *et al.* also assessed the statistical significance of the trial results included in the meta-analyses. They found that published trials (30 %) were more likely to have statistically significant results than grey trials (19 %).

The majority of trials (96 %) were published in English. This was similar for both grey and published trials and no language restrictions appear to have been imposed by the authors of the studies. Only three of the eight studies (Burdett *et al.*, 2003; Fergusson *et al.*, 2000; McAuley *et al.*, 2000) contained trials published in languages other than English. For all eight studies the most common type of grey literature analysed were abstracts (49 %). Unpublished data were the second largest type of grey literature (33 %). The definition of what constituted unpublished data varied since it was that used by the authors of the studies. However, it generally included data from trial registers, file-drawer data and data from individual trialists. Book chapters were the third largest type of grey literature (10 %), with unpublished reports, pharmaceutical company data, publications in press, letters and theses making up the small remainder.

All four of the studies containing multiple meta-analyses found that published trials showed an overall greater treatment effect than grey trials. The study by McAuley *et al.* (2000) found that grey trials were likely to produce smaller estimates of the intervention effect, on average by 15 %, than published trials (ratio of odds ratios 1.15; 95 % CI 1.04–1.28). In the study by Egger *et al.* (2003) the pooled estimates from grey trials were on average 7 % less beneficial than estimates from published trials (ratio of odds ratios 1.07; 95 % CI 0.98–1.15). However, in the Egger *et al.* study, there was notable heterogeneity between meta-analyses and there was a wide variation between different medical specialities. For example, in some areas there was little difference between the results of grey and published trials, but in obstetrics and gynaecology published trials showed a much larger treatment effect

Table 4.3   Systematic review of meta-analyses containing grey literature.

Study	No. meta-analyses	No. trials	No. participants	Area of health care
Bhandari *et al.* (2000)	1	Published: 4 Grey: 5	Published: 366 Grey: 348	Orthopaedic surgery
Burdett *et al.* (2003)	11	Published: 75 Grey: 45	Published: 12 156 Grey: 6 221	Cancer
Egger *et al.* (2003)	60	Published: 630 Grey: 153	Published: 146 160 Grey: 21 573	Various medical specialties
Fergusson *et al.* (2000)	10	Published: 108 Grey: 6	Published: 10 322 Grey: 820	Cardiac and orthopaedic surgery
Horn and Limburg (2002)	1	Published: 17 Grey: 4	Published: 5 626 Grey: 788	Stroke
Jeng *et al.* (1995)	1	Published: 4 Grey: 4	Published: 239 Grey: 140	Recurrent miscarriage
Man-Song-Hing *et al.* (1998)	1	Published: 4 Grey: 3	Published: 73 Grey: 336	Nocturnal leg cramps
McAuley *et al.* (2000)	41	Published: 365 Grey: 102	Published: 23 286 Grey: 194 141	Various medical specialties

Source: Hopewell *et al.* (2003c).

(ratio of odds ratios 1.34; 95 % CI 1.09–1.66). In the review of individual patient data meta-analyses by Burdett *et al.*, published trials also showed a significantly greater treatment effect in favour of the experimental intervention (ratio of hazard ratios for grey versus published trials was 1.04; 95 % CI 1.01–1.08). Three of the studies containing single meta-analyses (Bhandari *et al.*, 2000; Jeng *et al.*, 1995; Man-Son-Hing *et al.*, 1998) found that published trials showed an overall greater treatment effect than grey trials, although this difference was not found to be statistically significant. The remaining study, which was a single meta-analysis, found that published trials showed no treatment effect and that grey trials showed a negative treatment effect (Horn and Limburg, 2002).

Only one of the eight studies (McAuley *et al.*, 2000) assessed the type of grey literature and its impact on the overall results of the meta-analyses. McAuley *et al.* found that published trials showed an even greater treatment effect than grey trials if abstracts were excluded from the meta-analyses. The ratio of odds ratios for all grey trials versus published trials was 1.15 (95 % CI 1.04–1.28) compared to 1.33 (95 % CI 1.10–1.60) for grey trials, excluding abstracts, versus published trials.

Problems of searching for grey literature

The identification of relevant studies in the grey literature and their inclusion in systematic reviews can be particularly time-consuming and difficult. In an attempt to determine whether researchers search for, obtain and then include grey literature in meta-analyses, Cook *et al.* (1993) identified all articles indexed with the keyword 'meta-analysis' in MEDLINE from January 1989 to February 1991. One hundred and fifty meta-analyses were identified; of these 33 % did not search for unpublished information, 53 % searched for unpublished information and, of these, 31 % included unpublished information in the primary analysis. It was unclear whether unpublished information had been searched for in the remaining 13 % of the meta-analyses.

The Cochrane Collaboration recommends that those carrying out systematic reviews take practical steps to avoid publication bias by trying to identify and include data from trials found in the grey literature. To see how this recommendation was implemented in practice Stewart and Tierney (2000) assessed all new reviews published in Issue 1 2000 of *The Cochrane Library* to see whether Cochrane reviewers searched for grey and unpublished trials and whether that information was included in the reviews. They found that 71 % of reviews reported searching sources other than bibliographic databases, 75 % of reviews reported communicating with trialists and 52 % specifically mentioned seeking unpublished trials. Of the 52 new reviews assessed, 59 % included only published trials, 29 % included trials reported as abstracts and letters and 20 % included unpublished trials.

However, even when researchers do search for grey literature, there is some evidence to suggest that the considerable time, effort and expense may not always be worthwhile. In a massive attempt to obtain information about unpublished trials in perinatal medicine, Hetherington *et al.* (1989) sent letters to 42 000 obstetricians and paediatricians in 18 countries. They were notified of 395 unpublished randomized trials. Only 18 of these trials had been completed more than two years before the survey; 125 had ceased recruitment within the two years prior to the survey, 193 were actively recruiting at the time of the survey and 59 were about to begin recruitment. They concluded that obtaining information about trials retrospectively was unsuccessful; however, the response rate to requests for details about ongoing and planned trials was good.

Difficulties of including grey literature

There is some debate as to whether studies found in the grey literature should be included in meta-analyses because they might be incomplete and their methodological quality can be difficult to assess. A survey by Cook *et al.* (1993) showed that 78 % of meta-analysts and methodologists felt that unpublished material should definitely or probably be included in meta-analyses. Journal editors were less enthusiastic about the inclusion of unpublished studies: 47 % of editors felt that unpublished data should probably or certainly be included in meta-analyses. Interestingly, Cook *et al.* found that the editors' and methodologists' attitudes varied depending on the type of material included. Their attitudes were less favourable towards the inclusion of journal supplements, published abstracts, published symposia, book chapters and dissertations (none of which may have been peer-reviewed) and more favourable towards unpublished presentations or materials that have never been published or presented. This study raised concerns that studies found in the grey literature may be of poorer quality, as they have not been peer-reviewed. Traditionally, the peer review process has been thought to ensure the scientific quality of the studies it reviews and to help guarantee against scientific fraud. In reality, however, peer review still relies on the opinions of expert reviewers, which can be open to arbitration and bias (Jefferson *et al.*, 2003).

There is some empirical evidence to show that studies found in the grey literature may be of poorer methodological quality than those published as journal articles (Egger *et al.*, 2003; MacLean *et al.*, 2003). In an analysis of 60 meta-analyses of health-care interventions, Egger *et al.* (2003) found that trials published in the grey literature were less likely to report adequate concealment of allocation than published trials (33.8 % versus 40.7 %). Trials published in the grey literature were also less likely to report blinding of outcome assessment than published trials (45.1 % versus 65.8 %).

Studies presented at scientific meetings and reported only as abstracts in the proceedings of the meeting are a principal source of grey literature. However, concerns have been raised over the poor reporting and quality of these abstracts. Several studies have highlighted that many abstracts submitted to scientific meetings may have had inadequacies in the aims, methods, results and conclusions of the study or, indeed, have lacked numerical or statistical data (McIntosh, 1996; Panush *et al.*, 1989). A study of trials reported as abstracts and presented at the American Society of Clinical Oncology conferences in 1992 and 2002 found that only 2 % of trials reported in the abstracts stated the method of allocation concealment, 14 % described adequate methods of blinding and only 21 % reported, or suggested, intention to treat analysis. For the majority of trials it was not possible to determine aspects of trial quality, due to the limited amount of information available in the abstract. Information was also limited on the number of participants initially randomized and the number actually included in the analysis (Hopewell and Clarke, 2005).

Several studies have highlighted some significant differences between the content and quality of information reported about a study in a conference abstract compared to its subsequent full publication (Bhandari *et al.*, 2002; Chokkalingam *et al.*, 1998; Weintraub, 1987). One study compared the consistency of information reported in an abstract and presented at an annual orthopaedic surgeons' conference in 1996, with that of its subsequent full publication. Bhandari *et al.* (2002) identified significant inconsistencies in the study title and authorship between what was reported in the abstract and its subsequent full publication. They also identified differences in the number of participants included in the full publication compared to what was reported in the abstract.

Abstracts submitted to scientific meetings are often subject to peer review, which should provide some level of scientific control. However, even when this is the case there is still considerable variation in the evaluation of the abstracts (Vilstrup and Sorensen, 1998; Rubin *et al.*, 1993; Timmer *et al.*, 2003). What is needed are clearer guidelines for authors on submitting an abstract for presentation at a scientific meeting, which outline the key elements of the study to be described. Most of the work on improving the quality of trials reported as abstracts has concentrated on the development of structured abstracts in journal articles (Haynes *et al.*, 1990, 1996; Mulrow *et al.*, 1988). Structured abstracts in journal articles face similar constraints to those of abstracts submitted to scientific meetings, the most obvious being the limited amount of detail on the study's methodology and results that authors can include. However, the quality of trials reported as abstracts in a structured format in journal articles can also still be poor (Froom and Froom, 1993; Scherer and Crawley, 1998). Despite authors being provided with specific instructions about abstract accuracy, inconsistencies in the data presented have still occurred (Pitkin and Branagan, 1998). Further research is needed to address these important issues and ultimately improve the quality of abstracts presented at scientific meetings.

Perhaps the biggest problem in trying to identify and include studies found in the grey literature is obtaining a full and representative sample. As has been discussed earlier in this chapter, there is the potential for bias to occur when searching for and selecting studies for inclusion in a systematic review. To avoid introducing an additional bias into the review, it is important that the same rigorous criteria for searching for, and selecting, published studies should also apply to studies found in the grey literature. The problem, however, comes in trying to identify these studies in an unbiased way. Although some of these studies will be indexed in electronic resources which are relatively easy to access, such as CENTRAL, SIGLE, Biological Abstracts, ERIC, and BiomedCentral (Box 4.2), these are just a small number of the vast range of potential sources which currently exist.

The registration of studies at the time they start, and before their results become known, would help eliminate the risk of publication bias in both the grey and published literature. One such initiative in clinical trials is the development of the metaRegister of Controlled Trials and the introduction of an International Standard Randomized Controlled Trial Number (ISRCTN) for all clinical trials. This comprehensive register of initiated trials assigns a unique ISRCTN identifier to each trial. The aim of this initiative is to inform patients, clinicians, researchers and others about which trials have been started and therefore help to address directly the problems of publication bias. Initiatives such as this and the prospective registration of ongoing trials are discussed further in Chapter 3.

RECOMMENDATIONS

As has been shown in this chapter, there are a number of key sources to consider when carrying out a comprehensive search for a systematic review. These sources include searching electronic bibliographic databases, handsearching key journals, searching conference proceedings, contacting researchers, searching research registers and searching the Internet. These sources are likely to identify studies in both the published and grey literature. The optimal extent of the search is very much dependent on the subject matter and the resources available for the search. For example, those carrying out a systematic review need to consider when it is appropriate to supplement electronic searches with a search by hand of key journals or journal supplements not indexed by electronic databases or a search by hand of conference proceedings for a specific subject area. It is also crucial to assess the quality of the studies identified from the search, irrespective of whether they were found in the published or grey literature. Whatever is searched, it is important that the systematic reviewer is explicit and documents in their review exactly which sources they have searched and which strategies they have used so that the reader can assess the validity of the conclusions of the review based on the search that has been conducted.

ACKNOWLEDGEMENTS

We are grateful to Anne Eisinga for her help in identifying key electronic information resources for grey literature and for commenting on draft versions of this chapter.

DISCLAIMER

The views expressed in this paper represent those of the authors and are not necessarily the views or the official policy of The Cochrane Collaboration.

REFERENCES

Alberani, V., de Castro Pietrangeli, P. and Mazza, A.M. (1990). The use of grey literature in health sciences: a preliminary survey. *Bulletin of the Medical Library Association*, **78**, 358–363.

Auger, C.P. (1998). *Information Sources in Grey Literature (Guides to Information Sources)*, 4th edition. London: Bowker Saur.

Bhandari, M., Devereaux, P.J., Guyatt, G.H., Cook, D.J., Swiontkowski, M.F., Sprague, S. and Schemitsch, E.H. (2002). An observational study of orthopaedic abstracts and subsequent full-text publications. *Journal of Bone and Joint Surgery (American)*, **84**, 615–621.

Bhandari, M., Guyatt, G.H., Tong, D., Adili, A. and Shaughnessy, S.G. (2000). Reamed versus nonreamed intramedullary nailing of lower extremity long bone fractures: A systematic overview and meta-analysis. *Journal of Orthopaedic Trauma*, **14**, 2–9.

Boruch, R.F., McSweeney, A.J. and Soderstrom, E.J. (1978). Randomized field experiments for program planning, development, and evaluation: an illustrative bibliography. *Evaluation Quarterly*, **2**(4), 655–695.

Burdett, S., Stewart, L.A. and Tierney, J.F. (2003). Publication bias and meta-analyses: A practical example. *International Journal of Technology Assessment in Health Care*, **19**, 129–134.

Callaham, M.L., Wears, R.L., Weber, E.J., Barton, C. and Young, G. (1998). Positive-outcome bias and other limitations in the outcome of research abstracts submitted to a scientific meeting. *Journal of the American Medical Association*, **280**, 254–257.

Chokkalingam, A., Scherer, R. and Dickersin, K. (1998). Agreement of data in abstracts compared to full publications. *Controlled Clinical Trials*, **19**, 61–62S.

Clarke, M. and Clarke, T. (2000). A study of the references used in Cochrane protocols and reviews. Three bibles, three dictionaries, and nearly 25 000 other things. *International Journal of Technology Assessment in Health Care*, **16**, 907–909.

Clarke, M. and Oxman, A.D. (eds) (2003). Cochrane Reviewers' Handbook 4.2.0 [updated March 2003]. In *The Cochrane Library, Issue 2, 2003*. Oxford: Update Software.

Cook, D.J., Guyatt, G.H., Ryan, G., Clifton, J., Buckingham, L., Willan, A., McIlroy, W. and Oxman, A.D. (1993). Should unpublished data be included in meta-analyses? Current convictions and controversies. *Journal of the American Medical Association*, **269**, 2749–2753.

de Smidt, G.A. and Gorey, K.M. (1997). Unpublished social work research: Systematic replication of a recent meta-analysis of published intervention effectiveness research. *Social Work Research*, **21**, 58–62.

Detsky, A.S., Baker, J.P., O'Rourke, K. and Goel, V. (1987). Perioperative parenteral nutrition: A meta-analysis. *Annals of Internal Medicine*, **107**, 195–203.

Dickersin, K. (1997). How important is publication bias? A synthesis of available data. *AIDS Education and Prevention*, **9**(1 Suppl.), 15–21.

Dickersin, K., Manheimer, E., Wieland, S., Robinson, K.A., Lefebvre, C. and McDonald, S. (2002). Development of the Cochrane Collaboration's CENTRAL Register of controlled clinical trials. *Evaluation and the Health Professions*, **25**, 38–64.

Dickersin, K., Scherer, R. & Lefebvre, C. (1994). Identifying relevant studies for systematic reviews. *British Medical Journal*, **309**, 1286–1291.

Donaldson, I.J. and Cresswell, P.A. (1996). Dissemination of the work of public health medicine trainees in peer-reviewed publications: An unfulfilled potential. *Public Health*, **110**, 61–63.

Egger, M., Zellweger-Zahner, T., Schneider, M., Junker, C., Lengeler, C. and Antes, G. (1997). Language bias in randomised controlled trials published in English and German. *Lancet*, **350**, 326–329.

Egger, M., Davey Smith, G. and O'Rourke, K. (2001). Rationale, potentials, and promise of systematic reviews. In M. Egger, G. Davey Smith and D.G. Altman (eds), *Systematic Reviews in Health Care: Meta-analysis in Context* (pp. 3–19). London: BMJ Books.

Egger, M., Jüni, P., Bartlett, C., Holenstein, F. and Sterne, J. (2003). How important are comprehensive literature searches and the assessment of trial quality in systematic reviews? Empirical study. *Health Technology Assessment*, **7**, 1–76.

Eysenbach, G., Tuische, J. and Diepgen, T.L. (2001). Evaluation of the usefulness of Internet searches to identify unpublished clinical trials for systematic reviews. *Medical Informatics and the Internet in Medicine*, **26**, 203–218.

Fergusson, D., Laupacis, A., Salmi, L.R., McAlister, F.A. and Huet, C. (2000). What should be included in meta-analyses? An exploration of methodological issues using the ISPOT meta-analyses. *International Journal of Technology Assessment in Health Care*, **16**, 1109–1119.

Froom, P. & Froom, J. (1993). Deficiencies in structured medical abstracts. *Journal of Clinical Epidemiology*, **46**, 591–594.

Glass, G.V., McGaw, B. and Lee Smith, M. (1981). *Meta-analysis in Social Research*. London: Sage Publications.

Gotzsche, P.C. (1987). Reference bias in reports of drug trials. *British Medical Journal (Clinical Research Edition)*, **295**, 654–656.

Grayson, L. & Gomersall, A. (2003). A difficult business: Finding the evidence for social science reviews. Working Paper 19, ESRC UK Centre for Evidence Based Policy and Practice.

Hackshaw, A.K., Law, M.R. and Wald, N.J. (1997). The accumulated evidence on lung cancer and environmental tobacco smoke. *British Medical Journal*, **35**, 980–988.

Haynes, R.B., Mulrow, C.D., Huth, E.J., Altman, D.G. and Gardner, M.J. (1990). More informative abstracts revised. *Annals of Internal Medicine*, **11**, 69–76.

Haynes, R.B., Mulrow, C.D., Huth, E.J., Altman, D.G. and Gardner, M.J. (1996). More informative abstracts revised. *Cleft Palate and Craniofacial Journal*, **33**, 1–9.

Hetherington, J., Dickersin, K., Chalmers, I. and Meinert, C.L. (1989). Retrospective and prospective identification of unpublished controlled trials: Lessons from a survey of obstetricians and pediatricians. *Pediatrics*, **84**, 374–380.

Hopewell, S. and Clarke, M.. (2001). Methodologists and their methods. Do methodologists write up their conference presentations or is it just 15 minutes of fame? *International Journal of Technology Assessment in Health Care*, **17**, 601–603.

Hopewell, S. and Clarke, M. (2005). Abstracts presented at the American Society of Clinical Oncology conference: How completely are trials reported? *Clinical Trials*, **2**, 285–268.

Hopewell, S., Clarke, M., Stewart, L. and Tierney, L. (2003a). Time to publication for results of clinical trials (Cochrane Methodology Review). In *The Cochrane Library, Issue 4*, 2003. Chichester: John Wiley & Sons, Ltd.

Hopewell, S., Clarke, M., Lefebvre, C. and Scherer, R. (2003b). Handsearching versus electronic searching to identify reports of randomized trials (Cochrane Methodology Review). In *The Cochrane Library, Issue 4*, 2003. Chichester: John Wiley & Sons, Ltd.

Hopewell, S., McDonald, S., Clarke, M. and Egger, M. (2003c). Grey literature in meta-analyses of randomized trials of health care interventions (Cochrane Methodology Review). In *The Cochrane Library, Issue 4*, 2003. Chichester: John Wiley & Sons, Ltd.

Horn, J. and Limburg, M. (2002). Calcium antagonists for acute ischemic stroke. In *The Cochrane Library, Issue 1*, 2002. Oxford: Update Software.

Jefferson, T.O., Alderson, P., Davidoff, F. and Wager, E. (2003). Editorial peer-review for improving the quality of reports of biomedical studies (Cochrane Methodology Review). In *The Cochrane Library, Issue 4*, 2003. Chichester: John Wiley & Sons, Ltd.

Jeng, G.T., Scott, J.R. & Burmeister, L.F. (1995). A comparison of meta-analytic results using literature vs individual patient data. Paternal cell immunization for recurrent miscarriage. *Journal of the American Medical Association*, **274**, 830–836.

Jüni, P., Holenstein, F., Sterne, J., Bartlett, C. and Egger, M. (2002). Direction and impact of language bias in meta-analyses of controlled trials: empirical study. *International Journal of Epidemiology*, **31**(1), 115–123.

Lipsey, M.W. and Wilson, D.B. (1993). The efficacy of psychological, educational, and behavioral treatment: Confirmation from meta-analysis. *American Psychologist*, **48**, 1181–1209.

MacLean, C.H., Morton, S.C., Ofman, J.J., Roth, E.A. and Shekelle, P.G. (2003). How useful are unpublished data from the Food and Drug Administration in meta-analysis? *Journal of Clinical Epidemiology*, **56**, 44–51.

Mallett, S., Hopewell, S. and Clarke, M. (2000). The use of grey literature in the first 1000 Cochrane reviews. Paper presented at the Fourth Symposium on Systematic Reviews: Pushing the Boundaries, Oxford, 2–4 July.

Man-Song-Hing, M., Wells, G. and Lau, A. (1998). Quinine for noctural leg cramps: A meta-analysis including unpublished data. *Journal of General Internal Medicine*, **13**, 600–606.

McAuley, L., Pham, B., Tugwell, P. & Moher, D. (2000). Does the inclusion of grey literature influence estimates of intervention effectiveness reported in meta-analyses? *Lancet*, **356**, 1228–1231.

McDaniel, M.A., Whetzel, D., Schmidt, F.L. and Maurer, S. (1994). The validity of the employment interview: A comprehensive review and meta-analysis. *Journal of Applied Psychology*, **79**, 599–616.

McIntosh, N. (1996). Abstract information and structure at scientific meetings. *Lancet*, **347**, 544–545.

Mulrow, C.D. (1994). Rationale for systematic reviews. *British Medical Journals*, **309**, 597–599.

Mulrow, C.D., Thacker, S.B. and Pugh, J.A. (1988). A proposal for more informative abstracts of review articles. *Annals of Internal Medicine*, **108**, 613–615.

Panush, R.S., Delafuente, J.C., Connelly, C.S., Edwards, N.L., Greer, J.M., Longley, S. and Bennett, F. (1989). Profile of a meeting: How abstracts are written and reviewed. *Journal of Rheumatology*, **16**, 145–147.

Petrosino, A., Boruch, R.F., Rounding, C., McDonald, S. and Chalmers, I. (2000). The Campbell Collaboration Social, Psychological, Educational and Criminological Trials Register (C2-SPECTR) to facilitate the preparation and maintenance of systematic reviews of social and educational interventions. *Evaluation and Research in Education*, **14**, 206–218.

Pitkin, R.M. and Branagan, M.A. (1998). Can the accuracy of abstracts be improved by providing specific instructions? A randomized controlled trial. *Journal of the American Medical Association*, **280**, 267–269.

Raudenbush, S.W. (1984). Magnitude of teacher expectancy effects on pupil IQ as a function of the credibility of expectancy induction. *Journal of Educational Psychology*, **76**, 85–97.

Rubin, H.R., Redelmeier, D.A., Wu, A.W. and Steinberg, E.P. (1993). How reliable is peer review of scientific abstracts? Looking back at the 1991 Annual Meeting of the Society of General Internal Medicine. *Journal of General Internal Medicine*, **8**, 255–258.

Scherer, R.W. and Crawley, B. (1998). Reporting of randomized clinical trial descriptors and use of structured abstracts. *Journal of the American Medical Association*, **280**, 269–272.

Scherer, R.W., Langenberg, P., von Elm, E. (2005). Full publication of results initially presented in abstracts. *The Cochrane Database of Methodology Reviews, Issue 2*, 2005. Art. No.: MR000005. DOD: 10.1002/14651858.MR000005.pub2.

Schmidmaier, D. (1986) Ask no questions and you'll be told no lies: or how we can remove people's fear of 'Grey Literature'. *Librarian*, **36**, 98–112.

Song, F., Eastwood, A.J., Gilbody, S., Duley, L. and Sutton, A.J. (2000). Publication and related biases. *Health Technology Assessment*, **4**, 1–115.

Stewart, L. and Tierney, J. (2000). Publication bias is a serious threat to the validity of Cochrane reviews. Paper presented at the Eight International Cochrane Colloquium, Cape Town, 25–29 October.

Taus, C., Pucci, E., Giuliani, G., Telaro, E. and Pistotti, V. (1999). The use of 'grey literature' in a sub-set of neurological Cochrane reviews. Paper presented at the Seventh International Cochrane Colloquium; Rome, 5–9 October.

Timmer, A., Sutherland, L.R. and Hilsden, R.J. (2003). Development and evaluation of a quality score for abstracts. *BMC Medical Research Methodology*, **3**, 2.

The Cochrane Library, Issue 1, 2004. *EMBASE Release Notes*. Chichester: John Wiley & Sons, Ltd.

Turner, H., Boruch, R., Lavenberg, J., Schoeneberger, J. and de Moya, D. (2004). Electronic registers of trials. Paper Presented at the Fourth Annual Campbell Collaboration Colloquium: A First Look at the Evidence; Washington, DC, 18–20 February.

Van Loo, J. (1985). Medical and psychological effects of unemployment: A 'grey' literature search. *Health Libraries Review*, **2**, 55–62.

Vilstrup, H. & Sorensen, H.T. (1998). A comparative study of scientific evaluation of abstracts submitted to the 1995 European Association for the Study of the Liver Copenhagen meeting. *Danish Medical Bulletin*, **45**, 317–319.

Weber, E.J., Callaham, M.L., Wears, R.L., Barton, C. and Young, G. (1998). Unpublished research from a medical specialty meeting: Why investigators fail to publish. *Journal of the American Medical Association*, **280**, 257–259.

Weintraub, W.H. (1987). Are published manuscripts representative of the surgical meeting abstracts? An objective appraisal. *Journal of Pediatric Surgery*, **22**, 11–13.

White, K.R. (1982). The relation between socioeconomic status and academic achievement. *Psychological Bulletin*, **91**, 461–481.

Statistical Methods for Assessing Publication Bias

CHAPTER 5

The Funnel Plot

Jonathan A.C. Sterne
University of Bristol, UK

Betsy Jane Becker
Florida State University, USA

Matthias Egger
University of Bern, Switzerland

KEY POINTS

- Funnel plots are a primary visual tool for the investigation of publication and other bias in meta-analysis.
- Funnel plots are simple scatter plots of the treatment effects estimated from individual studies against a measure of study size.
- Publication bias may lead to asymmetrical funnel plots.
- For studies with binary outcomes, standard error is the best measure of study size, while risk ratios or odds ratios should be used as the measure of treatment effect.
- It is important to realize that publication bias is only one of a number of possible causes of funnel plot asymmetry. Funnel plots should be seen as a generic means of examining small-study effects (the tendency for the smaller studies in a meta-analysis to show larger treatment effects) rather than as a tool to diagnose specific types of bias.

DEFINITION AND RATIONALE

Funnel plots are a primary visual tool for the investigation of publication and other bias in meta-analysis. In this chapter we describe funnel plots and review their use in both the social science and medical literatures. We provide guidelines on

---

Publication Bias in Meta-Analysis – Prevention, Assessment and Adjustments   Edited by H.R. Rothstein, A.J. Sutton and M. Borenstein   © 2005 John Wiley & Sons, Ltd

choice of axis in funnel plots, explain the different possible causes of funnel plot asymmetry, and compare the use of funnel plots with other methods to examine bias in meta-analysis. Finally, we illustrate the use of funnel plots with a number of examples.

Funnel plots are simple scatter plots of the treatment effects estimated from individual studies against a measure of study size. The name 'funnel plot' is based on the fact that the precision in the estimation of the underlying treatment effect increases as the sample size of the studies in the review increases. Therefore, when a measure of study size is plotted on the vertical axis, results from small studies will scatter widely at the bottom of the graph, with the spread narrowing among larger studies. In the absence of bias (and when studies estimate the same underlying effect) the plot will resemble a symmetrical inverted funnel, as shown in Figure 5.1(a).

If there is bias, for example because smaller studies showing no statistically significant effects (open circles in Figure 5.1) remain unpublished, then the funnel plot will appear asymmetrical, with a gap in the bottom right-hand side of the graph (Figure 5.1(b)). In this situation the combined effect from meta-analysis will

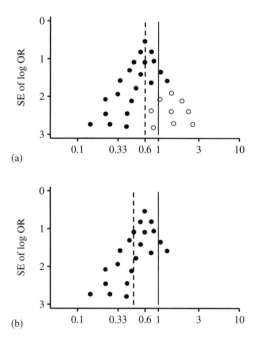

(a)

(b)

Figure 5.1 Hypothetical funnel plots: (a) symmetrical plot in the absence of bias (open circles indicate smaller studies showing no beneficial effects); (b) asymmetrical plot in the presence of publication bias (smaller studies showing no beneficial effects are missing); (c) asymmetrical plot in the presence of bias due to low methodological quality of smaller studies (open circles indicate small studies of inadequate quality whose results are biased towards larger beneficial effects). (Based on figure from Sterne *et al.*, 2001).

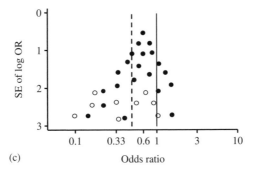

(c)   Odds ratio

Figure 5.1 (Continued).

overestimate the average effect or strength of relationship. The more pronounced the asymmetry, the more likely it is that the amount of bias will be substantial.

Empirical evidence supports the use of funnel plots to examine publication bias. For example, Egger *et al.* (2003) identified 58 meta-analyses that included at least one unpublished trial and used statistical methods (described in Chapter 6) to assess funnel plot asymmetry. After excluding the unpublished trials funnel plot asymmetry (smaller trials showing larger treatment effects) became more pronounced. Publication bias is, however, only one of a number of possible explanations for funnel plot asymmetry: other explanations are discussed in more detail in later on. For example, trials of lower quality may yield exaggerated estimates of intervention effects (Schulz *et al.*, 1995). Smaller studies are, on average, conducted and analysed with less methodological rigour than larger studies (Egger *et al.*, 2003), so that asymmetry may also result from the overestimation of intervention effects in smaller studies of lower methodological quality (Figure 5.1(c)). Conversely, in some cases interventions may be more beneficial in smaller studies, because they were more fully or carefully implemented than in the larger studies.

HISTORY OF FUNNEL PLOTS

Funnel plots were first used in educational research and psychology by Light and Pillemer (1984). For example, Figure 5.2 showed data from a meta-analysis of educational programmes for surgical patients drawn from the work of Devine and Cook (1983). However, in spite of the fact that funnel plots first appeared in the social science literature, they are still not widely used in that realm. Light *et al.* (1994) examined 74 meta-analyses published from 1985 to 1991 in the review journal *Psychological Bulletin* and found only one review using a funnel plot. More recently, Becker and Morton (2002) found only one funnel plot in the 28 meta-analyses that appeared from 1999 to 2002 in two prominent social science review journals (*Review of Educational Research* and *Psychological Bulletin*). Becker and Morton reported that, in contrast, 6 of 19 meta-analyses published in the *Journal of the American Medical Association* from 2001 to April 2002 discussed funnel plots for their data (though only two of those plots appeared in print).

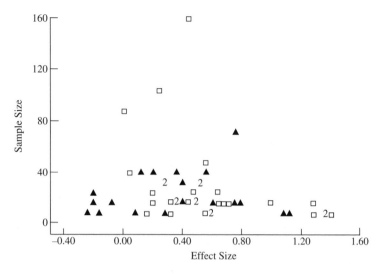

Figure 5.2   Funnel plot of published (□) and unpublished (▲) studies from a meta-analysis of educational programmes for surgical patients drawn from the work of Devine and Cook (1983). (Reproduced, with permission, from Light and Pillemer, 1984.)

The possibility of using funnel plots as a means of diagnosing the existence of publication bias in medical research gained widespread attention following an article by Begg and Berlin (1988). These authors noted that publication bias might be a particular problem in meta-analyses based only on published studies, and proposed methods, including funnel plots, for empirical assessments of publication bias. So far as we are aware, the first funnel plot published in the medical literature was in a paper, also published in 1988, by Vandenbrouke, on the subject of publication bias in studies of passive smoking and lung cancer (Vandenbrouke, 1988). This funnel plot, reproduced in Figure 5.3, has three features discussed in more detail in subsequent sections. First, the measure of association used is the relative risk (risk ratio, RR) and is plotted on a log scale so that, for example, the distance between RRs of 0.5 and 1 is the same as the distance between RRs of 1 and 2. Second, the 13 studies were a mixture of case–control and cohort studies, with the odds ratios from case–control studies assumed to estimate risk ratios because the outcome is rare. Because the sample size in a cohort study is not comparable to that in a case–control study, Vandenbrouke used the standard error of $\log_e(RR)$ as the measure of study size. Third, in this plot the intervention effect was plotted on the vertical rather than horizontal axis. Choice of axis in funnel plots is discussed in the next section.

In 1995, an influential editorial by Egger and Davey Smith (1995) proposed that use of funnel plots might have alerted investigators to the unreliability of a meta-analysis of small trials of the effect of magnesium treatment for myocardial infarction, whose results were subsequently called into question by larger studies that found no little or evidence that magnesium treatment reduced mortality. The funnel plots from this article are reproduced in Figure 5.4. The very large ISIS-4

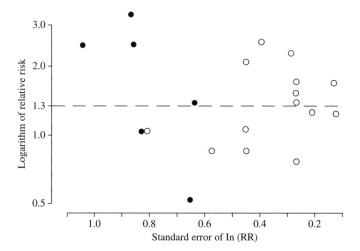

Figure 5.3 Funnel plot of the relative risk (logarithmic scale) according to standard error of $\log_e(RR)$ for men (closed circles) and women (open circles) from 13 studies of passive smoking and lung cancer. (Reproduced, with permission, from Vandenbroucke, 1988.)

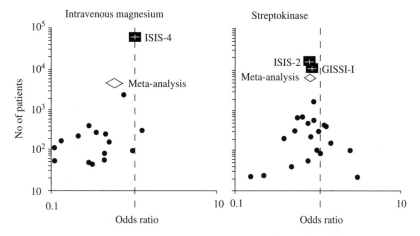

Figure 5.4 Funnel plots for meta-analyses refuted and confirmed by subsequent large trials: intravenous magnesium (left) and streptokinase (right). (Reproduced, with permission, from Egger and Davey Smith, 1995.)

(Fourth International Study of Infarct Survival) trial (58 000 participants) excluded any substantial beneficial effect of intravenous magnesium on mortality following myocardial infarction, since the 95 % confidence interval (CI) for the mortality odds ratio was 0.99 to 1.13) This was in sharp contrast to a previous meta-analysis of small trials, which suggested a substantial beneficial effect. Egger and Davey Smith contrasted the asymmetrical funnel plot from that meta-analysis with the symmetrical funnel plot seen in a meta-analysis of streptokinase trials, for which the results were consistent with those of the subsequent large ISIS-2 trial (ISIS-2 (Second International Study of Infarct Survival) Collaborative Group, 1988). Recently, evidence has emerged that publication bias may be a serious problem in the rapidly expanding field of genetic meta-analysis (Ioannidis *et al.*, 2001; Lohmueller *et al.*, 2003). Trikalinos and Ioannidis discuss bias in genetic meta-analyses in Chapter 13 in this volume.

A recent review of meta-analyses of at least five controlled clinical trials published in leading medical journals (*Annals of Internal Medicine, British Medical Journal, Lancet, Journal of the American Medical Association, American Journal of Cardiology, Cancer, Circulation,* and *Obstetrics and Gynaecology*) has found a dramatic increase in the use of funnel plots (Gerber *et al.*, 2005). Figure 5.5 shows that the proportion of meta-analyses that reported on funnel plots rose from 5 % or less during the years 1993–1996, to over 45 % in 2001 and 2002.

The limited application of funnel plots by reviewers in the social sciences may have resulted from the limited but somewhat critical attention these plots have received from social science methodologists. Two of the most popular books on meta-analysis for the social sciences do not mention funnel plots at all. Cooper (1998) discusses publication-bias issues, but focuses on how extensive literature-searching procedures can address the problem. Rosenthal (1991) also discusses

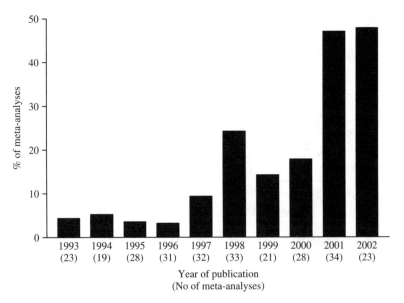

Figure 5.5   Use of funnel plots in meta-analyses published in four general and internal medicine journals and the four specialist journals, 1993–2002. The bars show the proportion of meta-analyses of at least five controlled clinical trials that reported on funnel plots and were published in *Annals of Internal Medicine, BMJ, Lancet, JAMA, American Journal of Cardiology, Cancer, Circulation,* or *Obstetrics and Gynaecology* during each of the publication years 1993–2001 inclusive. The figures in parentheses show the number of meta-analyses published in these journals, in each year.

publication bias, but favours his failsafe number (see Chapter 8 in this volume) and analyses of probability values as ways of dealing with the issue. Hedges and Olkin (1985) also do not mention funnel plots, but their book was a treatment of statistical methods in meta-analysis and when it was published the tests to accompany funnel plots had not yet been developed.

Social science authors who have discussed funnel plots are not always enthusiastic. Mullen (1989) gave some detail on funnel plots, but noted that (at the time) no numerical method was available to assess 'the deviation of a funnel plot from "funnelness"' (p. 39). Light *et al.*'s (1994) chapter in the *Handbook of Research Synthesis* also briefly discussed funnel plots. However, in their conclusions they did not mention the funnel plot as a specific element to be included in a report, recommending instead that stem-and-leaf and 'schematic plots' (box-and-whisker plots) be included. The discussion by Wang and Bushman (1998) appears in the context of an article promoting the use of normal quantile plots, which are described by the authors as 'less ambiguous to interpret than funnel plots'. They list three problems with funnel plots, the primary issue being the difficulty in determining whether the plot actually resembles a funnel, especially for small reviews. Once again the funnel plot was not strongly recommended by these authors.

Two more recent sources are more positive regarding use of funnel plots. One is *Practical Meta-analysis* (Lipsey and Wilson, 2001). Lipsey and Wilson discuss funnel plots in a section on graphical techniques. They argue again that, for small reviews, the plots can be hard to interpret and caution reviewers against overinterpreting funnel plots. Rothstein *et al.* (2002) also briefly cover funnel plots as a useful 'means of alerting the researcher that there may be a problem that needs to be explored' (p. 554).

In contrast, recent books on meta-analysis in medical research have dealt in detail with the use of funnel plots to detect publication and other bias. These include the books by Egger *et al.* (2001) and Sutton *et al.* (2002). On the other hand, Glasziou *et al.* (2001) were less enthusiastic. Empirical evidence for the existence of publication bias is discussed in Chapter 10 of this book.

GUIDELINES ON CHOICE OF AXIS FOR FUNNEL PLOTS

Figures 5.1, 5.3 and 5.4 illustrate the diverse choices of axis used in funnel plots. The array of possible choices leads to a concern that the visual impression given by funnel plots may be influenced by choice of axis, that investigators might choose their axes in order to convey the desired impression, or that the chosen axes may not be appropriate for detecting bias.

In the social sciences (from where, as we have noted, funnel plots originated) outcomes are typically quantitative, so that intervention effects are measured as mean differences or, more commonly, as standardized mean differences (effect sizes). Correlations also are common in meta-analyses of relationships. In contrast, the majority of endpoints in randomized trials of medical interventions are binary, with intervention effects most commonly expressed as ratio measures (odds ratio, risk ratio or hazard ratio). The use of ratio measures is justified by empirical evidence that there is less between-trial heterogeneity in intervention effects based on ratio measures than difference measures (Deeks and Altman, 2001; Engels *et al.*, 2000). Because considerations for choice of axis in funnel plots are different for ratio measures than for difference measures, we will deal with them separately.

In this chapter, we will adopt the convention of plotting intervention effects on the horizontal axis and the measure of study size on the vertical axis, since this is the most common convention in meta-analysis. It should be noted that it is not an error to plot the axes the other way round (as in Figure 5.3). Indeed, such a choice is arguably more consistent with standard practice, in that the variable on the vertical axis is usually hypothesized to depend on the variable on the horizontal axis.

FUNNEL PLOTS FOR STUDIES WITH BINARY OUTCOMES

Choice of vertical axis (measure of study size)

Sterne and Egger (2001) considered choice of axis in funnel plots of meta-analyses with binary outcomes. They illustrated issues in the choice of vertical axis (measure of study size) in the context of the randomized controlled trials of magnesium treatment in the prevention of death following myocardial infarction discussed earlier. Figure 5.6 shows funnel plots for these 16 trials, using six different choices of vertical axis. In each case, the horizontal axis is the log odds ratio, with a log odds ratio of 0 (odds ratio equal to 1) corresponding to no intervention effect, and negative log odds ratios (odds ratios less than 1) corresponding to a beneficial treatment effect (mortality lower in individuals treated with magnesium than in individuals treated with placebo). The vertical line shows the summary log odds ratio calculated using fixed-effect meta-analysis, which is 0.015 (95 % CI −0.045 to 0.075) since it is dominated by the ISIS-4 study, in which magnesium had no effect on mortality. For the purposes of displaying the centre of the plot in the absence of bias, calculation of the summary log odds ratio using fixed-rather than random-effects meta-analysis is preferable because the random-effects estimate gives greater relative weight to smaller studies, and will therefore be more affected if publication bias is present (Poole and Greenland, 1999).

Plot (a) of Figure 5.6 uses standard error as the vertical axis. The largest studies have the smallest standard errors, so to place the largest studies at the top of the graph, the axis has to be inverted (with standard error 0 at the top). The diagonal lines show the expected 95 % confidence intervals around the summary estimate, that is, [summary effect estimate − (1.96 × standard error)] and [summary effect estimate + (1.96 × standard error)] for each standard error on the vertical axis. They indicate the extent of between-study heterogeneity: in the absence of heterogeneity 95 % of the studies should lie within the funnel defined by these straight lines. In this plot, several points fall to the left of the lines, reflecting odds ratios lower than expected under the fixed-effect assumption (more beneficial effects of magnesium).

Plot (b) displays the results of the same 16 studies, using precision (defined as 1/standard error) as the vertical axis. No inversion of the axis is required, since the larger the study the smaller the standard error and hence the larger the precision. The lines showing expected 95 % confidence intervals are derived by calculating

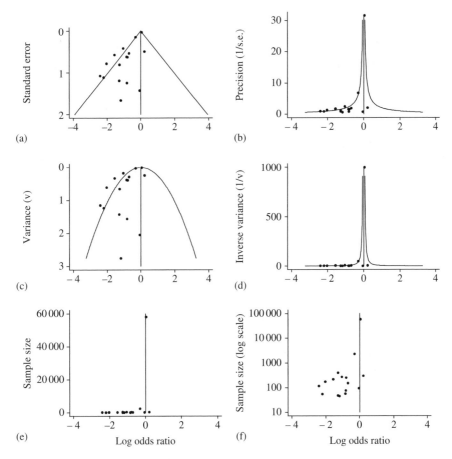

Figure 5.6   Funnel plots for the magnesium trials with different vertical axes. The points correspond to the treatment effects from individual trials, and the diagonal or curved lines show the expected 95 % confidence intervals around the summary estimate. Odds ratios are plotted on a logarithmic scale, to ensure that effects of the same magnitude but in opposite directions are equidistant from 1.0. (Based on a figure published in Sterne and Egger, 2001.)

the standard error corresponding to the value of the precision at any point on the vertical axis. For example, if the precision is 10 then the standard error is $1/10 = 0.1$, and so the lines are at (summary effect estimate $- 1.96 \times 0.1$) and (summary effect estimate $+ 1.96 \times 0.1$). The lines are sharply curved and the plot emphasizes the difference between the largest studies while smaller studies are compressed at the bottom. Plots (c) and (d) use variance (squared standard error) and inverse variance (1/variance) respectively for the vertical axis. In plot (c), the smaller studies are further apart and the larger studies closer together than for standard error, so that the funnel is flattened at the top. Plot (d) might be thought informative since the weights in fixed-effects meta-analysis are proportional to the inverse variance. However, this plot compresses the smaller studies together even more than plot (b), so that the expected 95 % confidence intervals are even more sharply curved. The degree of compression will depend on the range of standard errors in the review. However, some compression will always occur relative to the plot using precision because the range of the *squared* precision values will always be greater than that of the precision (1/standard error) values.

Plots (e) and (f) use log sample size and sample size respectively for the vertical axis. While plot (f) gives an impression of asymmetry similar to that in plot (a), plot (e) compresses the smaller studies at the bottom to an even greater degree than plot (b) and in a manner similar to plot (d). Because the standard error of the log odds ratio depends on the number of events in each group as well as on the number of subjects, it is not possible to calculate expected 95 % confidence intervals without making assumptions about the numbers of events that could have occurred at each hypothetical sample size (for instance, making the unrealistic assumption that event rates are the same in each study).

Interestingly, omission of the very large ISIS-2 trial from these funnel plots did not alter these patterns greatly – for example, the plots using precision or inverse variance as vertical axis still emphasized the differences between the next largest trial (the second Leicester Intravenous Magnesium Intervention Trial (LIMIT-2), with 2316 subjects; Woods and Fletcher, 1994) and the remaining trials, and the lines denoting expected 95 % confidence limits were still curved.

Clearly, the choice of variables used in funnel plots should be consistent with the fundamental assumption underlying the method: that in the absence of bias a plot from studies which estimate the same intervention effect will be symmetrical and bear some resemblance to a funnel. This is the case for standard error and variance (and functions thereof) but not necessarily for sample size or functions of sample size. Further, the major factor affecting the probability of publication of a study is its *p*-value (Dickersin, 1997; Dickersin *et al.*, 1992; Easterbrook *et al.*, 1991; Ioannidis, 1998; Stern and Simes, 1997), which is more directly a function of standard error or variance rather than of sample size.

For binary outcomes, the power of a study, and the precision of the estimate of a intervention's effect, are determined both by the total sample size and by the number of participants developing the event of interest. A study with 100 000 subjects and 10 events is less powerful and will produce less precise estimates than a study with 1000 subjects and 100 events. In other words, studies with very different sample sizes may have the same standard error and precision, and vice versa. In the absence of bias the shape of plots using sample size on the vertical axis will therefore not necessarily correspond to a funnel. A predictable shape in the absence of bias is required in order to use any plot to assess evidence for bias. For the same reason it is not possible to calculate expected 95 % confidence limits for funnel plots of sample size. Sterne and Egger (2001) concluded that sample size, and direct functions of sample size, are therefore not a good choice for the vertical axis in funnel plots.

Standard error, precision, variance and inverse variance are four valid choices for the vertical axis. However, there are important differences in the shapes of plots using these measures. The plots discussed above suggested that the use of standard error is likely to be preferable in many situations. Interpretation of funnel plots is facilitated by inclusion of lines representing the 95 % confidence limits around the summary intervention effect, which show the expected distribution of studies in the absence of heterogeneity or selection biases. Such lines can be drawn for all four variables; however, the lines will be straight and thus represent the shape of an inverted funnel only for standard error. For all other measures lines will be curved and, in the case of precision and inverse variance, smaller studies will often be compressed at the bottom of the graph. This makes the visual assessment of asymmetry more difficult and gives emphasis to the larger studies, which on average are less prone to bias. Because use of precision or the inverse variance for the vertical axis increases the distance between the largest study or studies and the rest, this choice might be preferable when investigators wish to focus on a comparison of meta-analyses of small studies with subsequent large studies (Cappelleri *et al.*, 1996).

An important note of caution, discussed in more detail in Chapter 6, applies to funnel plots for studies with binary outcomes such as odds ratios. When the true odds ratio differs from 1, there will be an association between the sampling distributions of the log odds ratio and its standard error, even in the absence of any real association between study size and the intervention effect. This was investigated in simulation analyses by Sterne *et al.* (2000), who found that substantial associations occurred when there was a substantial treatment effect, or the average number of events per trial in one or both groups was small, or all trials in the meta-analysis had similar sample sizes. In such circumstances it would be unwise to investigate bias using funnel plots.

Choice of horizontal axis (treatment effect estimate)

Sterne and Egger (2001) empirically investigated whether the prevalence of funnel plot asymmetry in published meta-analyses depends on the choice of treatment effect. A hand search of four general medicine and four specialist journals yielded 78 meta-analyses (containing 1145 trials) based on at least five trials with binary endpoints, and which randomized individuals and reported the number of subjects and events in each trial.

For each meta-analysis, the evidence for funnel plot asymmetry was investigated using the regression method proposed by Egger *et al.* (1997). This is described in more detail later in this section. There was evidence of funnel plot asymmetry for one or more of the treatment effect measures in 32 of the 78 meta-analyses. Consistent with the evidence cited earlier, treatment effects measured using risk differences showed more between-trial heterogeneity than effects measured with the log odds ratio or log risk ratio. The number of meta-analyses with evidence of funnel plot asymmetry was 21 (27 %), 22 (28 %) and 17 (22 %) for log odds ratio, log risk ratio and risk difference, respectively. While there was substantial concordance between log odds ratio and log risk ratio, use of risk difference suggested asymmetry in eight meta-analyses not identified as biased using the other treatment effect measures.

Engels *et al.* (2000) pointed out that use of risk differences can lead to studies with very few events in either treatment or control groups being given more weight

Table 5.1 Possible choices of axis in funnel plots for binary outcomes: advantages, disadvantages and recommendations.

Axis / measure	Advantages and disadvantages	Recommendations
Vertical axis		
Standard error	Funnel shape with straight 95 % confidence lines. Emphasis of the plot is on smaller studies where bias is more likely. Axis has to be inverted to place the largest studies at the top of the graph.	The best choice in most cases.
Precision (inverse standard error)	Plot is not funnel-shaped; 95 % confidence lines are curved. Emphasis of the plot is on larger studies; smaller studies are compressed at the bottom.	An option in studies which focus on a comparison of meta-analyses of small studies with subsequent large studies.
Variance (squared standard error)	Plot is not funnel-shaped; 95 % confidence lines are curved. Emphasis of the plot is on smaller studies where bias is more likely.	Not recommended.
Inverse variance	Plot is not funnel-shaped; 95 % confidence lines are curved. Emphasis of plot is on larger studies, smaller studies are compressed at the bottom.	An option in studies which focus on a comparison of meta-analyses of small studies with subsequent large studies, but precision would usually be better.
Sample size or log sample size	Expected shape of plot in absence of bias is unpredictable.	Invalid choice.
Horizontal axis		
Log odds ratio	Scale is not constrained and plots have the same shape whether the outcome is defined as occurrence or non-occurrence of the disease. Odds ratios may be misinterpreted as risk ratios.	The best choice in most cases.
Log risk ratio	Readily understood measure. Scale is naturally constrained so that heterogeneity may be introduced if the event rate is high.	Valid choice in many cases, but not recommended if the event rate is high. Can give different conclusions depending on outcome definition.
Risk difference	Readily understood measure. Often associated with increased heterogeneity which may result in funnel plot asymmetry which is not apparent when ratio measures are used.	Not recommended in most cases.

than studies with many events. Sterne and Egger (2001) concluded that this, and the greater heterogeneity of the risk difference between studies, contributed to the discrepant results of tests for funnel plot asymmetry. They concluded that risk differences will not usually be the best choice of horizontal axis in funnel plots. Independently of the reason for funnel plot asymmetry, such asymmetry calls into question the wisdom of statistically combining studies in meta-analysis.

Conclusions from tests of funnel plot asymmetry were similar whether log odds ratio or log risk ratio was chosen. This might be expected, as odds ratio and risk ratio give similar results unless event rates are high. However, funnel plots based on the odds ratio have the same shape whether the outcome is defined as occurrence or non-occurrence of the event of interest (for example, smoking cessation or continued smoking). In contrast, the shape of funnel plots based on risk ratios will differ depending on the definition of the outcome as occurrence or non-occurrence. The recommendations for the choice of axis in funnel plots made by Sterne and Egger (2001) are summarized in Table 5.1. In summary, standard error is likely to be the best choice of the vertical axis in funnel plots, unless the aim is to emphasize the difference between the largest studies and all others, when precision is more appropriate. A log ratio measure, preferably the log odds ratio, should generally be used for the horizontal axis.

FUNNEL PLOTS FOR STUDIES WITH NUMERICAL OUTCOMES

Choice of vertical axis (measure of study size)

So far as we are aware, no empirical investigations have examined the best choice of axes for funnel plots where the outcomes for subjects in the studies being displayed are measured on a numerical scale (such as depression score, blood pressure). The study outcomes in such cases are typically standardized numerical indices such as the standardized mean difference or correlation. Issues regarding choice of horizontal axis (treatment effect measure) seem less likely to be controversial when the treatment effects are essentially difference measures. While there are many reasons for considering treatment effects measured as mean differences to be more readily interpretable than standardized mean differences or correlation coefficients (see Bond $et\,al.$, 2003), these choices do not have obvious implications for the shape of funnel plots.

With at least two of the typical indices of effect used in the social sciences – the standardized mean difference d and the correlation r – the standard errors of the study outcomes depend on both the size of the study (n_{obs}) and the study result. For example, the standard error of the correlation is $SE(r) = (1 - \rho^2)/\sqrt{n_{obs}}$, which depends on the size of the population correlation as well as n_{obs}. (Typically this quantity is estimated by substituting the sample r for ρ.) Thus just as the size of the standard error of the log odds ratio depends on the number of events, the size of the standard error of r depends on the correlation observed in the sample. For instance, the standard error of a correlation of $r = 0.2$ from a sample of $n_{obs} = 178$ subjects is 0.072. The same standard error is obtained with a correlation of $r = 0.8$ and $n_{obs} = 25$. Thus it is difficult to compute stable confidence bands for funnel plots of correlations whether the rs are plotted against n_{obs} or their standard errors. Of course when the population correlation is constant then the standard error is a simple function of sample size and stable bands can be obtained. To compute error bands for a funnel plot, one can use the summary effect estimate in the computation of standard errors. However, again this makes clear that asymmetry in a correlation funnel plot may result from the presence of different population correlations in some studies, rather than from publication bias. Similar statements can be made about the standardized mean difference d, the variance of which depends on the population standardized mean difference.

An alternative for both the correlation and d is to make plots using variance-stabilizing transformations. These can be used to obtain indices with standard errors that are relatively simple functions of n_{obs}. For the correlation, we may use the familiar Fisher (1921) z transformation $z = \frac{1}{2}\ln[(1 + r)/(1 - r)]$. The value z has a standard error $1/\sqrt{(n_{obs} - 3)}$, which can be used to plot a confidence band for the z values against either n_{obs} or their standard errors. Similarly, standardized mean differences can be transformed using the inverse hyperbolic sine function described by Hedges and Olkin (1985). However, these indices are not as easily interpreted as the original study outcomes r and d. The scale of Fisher's z ranges from negative to positive infinity, thus it is not bounded at ± 1. To the extent that readers attach meaning to the values of study outcomes appearing in funnel plots, the variance-stabilized versions of r and d may not be as informative as the original values.

Choice of horizontal axis (treatment effect estimate)

Within the social sciences little has been done to investigate the use of different indices of study outcome in the funnel plot. However, some variation is seen in the indices that are used. Light *et al.* (1994) reproduced a funnel plot from Booth-Kewley and Friedman (1987) that plots correlations against sample size. But Mullen (1989, p. 76) plotted Fisher z values rather than correlations for a set of hypothetical correlational results. Mullen unfortunately did not give a rationale for this choice, but that may simply illustrate that variations in the outcome index used are not unusual enough to require justification.

In summary, the relative merits of the different possible choices of vertical axis in funnel plots (standard error, variance, precision, etc.) in studies with numerical outcome measures seem likely to be similar to those for studies with binary outcome measures. For studies with outcomes measured as correlations or as standardized mean differences, variance-stabilizing transformation of the outcome may be necessary.

INTERPRETATION – SOURCES OF FUNNEL PLOT ASYMMETRY

As discussed in the first section of this chapter, and illustrated in Figure 5.2(b), funnel plots were first proposed as a means of detecting a specific form of bias – publication bias. We have seen (Figure 5.2(c)) that the exaggeration of treatment

Table 5.2   Potential sources of asymmetry in funnel plots.

1. Selection biases
   Publication bias
   Location biases
      Language bias
      Citation bias
      Multiple publication bias

2. True heterogeneity
   Size of effect differs according to study size:
      Intensity of intervention
      Differences in underlying risk
      Other characteristics confounded with size

3. Data irregularities
   Poor methodological design of small studies
   Inadequate analysis
   Fraud

4. Artefact
   Heterogeneity due to poor choice of effect measure

5. Chance

Source: based on Egger *et al.* (1997).

effects in small studies of low quality provides a plausible alternative mechanism for funnel plot asymmetry. Egger *et al.* (1997) listed different possible reasons for funnel plot asymmetry; these are summarized in Table 5.2.

It is important to realize that *funnel plot asymmetry need not result from bias*. The studies displayed in a funnel plot may not always estimate the same underlying effect, and heterogeneity between results may lead to asymmetry in funnel plots if the true effect is larger in the smaller studies. For example, if a combined outcome is considered (for example, the outcome 'cardiovascular event' might include both stroke and myocardial infarction) then substantial benefit may be seen only in subjects at high risk for the component of the combined outcome that is affected by the intervention (Davey Smith and Egger, 1994; Glasziou and Irwig, 1995). A cholesterol-lowering drug which reduces coronary heart disease (CHD) mortality will have a greater effect on all-cause mortality in high-risk subjects with established cardiovascular disease than in young, asymptomatic subjects with isolated hypercholesterolaemia (Davey Smith *et al.*, 1993). This is because a consistent relative reduction in CHD mortality will translate into a greater relative reduction in all-cause mortality in high-risk subjects in whom a greater proportion of all deaths will be from CHD. Studies conducted in high-risk subjects will also tend to be smaller, because of the difficulty in recruiting such subjects and because increased event rates mean that smaller sample sizes are required to detect a given effect. Improvement of standard treatments over time may also lead to small-study effects, since small studies are generally conducted before larger ones, and improvement of standard (control) treatments will tend to reduce the relative efficacy of the experimental treatment.

Some interventions may also have been implemented less thoroughly in larger studies, thus explaining the more positive results in smaller studies. This is particularly likely in studies of complex interventions in chronic diseases, such as rehabilitation after stroke or multifaceted interventions in diabetes mellitus. For example, an asymmetrical funnel plot was found in a meta-analysis of trials examining the effect of inpatient comprehensive geriatric assessment programmes on mortality (Egger *et al.*, 1997; Stuck *et al.*, 1993) An experienced consultant geriatrician was more likely to be actively involved in the smaller trials, and this may explain the larger treatment effects observed in these trials (Egger *et al.*, 1997; Stuck *et al.*, 1993).

Odds ratios are more extreme (further from 1) than the corresponding risk ratio if the event rate is high. Because of this, a funnel plot that shows no asymmetry when plotted using risk ratios could still be asymmetric when plotted using odds ratios. This would happen if the smaller studies were consistently conducted in high-risk subjects, and the large studies in subjects at lower risk, although differences in underlying risk would need to be substantial.

Finally, it is of course possible that an asymmetrical funnel plot arises merely by the play of chance. Funnel plot asymmetry thus raises the possibility of bias but it is not proof of bias. However, asymmetry (unless produced by chance alone) will always lead us to question the interpretation of the overall estimate of effect when studies are combined in a meta-analysis – for example, if the study size predicts the treatment effect, then what treatment effect will apply if the treatment is adopted in routine practice? Sterne *et al.* (2000, 2001) have suggested that the funnel plot should be seen as a generic means of examining 'small-study effects' (the tendency for the smaller studies in a meta-analysis to show larger treatment effects) rather than as a tool to diagnose specific types of bias.

HOW MANY STUDIES ARE REQUIRED TO EXAMINE FUNNEL PLOT ASYMMETRY?

Because the assessment of funnel plots is of necessity subjective, it is not possible to provide formal guidance as to the number of studies in a meta-analysis that are required before funnel plots may be used to assess evidence for small-study effects in a meta-analysis. Informally, it seems unlikely that funnel plots will be useful in meta-analyses containing a small number of studies (fewer than five, say). The issue of power in examining funnel plot asymmetry is addressed in Chapter 6, on statistical tests for funnel plot asymmetry.

ADVANTAGES AND DISADVANTAGES COMPARED TO OTHER METHODS FOR EXAMINING BIAS

Other chapters in this book focus mainly on statistical methods for detecting bias, particularly publication bias, in meta-analysis. Visual assessment of funnel plots will of necessity lead to a subjective appraisal of the evidence for small-study effects, but will be of use in assisting a meta-analyst in gaining an understanding of the nature of the data. The desire for a less subjective appraisal of the evidence for funnel plot asymmetry led to the development of the statistical tests for funnel plot asymmetry described in Chapter 6.

The many possible causes of funnel plot asymmetry can be seen as both a weakness and a strength of this means of probing for bias in meta-analysis, compared to methods that look for specific types of bias, in particular publication bias. While it is natural to wish to reach definitive conclusions when funnel plot asymmetry is found, it will be clear from the discussion in the previous section that we believe investigators should consider instead a range of possible explanations, and will therefore need to be cautious in their interpretations. In contrast, formal methods to detect selection bias may seem more attractive because they can lead to the conclusion that a particular form of bias (publication bias) has led to the observed pattern of study results. For example, it has been proposed that we may use the available data in a meta-analysis to conduct sensitivity analyses in which the overall treatment effect is estimated after allowing for varying amounts of publication bias (Copas, 1999; Copas and Shi, 2000; Hedges and Vevea, 1994; Chapter 9, this volume). However, arguments similar to those earlier in this chapter imply that evidence of selection bias might also arise for other reasons. Given the typically low power of tests for funnel plot asymmetry (see Chapter 6) we think it unlikely that it will be possible to distinguish statistically between different forms of bias unless the meta-analysis contains a large number of studies (say, more than 50), in which case multivariable meta-regression analyses might assess the relative impact of study size, trial quality and other trial characteristics on the effect of the intervention. Such meta-analyses are very unusual in medical research (Sterne *et al.*, 2000).

ILLUSTRATIVE EXAMPLES

In this section we show three examples of the application of funnel plots to the data sets used throughout the book. We illustrate how displaying study characteristics (moderator variables) may change the interpretation of funnel plots, and highlight differences between the data sets where possible.

The effects of environmental tobacco smoke

Our first example is a meta-analysis of 37 studies of the effect of environmental tobacco smoke on the risk of lung cancer in lifetime non-smokers (Hackshaw *et al.*, 1997). Whether this meta-analysis was affected by publication bias has been a matter of ongoing controversy – see Copas and Shi (2000) and the subsequent correspondence (http://www.bmj.com). Odds ratios comparing spouses of smokers with spouses of non-smokers were derived from five cohort studies and 34 case–control studies. Each study provided an estimate of the risk ratio comparing the risk of lung cancer for non-smoking spouses of smokers relative to the risk of lung cancer in non-smoking spouses of nonsmokers. Figure 5.7 is a funnel plot of these

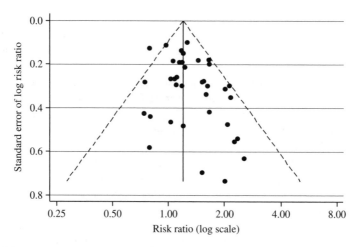

Figure 5.7   Funnel plot of data from the meta-analysis of the effects of environmental tobacco smoke on lung cancer, reported by Hackshaw *et al.* (1997).

data in which the standard error of the log risk ratio is plotted against the log risk ratio. The horizontal axis displays the risk ratio on a log scale; this gives a plot that is identical in shape to a plot of standard error of log RR against log RR. The plot shows asymmetry because smaller studies tend to show greater adverse effects of passive smoking (risk ratios greater than 1). This could, however, be due to the play of chance. Visual inspection alone does not allow us to say definitively whether the funnel plot is asymmetric (see also Chapter 6).

Teacher expectancy effects

Raudenbush (1984) reviewed 18 studies of teacher expectancy effects. In these studies students had been randomly assigned to either a control or 'expectancy induction' group where teachers were told to expect superior performance by their students. The work grew out of Robert Rosenthal's interest (see Rosenthal and Jacobson, 1968) in the effects of teachers' expectations of students' ability on students' performance – the so-called 'expectancy effect'. The effect size reflects the difference between these two groups on scores on an intelligence test.

One key difference between studies was the time at which the expectancy message was delivered. In some studies teachers were given the message before they had any contact with their students and in others the message was delivered after teachers had known their students for one or more weeks. In our analyses below, prior contact of more than one week is defined as high contact. Thus we are able to test the hypothesis that when teachers know their students better at the time of an expectancy induction, the expectancy effect will be smaller.

The results of each study were expressed as a standardized mean difference. Figure 5.8(a) is a funnel plot of these data, which appears asymmetric. However, Figure 5.8(b) shows that when the data are subdivided according to the amount of prior contact between the teacher and the students, it is apparent both that the

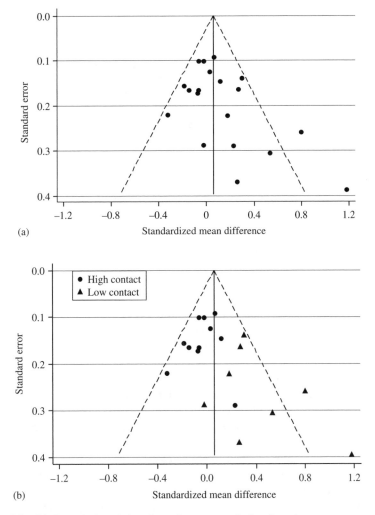

Figure 5.8 (a) Funnel plot of data from the meta-analysis of teacher expectancy effects, reported by Raudenbush (1984). (b) Funnel plot indicating whether the study had high or low prior contact.

standardized mean differences tended to be higher in the studies with low prior contact, and that these studies tended to be smaller. Little evidence of asymmetry is seen within studies of high or low contact.

Validity of employment interviews

McDaniel *et al.* (1994) examined correlations between employment interview scores and job performance in a meta-analysis of 160 studies on 25 244 people. The goal of the research was to see whether interviews are a valid means of selecting employees

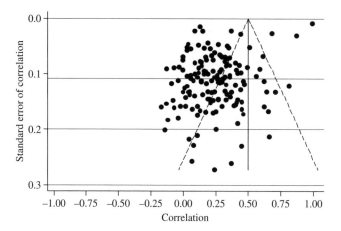

Figure 5.9 Funnel plot of data from the meta-analysis of employment interview validity, reported by McDaniel *et al.* (1994). This plot uses the untransformed correlation as the measure of intervention effect.

who will do well on the job, with a specific hypothesis that interview score is linearly related to job performance. Subsidiary hypotheses concerned whether the degree of structure of an interview (which was coded as either high or low in the meta-analysis) moderates its validity as an employment selection device.

Figure 5.9 shows a funnel plot of the correlation (horizontal axis) versus its standard error (vertical axis) with the standard error axis inverted so that larger standard errors are near the bottom of the plot. This plot shows considerably greater heterogeneity among studies with small standard errors (near the top of the funnel) than the plot's confidence bands suggest would occur if all studies came from a single population. Indeed, there is clear evidence of heterogeneity ($Q = 789.73$, $df = 159$, $p < 0.001$). Additionally, the fixed-effect estimate plotted appears to be some distance from the majority of the points. This is because of the influential points in the top right-hand corner (precise studies with large correlations). If a random-effects model is used to combine the data a pooled correlation of 0.24 is obtained (compared to 0.50) which is closer to the centre of the main body of studies since, in the presence of heterogeneity, studies get a more equal weighting using a random-effects model and hence less precise studies are more influential. Such large discrepancies between fixed- and random-effect pooled estimates are an indicator of asymmetry in the distribution of studies. As explained earlier in this chapter, plots such as this are difficult to interpret because the standard error of a correlation depends on the sample correlation.

Also as described earlier, these correlations were transformed to Fisher's z values for another plot. Figure 5.10(a) is a funnel plot of standard error of the transformed correlation (vertical axis) against the transformed correlation (horizontal axis). There is a clear impression of asymmetry, with smaller studies giving larger transformed correlations (however, note that the appearance of the plot has changed considerably from Figure 5.9). In Figure 5.10(b) these data are subdivided according to whether

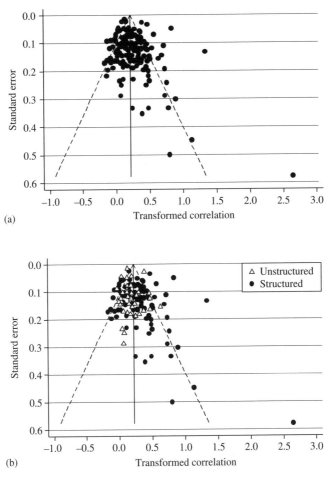

Figure 5.10   (a) Funnel plot of data from the meta-analysis of employment interview validity, reported by McDaniel *et al.* (1994). (b) Funnel plot indicating whether the study used structured or unstructured interviews. These plots use transformed correlations as the measure of intervention effect, because the standard error of the transformed correlation depends only on the sample size.

the study used structured or unstructured employment interviews. It appears that the asymmetry occurs only in the studies that used structured interviews.

CONCLUSIONS

Funnel plots are a useful graphical means of examining the evidence for small-study effects in meta-analyses. Statistical methods to examine bias that are based on funnel plots are described in Chapters 6 and 7 of this book.

REFERENCES

Becker, B.J. and Morton, S.C. (2002). Publication bias: Methods and practice in evidence-based medicine and social science. Mathematical Sciences Research Institute, Berkeley, California.

Begg, C.B. and Berlin, J.A. (1988). Publication bias: a problem in interpreting medical data. *Journal of the Royal Statistical Society (Series A)*, **151**, 419–463.

Bond, C.F., Wiitala, W.L. and Richard, F.D. (2003). Meta-analysis of raw mean differences. *Psychological Methods*, **8**, 406–418.

Booth-Kewley, S. and Friedman, H.S. (1987). Psychological predictors of heart disease: A quantitative review. *Psychological Bulletin*, **101**, 343–362.

Cappelleri, J.C., Ioannidis, J.P., Schmid, C.H., de Ferranti, S.D., Aubert, M., Chalmers, T.C. and Lau, J. (1996). Large trials vs meta-analysis of smaller trials: How do their results compare? *Journal of the American Medical Association*, **276**, 1332–1338.

Cooper, H.M. (1998). *Synthesizing Research: A Guide for Literature Reviews* (3rd edition). Thousand Oaks, CA: Sage.

Copas, J. (1999). What works? Selectivity models and meta-analysis. *Journal of the Royal Statistical Society (Series A)*, **162**, 95–109.

Copas, J.B. and Shi, J.Q. (2000). Reanalysis of epidemiological evidence on lung cancer and passive smoking. *British Medical Journal*, **320**, 417–418.

Davey Smith, G. and Egger, M. (1994). Who benefits from medical interventions? Treating low risk patients can be a high risk strategy. *British Medical Journal*, **308**, 72–74.

Davey Smith, G., Song, F. and Sheldon, T.A. (1993). Cholesterol lowering and mortality: The importance of considering initial level of risk. *British Medical Journal*, **306**, 1367–1373.

Deeks, J.J. and Altman, D.G. (2001). Effect measures for meta-analysis of trials with binary outcomes. In M. Egger, G. Davey Smith and D.G. Altman (eds), *Systematic Reviews in Health Care: Meta-Analysis in Context*, 2nd edition, pp. 313–335. London: BMJ Books.

Devine, E. and Cook, T.D. (1983). A meta-analytic analysis of effects of psychoeducational interventions on length of postsurgical hospital stay. *Nursing Research*, **32**, 267–274.

Dickersin, K. (1997). How important is publication bias? A synthesis of available data. *AIDS Education & Prevention*, **9**, S15–21.

Dickersin, K., Min, Y.I. and Meinert, C.L. (1992). Factors influencing publication of research results: follow-up of applications submitted to two institutional review boards. *Journal of the American Medical Association*, **263**, 374–378.

Easterbrook, P.J., Berlin, J.A., Gopalan, R. and Matthews, D.R. (1991). Publication bias in clinical research. *Lancet*, **337**, 867–872.

Egger, M. and Davey Smith, G. (1995). Misleading meta-analysis. Lessons from an 'effective, safe, simple' intervention that wasn't. *British Medical Journal*, **310**, 752–754.

Egger, M., Davey Smith, G., Schneider, M. and Minder, C. (1997). Bias in meta-analysis detected by a simple, graphical test. *British Medical Journal*, **315**, 629–634.

Egger, M., Davey Smith, G. and Altman, D. (2001). *Systematic Reviews in Health care: Meta-analysis in Context*. London: BMJ Books.

Egger, M., Jüni, P., Bartlett, C., Holenstein, F. and Sterne, J. (2003). How important are comprehensive literature searches and the assessment of trial quality in systematic reviews? Empirical study. *Health Technology Assessment*, **7**, 1–68.

Engels, E.A., Schmid, C.H., Terrin, N.T., Olkin, I. and Lau, J. (2000). Heterogeneity and statistical significance in meta-analysis: An empirical study of 125 meta-analyses. *Statistics in Medicine*, **19**, 1707–1728.

Fisher, R.A. (1921). On the 'probable error' of a coefficient of correlation deduced from a small sample. *Metron*, **1**, 1–32.

Gerber, S., Tallon, D., Schneider, M. and Egger, M. (2005). Characteristics of meta-analyses published in leading medical journals 1993 to 2002: Bibliographic study. Manuscript in preparation.

Glasziou, P.P. and Irwig, L.M. (1995). An evidence based approach to individualising treatment. *British Medical Journal*, **311**, 1356–1359.

Glasziou, P., Irwig, L.M., Bain, C. and Colditz, G. (2001). *Systematic Reviews in Health Care: A Practical Guide*. Cambridge: Cambridge University Press.

Hackshaw, A.K., Law, M.R. and Wald, N.J. (1997). The accumulated evidence on lung cancer and environmental tobacco smoke. *British Medical Journal*, **315**, 980–988.

Hedges, L.V. and Olkin, I. (1985). *Statistical Methods for Meta-analysis*. Boston: Academic Press.

Hedges, L.V. and Vevea, J.L. (1994). Estimating effect size under publication bias: small sample properties and robustness of a random effects selection model. *Journal of Educational & Behavioral Statistics*, **21**, 299–332.

Ioannidis, J.P. (1998). Effect of the statistical significance of results on the time to completion and publication of randomized efficacy trials. *Journal of the American Medical Association*, **279**, 281–286.

Ioannidis, J.P.A., Ntzani, E.E., Trikalinos, T.A. and Contopoulos-Ioannidis, D.G. (2001). Replication validity of genetic association studies. *Nature Genetics*, **29**, 306–309.

ISIS-2 (Second International Study of Infarct Survival) Collaborative Group (1988). Randomised trial of intravenous streptokinase, oral aspirin, both, or neither among 17 187 cases of suspected acute myocardial infarction: ISIS-2. *Lancet*, **ii**, 349–360.

Light, R.J. and Pillemer, D.B. (1984). *Summing Up: The Science of Reviewing Research*. Cambridge, MA: Harvard University Press.

Light, R.J., Singer, J.D. and Willett, J.B. (1994). The visual presentation and interpretation of meta-analyses. In H.M. Cooper and L.V. Hedges (eds), *The Handbook of Research Synthesis*. New York: Russell Sage Foundation.

Lipsey, M.W. and Wilson, D.B. (2001). *Practical Meta-analysis*. Thousand Oaks, CA: Sage.

Lohmueller, K.E., Pearce, C.L., Pike, M., Lander, E.S. and Hirschhorn, J.N. (2003). Meta-analysis of genetic association studies supports a contribution of common variants to susceptibility to common disease. *Nature Genetics*, **33**, 177–182.

McDaniel, M.A., Whetzel, D., Schmidt, F.L. and Maurer, S. (1994). The validity of the employment interview: A comprehensive review and meta-analysis. *Journal of Applied Psychology*, **79**, 599–616.

Mullen, B. (1989). *Advanced BASIC Meta-analysis*. Hillsdale, NJ: Lawrence Erlbaum.

Poole, C. and Greenland, S. (1999). Random-effects meta-analyses are not always conservative. *American Journal of Epidemiology*, **150**, 469–475.

Raudenbush, S.W. (1984). Magnitude of teacher expectancy effects on pupil IQ as a function of the credibility of expectancy induction. *Journal of Educational Psychology*, **76**, 85–97.

Rosenthal, R. (1991) *Meta-analytic Procedures for Social Research*, revised edition. Newbury Park, CA: Sage.

Rosenthal, R. and Jacobson, L. (1968). *Pygmalion in the Classroom: Teacher Expectation and Pupils' Intellectual Development*. New York: Holt, Rinehart and Winston.

Rothstein, H.R., McDaniel, M.A. and Borenstein, M. (2002). Meta-analysis: A review of quantitative cumulation methods. In N. Schmitt and F. Drasgow (eds), *Measuring and Analyzing Behavior in Organizations*, pp. 534–570. San Francisco: Jossey-Bass.

Schulz, K.F., Chalmers, I., Hayes, R.J. and Altman, D.G. (1995). Empirical evidence of bias. Dimensions of methodological quality associated with estimates of treatment effects in controlled trials. *Journal of the American Medical Association*, **273**, 408–412.

Stern, J.M. and Simes, R.J. (1997). Publication bias: Evidence of delayed publication in a cohort study of clinical research projects. *British Medical Journal*, **315**, 640–645.

Sterne, J.A. and Egger, M. (2001). Funnel plots for detecting bias in meta-analysis: Guidelines on choice of axis. *Journal of Clinical Epidemiology*, **54**, 1046–1055.

Sterne, J.A.C., Gavaghan, D. and Egger, M. (2000). Publication and related bias in meta-analysis: Power of statistical tests and prevalence in the literature. *Journal of Clinical Epidemiology*, **53**, 1119–1129.

Sterne, J.A.C., Egger, M. & Davey Smith, G. (2001). Investigating and dealing with publication and other biases. In M. Egger, D.G. Altman and G. Davey Smith (eds), *Systematic Reviews in Health Care: Meta-analysis in Context*, 2nd edition. London: BMJ Books.

Stuck, A.E., Siu, A.L., Wieland, G.D., Adams, J. and Rubenstein, L.Z. (1993). Comprehensive geriatric assessment: A meta-analysis of controlled trials. *Lancet*, **342**, 1032–1036.

Sutton, A.J., Abrams, K., Jones, D.R., Sheldon, T.A. and Song, F. (2002) *Methods for Meta-analysis in Medical Research*. Chichester: John Wiley & Sons, Ltd.

Vandenbroucke, J.P. (1988). Passive smoking and lung cancer: a publication bias? *British Medical Journal of Clinical Research Education*, **296**, 391–392.

Wang, M.C. and Bushman, B.J. (1998). Using the normal quantile plot to explore meta-analytic data sets. *Psychological Methods*, **3**, 46–54.

Woods, K.L. and Fletcher, S. (1994). Long term outcome after magnesium sulphate in suspected acute myocardial infarction: The Second Leicester Intravenous Magnesium Intervention Trial (LIMIT-2). *Lancet*, **343**, 816–819.

CHAPTER 6

Regression Methods to Detect Publication and Other Bias in Meta-Analysis

Jonathan A.C. Sterne
University of Bristol, UK

Matthias Egger
University of Bern, Switzerland

KEY POINTS

- This chapter describes two statistical tests for funnel plot asymmetry.
- Begg and Mazumdar (1994) have proposed a non-parametric test based on the rank correlation between intervention effect estimates and their sampling variances.
- Egger *et al.* (1997) have proposed a regression method which tests for a linear association between the intervention effect and its standard error.
- Simulation analyses have shown that the power of the methods is low unless there is severe bias. The regression method is generally more powerful than the rank correlation method.
- To avoid the potential inflation of the Type I error rate, the methods should only be used if there is clear variation in study sizes, with one or more trials of medium or large size.

Publication Bias in Meta-Analysis – Prevention, Assessment and Adjustments Edited by H.R. Rothstein, A.J. Sutton
and M. Borenstein © 2005 John Wiley & Sons, Ltd

INTRODUCTION – RATIONALE FOR REGRESSION METHODS

As discussed in Chapter 5, funnel plots are a useful tool for diagnosing small-study effects in meta-analysis. Possible causes of funnel plot asymmetry are shown in Table 5.2. While the visual appraisal of funnel plots will assist a meta-analyst in gaining an understanding of the nature of the data, the desire for a less subjective appraisal of the evidence for funnel plot asymmetry led to the development of statistical tests for funnel plot asymmetry, which are described in this chapter.

Throughout this chapter, we will denote the intervention effect estimate (e.g. standardized mean difference or log odds ratio) from study i as θ_i, and its corresponding variance and standard error as v_i and s_i respectively, where $v_i = s_i^2$.

BEGG AND MAZUMDAR'S RANK CORRELATION METHOD

Begg and Mazumdar (1994) proposed an adjusted rank correlation method to examine the association between the effect estimates and their sampling variances. The first stage in conducting this test is to stabilize the variances by deriving standardized effect sizes $\{\theta_i^*\}$, where

$$\theta_i^* = (\theta_i - \bar{\theta})/\sqrt{v_i^*}$$

where $\bar{\theta} = \left(\sum \theta_i v_i^{-1}\right)\big/\sum v_i^{-1}$ is the usual fixed-effect estimate of the summary effect, and $v_i^* = v_i - (\sum v_i^{-1})^{-1}$ is the variance of $(\theta_i - \bar{\theta})$.

The test is based on deriving Kendall's rank correlation between θ_i^* and v_i^*, which is based on comparing the ranks of the two quantities. The ranks are the ordered values: for example, the largest value of θ_i^* will have rank 1, the next largest will have rank 2, and so on. If there are k possible studies then there are $k(k-1)/2$ possible pairs of studies. For studies i and j, the pairs (θ_i^*, v_i^*) and (θ_j^*, v_j^*) are said to be concordant if their ranks differ in the same directions, that is, if both the θ_i^* and v_i^* ranks for study i are lower than the corresponding ranks for study j, or both are higher. The ranks are said to be discordant if the comparison of the ranks of the two variables is in opposite directions. If there are x concordant ranks and y discordant ranks, the normalized test statistic is

$$z = \frac{x - y}{\sqrt{k(k-1)(2k+5)/18}}.$$

It is conventional to report the two-sided *p*-value corresponding to this test statistic. Note that because the ranks of the standard errors $\{s_i\}$ are the same as the ranks of the variances $\{v_i\}$, Begg and Mazumdar's test is equivalent to deriving the rank correlation between the standardized effect sizes and their standard errors.

Example – magnesium to prevent mortality following myocardial infarction

The trials of magnesium for prevention of death following myocardial infarction were discussed in Chapter 5. Application of the rank correlation method showed that Kendall's rank correlation was -0.077 ($p = 0.702$) when the very large ISIS-4 trial was excluded, and 0.105 ($p = 0.586$) when the ISIS-4 trial was included. The rank correlation method thus provided no evidence of funnel plot asymmetry for these data.

EGGER'S LINEAR REGRESSION METHOD

Egger *et al.* (1997) introduced a linear regression approach in which the standard normal deviate z_i (defined as $z_i = \theta_i/s_i$) is regressed against its precision $prec_i$ (defined as $prec_i = 1/s_i$):

$$\mathrm{E}[z_i] = \beta_0 + \beta_1 prec_i.$$

The smaller the trial, the smaller will be its precision. In the absence of funnel plot asymmetry, the points in a plot of z_i against $prec_i$ – a plot that corresponds to Galbraith's (1988) radial plot – will scatter about a line that runs through the origin at standard normal deviate zero (intercept $\beta_0 = 0$), with the slope β_1 indicating the size and direction of effect. This situation corresponds to a symmetrical funnel plot. If there is funnel plot asymmetry, the regression line will not run through the origin, so that the intercept β_0 provides a measure of asymmetry – the larger its deviation from zero the more pronounced the asymmetry. A test of the null hypothesis that $\beta_0 = 0$ (no funnel plot asymmetry) can be derived from the usual regression output produced by statistical packages: the two-sided p-value should be reported. This test is often referred to as the 'Egger test' and has been widely used as a test for funnel plot asymmetry.

It should be noted that in their original report, Egger *et al.* (1997) suggested an alternative version of this test in which the regression of z on p is weighted by the inverse of the variance of the effect estimate, that is, by $1/v$. This weighted version of the test lacked a theoretical justification, and is no longer advocated by Egger *et al.* We strongly suggest that users conduct only the unweighted regression of z on $prec$.

The *unweighted* regression of z on $prec$ can be shown to be identical to a *weighted* regression of the intervention effect θ on its standard error s ($\mathrm{E}[\theta_i] = \beta_1 + \beta_0 s_i$), where the weights $\{1/v_i\}$ are inversely proportional to the variance of the estimated intervention effect in each study. Note that the *intercept* β_0 in the regression of z_i on $prec_i$ corresponds to the *slope* in the weighted regression of θ_i on s_i. Therefore, the regression test for funnel plot asymmetry corresponds to a test of linear trend in a funnel plot of intervention effect against its standard error. The inverse-variance weighting allows for between-study heterogeneity by including a multiplicative overdispersion parameter ϕ by which the variance v_i in each study is multiplied (Thompson and Sharp, 1999). We will use the notation $\theta_i \sim N(\theta_i, \sigma_i^2)$ to indicate that the effect of intervention in study i is normally distributed with mean θ_i and variance σ_i^2. The statistical model for the regression test is

$$\theta_i \sim N(\beta_0 s_i + \beta_1, v_i\phi).$$

META-REGRESSION TO EXAMINE SMALL-STUDY EFFECTS

The regression test for funnel plot asymmetry is closely related to a random-effects meta-regression of θ_i on s_i. In random-effects meta-regression, the between-study heterogeneity is usually modelled using an additive between-study component of variance, τ^2, so that the total variance for study i is assumed to equal $v_i + \tau^2$ rather than $v_i \phi$ (Thompson and Sharp, 1999). The statistical model is therefore

$$\theta_i \sim N(\beta_0 s_i + \beta_1, v_i + \tau^2).$$

Using an additive between-study component of variance has the advantage that in the absence of between-study heterogeneity τ^2 will be estimated to be zero, so that the model reduces to a fixed-effect meta-regression. In contrast, under a multiplicative model it is perfectly possible for the estimated value of ϕ to be less than 1, in which case the model implies that there is less between-study heterogeneity than would be expected by chance alone. Since it is implausible that such underdispersion will arise other than by chance, this is an undesirable property of models with multiplicative overdispersion.

For both the rank correlation and regression methods, two-sided rather than one-sided p-values should be reported. Although the expectation is that the direction of small-study effects will be that small studies show larger treatment effects, it is always possible that the pattern may be in the opposite direction.

Example – magnesium to prevent myocardial infarction

For the meta-analysis of 16 trials of magnesium to prevent death following myocardial infarction, p-values from the regression method were 0.009 when the very large ISIS-4 trial was excluded, and less than 0.001 when the ISIS-4 trial was included. The regression method thus provided clear evidence of funnel plot asymmetry for these data. The bias coefficient (the coefficient of standard error in the weighted regression of log odds ratio on standard error) was -1.27 (95 % confidence interval (CI) -2.15 to -0.39) when ISIS-4 was excluded, and -1.67 (95 % CI -2.29 to -1.03) when ISIS-4 was included. Figure 6.1 shows the funnel plots for these trials (a) excluding and (b) including the ISIS-4 trial. The dotted lines are the fitted regression lines corresponding to the regression test for funnel plot asymmetry. Note that the outcome (log odds ratio) is plotted on the horizontal rather than the vertical axis and so the regression line would be vertical in the absence of bias. The slope of the regression line in the lower panel is further from vertical, corresponding to the absolute value of the bias coefficient being greater when the ISIS-4 trial is included.

It is noteworthy that in this example the rank correlation and regression methods led to very different conclusions about the evidence for funnel plot asymmetry. This is probably because the regression method allows for the very high precision of the treatment effect estimate in the ISIS-4 trial: by definition, a method based only on ranks cannot do this.

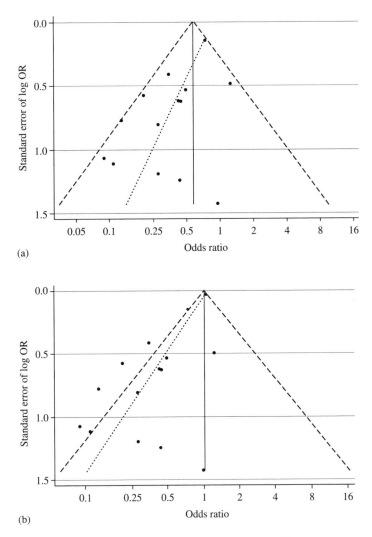

(a)

(b)

Figure 6.1 Funnel plots for magnesium trials (a) excluding and (b) including the very large ISIS-4 trial. The dotted line is the fitted regression line corresponding to the regression (Egger) test for funnel plot asymmetry. The dashed lines show the expected 95 % confidence intervals around the fixed-effect summary estimate. Odds ratios are plotted on a logarithmic scale, to ensure that effects of the same magnitude but in opposite directions are equidistant from 1.0.

CHARACTERISTICS OF THE METHODS

Simulation studies have been used to assess the characteristics of the rank correlation and regression tests for funnel plot asymmetry, and to compare them. Begg and Mazumdar (1994) evaluated their rank correlation method by using simulations based on selection models in which the probability that a study was selected for a meta-analysis was assumed to depend either on the observed p-value (one-sided or two-sided), or on the size of the observed treatment effect. These simulations were based either on 'small' meta-analyses (25 studies) or 'large' meta-analyses (75); the numbers being chosen based on the mean number of studies in samples of meta-analyses from medical research and social sciences, respectively. They concluded that the rank correlation test was 'fairly powerful' for large meta-analyses with 75 component studies, but had only moderate power for meta-analyses with 25 component studies. In simulations in which the selection process was assumed to be based on the study p-value, the power of the test ranged from 3 % to 58 % in small studies, and from 5 % to 99 % in large studies.

Sterne *et al.* (2000) compared the characteristics of the rank correlation and regression methods, in simulations that assumed that there was a linear association (or no association, assumed equivalent to no bias) between the treatment effect (log odds ratio) and its standard error. Based on a sample of 78 meta-analyses, they derived 'typical' meta-analyses containing 5, 10, 20 and 30 trials. Among the 78 meta-analyses the median number of trials was 10, despite the fact that meta-analyses containing fewer than 5 trials were excluded from the sample. These authors concluded that the regression method was more powerful than the rank correlation method, in situations typical of meta-analyses in medical research. For example, Figure 6.2 shows the power of the rank correlation and regression methods to detect small-study effects ($p < 0.1$), for amounts of underlying bias defined as no, moderate or severe. It can be seen that in the absence of bias the proportion of false positive results was close to the correct value of 10 %, while power, which increased with increasing numbers of trials, was low for meta-analyses with 10 or fewer trials and for moderate bias. In general, the rank correlation method was less powerful than the regression method, though it should be noted that these simulations assumed a linear association between log odds ratio and standard error, as is also assumed by the regression method. In further simulations in which the true treatment odds ratio differed from 1, the Type I error rate for the rank correlation test was generally too low.

Concerns about the validity of the rank correlation and regression methods have been raised by a number of authors, including Irwig *et al.* (1998), Macaskill *et al.* (2001) and Schwarzer *et al.* (2002). Problems with these methods result from the fact that when the true treatment effect is substantial (i.e. the odds ratio is a long way from 1) the measurement error in the log odds ratio will be correlated with measurement error in its standard error. In further simulations based on the characteristics of 78 meta-analyses, Sterne *et al.* (2000) concluded that the Type I error rate was inflated when there was a substantial treatment effect, or the average number of events per trial in one or both groups was small, or all trials in the meta-analysis had similar sample sizes. They suggested that use of the regression and rank correlation methods should be avoided in these circumstances, but noted that the main use of meta-analysis is to estimate moderate, rather than substantial, treatment effects.

It would clearly be desirable to develop a test for funnel plot asymmetry that retained the simplicity of the regression method, and was of comparable power, but which avoided the problem of inflated Type I error rates when the treatment effect is substantial. Macaskill *et al.* (2001) proposed an alternative test based on

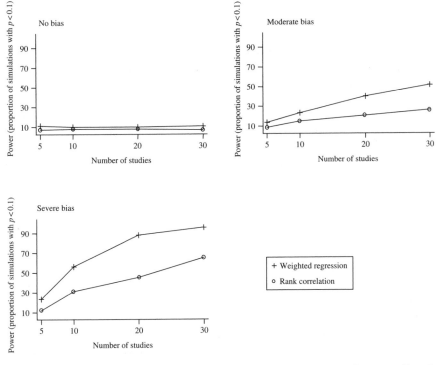

Figure 6.2 Power of the rank correlation and regression methods to detect small-study effects $(p < 0.1)$, for different amounts of underlying bias, from simulations reported in Sterne *et al.* (2000). Simulations were based on four hypothetical meta-analyses which are typical for those published in leading journals 1993–1997.

a regression of the standardized intervention effect on sample size. While this test does not suffer from inflated Type I error rates in the presence of substantial treatment effects, its power was lower than that of the regression method proposed by Egger *et al.* (1997). Further, as noted by Sterne and Egger (2001) and discussed in more detail in Chapter 5, the major factor affecting the probability of publication of a study is its *p*-value (Dickersin, 1997; Dickersin *et al.*, 1992; Easterbrook *et al.*, 1991; Ioannidis, 1998; Stern & Simes, 1997), which is a function of standard error or variance rather than of sample size. Thus, for binary outcomes, it seems to us that there is little rationale for a regression of intervention effect on sample size, and the method proposed by Macaskill *et al.* has not been widely used in practice.

WHEN ARE TESTS FOR FUNNEL PLOT ASYMMETRY APPROPRIATE?

The simulations reported by Begg and Mazumdar (1994) and by Sterne *et al.* (2000) lead to similar conclusions – that the power of the methods is low if the number of studies included in a meta-analysis is 10 or fewer, or in the absence of severe bias.

In medical research, the number of studies in a meta-analysis is often less than or equal to 10: in such situations investigators should be aware that the methods can detect only substantial small-study effects. As is generally true when parametric methods are compared to non-parametric methods (providing that the parametric model is approximately correct), the regression test for funnel plot asymmetry has greater power than the rank correlation test. Given that the tests will typically be used when power is at best moderate, this seems a major advantage of the regression method. The regression method is appropriate in situations in which meta-analysis generally makes sense; in estimating moderate treatment effects, based on a reasonable number of studies. However, to avoid the potential inflation of the Type I error rate described earlier, it should only be used if there is clear variation in study sizes, with one or more trials of medium or large size.

EXTENSIONS

It would clearly be desirable to extend the methods to allow for potential explanations for the funnel plot asymmetry, by examining whether the association between the intervention effect and study size is affected when we control for the association of other study characteristics with the intervention effect. We are not aware of any attempts to extend the rank correlation method to allow for other study characteristics. However, the regression method (in its weighted form) can be extended readily to include the effects of $p - 1$ additional study characteristics x_2 to x_p. For the reasons given earlier, it will generally be preferable to use random-effects meta-regression to examine the joint effect of a number of study characteristics on the estimated effect of intervention. The extended version of the meta-regression model described in section above is

$$\theta_i \sim N(\beta_0 s_i + \beta_1 + \beta_2 x_{2i} + \cdots + \beta_p x_{pi}, v_i + \tau^2).$$

Note that the regression coefficients from these models correspond to *differences between mean differences* when the outcome variable is numerical, and to *differences between log odds ratios* when outcomes are binary and the effect of the intervention is expressed as a log odds ratio. For binary outcomes, results of meta-regression analyses are often reported as the exponentiated regression coefficient and corresponding CI: these correspond to *ratios of odds ratios*.

For example, Sterne *et al.* (2001) reported the results of a reanalysis of a meta-analysis of randomized controlled trials of homeopathy (Linde *et al.*, 1997). The funnel plot of these trials was clearly asymmetric, and the estimated bias coefficient $\hat{\beta}_0$ from the regression test for funnel plot asymmetry was -1.72 (95 % CI -2.42 to -1.01). A meta-regression of each intervention log odds ratio on its standard error gave similar results ($\hat{\beta}_0 = -1.71$, 95 % CI -2.35 to -1.07).

The middle column of Table 6.1 shows results from univariable meta-regression analyses of a number of study characteristics, expressed as ratios of odds ratios. For example, the estimated ratio of odds ratios per unit increase in standard error is $e^{-1.71} = 0.18$, implying that the effects of homeopathy were substantially more beneficial in smaller studies. The univariable analyses also suggested that publication

Table 6.1 Meta-regression analysis of the results of 89 homoeopathy trials from the meta-analysis by Linde *et al.* (1997). Data kindly provided by Dr Klaus Linde.

Study characteristic	Univariable analysis		Controlling for all variables	
	Ratio of odds ratios* (95 % CI)	*p*-value	Ratio of odds ratios[a] (95 % CI)	*p*-value
Standard error of log OR	0.18 (0.10,0.34)	<0.001	0.20 (0.11,0.37)	<0.001
Language (Non-English vs. English)	0.73 (0.51,1.06)	0.097	0.73 (0.55,0.98)	0.038
Study quality				
Allocation concealment (not adequate vs. adequate)	0.70 (0.49,1.01)	0.054	0.98 (0.73,1.30)	0.87
Blinding (not double-blind vs. double-blind)	0.24 (0.12,0.46)	<0.001	0.35 (0.20,0.60)	<0.001
Handling of withdrawals (not adequate vs. adequate)	1.32 (0.87,1.99)	0.19	1.10 (0.80,1.51)	0.56
Publication type (Not MEDLINE-indexed vs. MEDLINE-indexed)	0.61 (0.42,0.90)	0.013	0.91 (0.67,1.25)	0.57

[a] Odds ratio with characteristic divided by odds ratio without characteristic. Ratios below 1 correspond to a smaller treatment odds ratio for trials with the characteristic, and hence a larger apparent benefit of homoeopathic treatment.

in a non-English language, inadequate allocation concealment, absence of double blinding and publication in a non MEDLINE-indexed journal were associated with increased (more beneficial) intervention effects. The right-hand column of Table 6.1 shows the results of a multivariable meta-regression analysis including all study characteristics. A strong association between intervention effect and its standard error remained even after controlling for the other variables, and publication in a non-English language and absence of double-blinding were still associated with more beneficial intervention effects. In contrast, there was no longer any evidence that allocation concealment was associated with the intervention effect.

ADVANTAGES AND DISADVANTAGES COMPARED TO OTHER METHODS FOR EXAMINING BIAS

It was explained in Chapter 5 that publication bias is only one of a number of possible reasons for funnel plot asymmetry. The tests for funnel plot asymmetry described in this chapter should therefore not be seen as specifically aimed at detecting publication bias, but rather as examining the evidence for 'small-study effects' (see Chapter 5) in a meta-analysis. This distinguishes them from many of the other methods described in this book – failsafe *N* (Chapter 7), trim and fill (Chapter 8), and selection model approaches (Chapter 9) – which examine the evidence that publication or selection biases have affected the results of a meta-analysis.

From different viewpoints, this may be seen as either a strength or a weakness of the tests for funnel plot asymmetry. In so far as they do not make specific assumptions about the processes leading to funnel plot asymmetry, they may be seen as valid in a wide range of circumstances. It seems likely that tests specifically aimed at detecting publication bias could give misleading results in situations where small studies differ from larger ones for reasons other than selection effects. On the other hand, a diagnosis of funnel plot asymmetry made as a result of one of the tests does not allow the investigator to conclude what factor or factors caused the asymmetry, to say whether it is results from the smaller or larger studies that are more valid, or to estimate the treatment effect 'corrected' for bias. They can only lead investigators to question the overall effect estimate and to consider further the possible causes of the small-study effects.

ILLUSTRATION USING THE THREE EXAMPLE DATA SETS

Table 6.2 shows results from the rank correlation and regression tests for funnel plot asymmetry, for the three example meta-analyses being considered throughout this book. Consistent with the results of the simulations discussed earlier in this chapter, the p-value from the regression test was smaller than the p-value from the rank correlation test in the passive smoking meta-analysis, while the p-values from the two methods were similar in the teacher expectancy and interview validity meta-analyses. There was some evidence for funnel plot asymmetry in each meta-analysis.

As discussed earlier in this chapter, the regression test for funnel plot asymmetry is closely related to a meta-regression of the intervention effect on its standard error. Table 6.3 (middle column) shows that for the teacher expectancy meta-analysis, fitting a meta-regression of the standardized mean difference on its standard error gave a regression coefficient of 1.75 (95 % CI 0.40 to 3.11, $p = 0.01$). This is similar to the results from the regression test shown in Table 6.2 (regression coefficient 1.58, 95 % CI -0.04 to 3.20, $p = 0.055$). As shown in Figure 5.8(b) in Chapter 5, the funnel plot asymmetry in the teacher expectancy meta-analysis seemed to be explained by differences (moderator effects) between the studies in which the teachers had high and low prior contact with the students. Table 6.3 (middle column) shows that the regression coefficient for low versus high contact was 0.37 (95 % CI

Table 6.2 Summary of results from the rank correlation and regression tests for funnel plot symmetry, for the three example meta-analyses considered in this book.

Meta-analysis	Rank correlation method		Regression method	
	p-value	Rank correlation	p-value	Coefficient of s.e. (95 % CI)
Passive smoking	0.21	0.14	0.024	0.89 (0.13, 1.66)
Teacher expectancy	0.086	0.29	0.055	1.58 (-0.04, 3.20)
Interview validity	0.015	0.13	0.031	0.63 (0.06, 1.19)

Table 6.3 Results from meta-regression analyses of the association between intervention effects and study characteristics, in the teacher expectancy meta-analysis.

Study characteristic	Univariable analysis		Controlling for the other variable	
	Coefficient (95 % CI)	*p*-value	Coefficient (95 % CI)	*p*-value
Standard error of intervention effect	1.75 (0.40, 3.11)	0.01	0.31 (−1.10, 1.72)	0.663
Low versus high contact	0.37 (0.20, 0.55)	<0.001	0.35 (0.14, 0.55)	0.001

0.20 to 0.55). Thus, on average, the standardized mean difference was 0.37 greater in low-contact studies than in high-contact studies. To investigate whether the small-study effects seen in this study can be explained by the differences between low- and high-contact studies, we fitted a regression model including both standard error of the intervention effect and an indicator variable for low contact studies. Results from this model are shown in the right-hand column of Table 6.3. After adjusting for contact, there is no longer any evidence that the intervention effect is related to its standard error in this meta-analysis. The estimated difference between low- and high-contact studies is little altered after adjusting for standard error.

REFERENCES

Begg, C.B. and Mazumdar, M. (1994). Operating characteristics of a rank correlation test for publication bias. *Biometrics*, **50**, 1088–1101.

Dickersin, K. (1997). How important is publication bias? A synthesis of available data. *AIDS Education & Prevention*, **9**, 15–21.

Dickersin, K., Min, Y.I. and Meinert, C.L. (1992). Factors influencing publication of research results: Follow-up of applications submitted to two institutional review boards. *Journal of the American Medical Association*, **263**, 374–378.

Easterbrook, P.J., Berlin, J.A., Gopalan, R. and Matthews, D.R. (1991). Publication bias in clinical research. *Lancet*, **337**, 867–872.

Egger, M., Davey Smith, G., Schneider, M. and Minder, C. (1997). Bias in meta-analysis detected by a simple, graphical test. *British Medical Journal*, **315**, 629–634.

Galbraith, R.F. (1988). A note on graphical presentation of estimated odds ratios from several clinical trials. *Statistics in Medicine*, **7**, 889–894.

Ioannidis, J.P. (1998). Effect of the statistical significance of results on the time to completion and publication of randomized efficacy trials. *Journal of the American Medical Association*, **279**, 281–286.

Irwig, L., Macaskill, P., Berry, G. and Glasziou, P. (1998). Bias in meta-analysis detected by a simple, graphical test. Graphical test is itself biased. *British Medical Journal*, **316**, 470–471.

Linde, K., Clausius, N., Ramirez, G., Melchart, D., Eitel, F., Hedges, L.V. and Jonas, W.B. (1997). Are the clinical effects of homeopathy placebo effects? A meta-analysis of placebo-controlled trials. *Lancet*, **350**, 834–843.

Macaskill, P., Walter, S.D. and Irwig, L. (2001). A comparison of methods to detect publication bias in meta-analysis. *Statistics in Medicine*, **20**, 641–654.

Schwarzer, G., Antes, G. and Schumacher, M. (2002). Inflation of type I error rate in two statistical tests for the detection of publication bias in meta-analyses with binary outcomes. *Statistics in Medicine*, **21**, 2465–2477.

Stern, J.M. and Simes, R.J. (1997). Publication bias: Evidence of delayed publication in a cohort study of clinical research projects. *British Medical Journal*, **315**, 640–645.

Sterne, J.A. and Egger, M. (2001). Funnel plots for detecting bias in meta-analysis: Guidelines on choice of axis. *Journal of Clinical Epidemiology*, **54**, 1046–1055.

Sterne, J.A.C., Gavaghan, D. and Egger, M. (2000). Publication and related bias in meta-analysis: Power of statistical tests and prevalence in the literature. *Journal of Clinical Epidemiology*, **53**, 1119–1129.

Sterne, J.A., Egger, M. and Davey Smith, G. (2001). Systematic reviews in health care: Investigating and dealing with publication and other biases in meta-analysis. *British Medical Journal*, **323**, 101–105.

Thompson, S.G. and Sharp, S.J. (1999). Explaining heterogeneity in meta-analysis: A comparison of methods. *Statistics in Medicine*, **18**, 2693–2708.

CHAPTER 7

Failsafe N or File-Drawer Number

Betsy Jane Becker

College of Education, Florida State University, USA

KEY POINTS

- Rosenthal (1979) introduced 'file-drawer' analysis and a formula for an estimate of the number of non-significant articles remaining in researchers' file drawers which became known as failsafe N.
- Other approaches to file-drawer analysis include Orwin's (1983) failsafe N for effect sizes and an application of Fisher's (1932) combined significance test. None of the failsafe methods incorporates information about study characteristics (e.g., moderator variables) into the computations.
- Available failsafe N methods lead to widely varying estimates of the number of additional studies, and thus to very different assessments of whether publication bias is a threat to the results of a review.
- Though it is conceptually attractive and relatively simple to use, Rosenthal's failsafe N is prone to misinterpretation and misuse and no statistical criterion is available for interpretation of its values.
- The failsafe N should be abandoned in favor of other more informative analyses such as those covered in other chapters of this book.

INTRODUCTION

The failsafe N or file-drawer number is a clever device suggested by Rosenthal (1979). It was one of the earliest approaches for dealing with the problem of publication bias and, in the social sciences, is still one of the most popular. Becker

---

Publication Bias in Meta-Analysis – Prevention, Assessment and Adjustments Edited by H.R. Rothstein, A.J. Sutton and M. Borenstein © 2005 John Wiley & Sons, Ltd

and Morton (2002) found that the failsafe N was presented in about one-third of a set of 28 meta-analyses published between 1999 and 2002 in two review journals in psychology and education.

Rosenthal's premise was that for any given topic, studies with statistically significant results are more likely to be published than those with non-significant results. Researchers may choose not to publish results that are not significant, instead storing them in file drawers (thus Rosenthal's label 'the file-drawer problem'). If this has happened, the published literature is not representative of the full set of studies that have been conducted, and conclusions drawn based on reviews of only published studies are likely to be biased. Rosenthal argued that if a large number of non-significant (and presumably unpublished) results were found and added to a meta-analysis, they could change the conclusions of the synthesis. Rosenthal thus asked, given a significant result for an overall test of combined significance in a meta-analysis, how many additional results it would take to reduce the overall test to non-significance.

DEFINITION OF THE FAILSAFE N

Before proceeding further, I will be more explicit about the methods Rosenthal used and the assumptions being made when computing the failsafe N. First, suppose we have a series of k independent studies, and study i examines the null hypothesis $H_0 : \theta_i = 0$. The parameter θ_i might represent a correlation between two variables or a standardized mean difference contrasting two groups, or an odds ratio. Each study also provides an observed one-tailed probability value p_i from a test of the directional null hypothesis that $\theta_i \leq 0$ (i.e., the alternative is that $\theta_i > 0$).

Rosenthal began with the test of combined significance (p-value summary) often called the 'sum of Zs', suggested by Stouffer *et al.* (1949). The test is

$$Z_S = \frac{\sum_{i=1}^{k} z_i}{\sqrt{k}},$$

where z_i is the normal deviate associated with the one-tailed p-value p_i in study i. The null hypothesis for the series of studies is that all k values of θ_i are zero (or less). If this null model is true, Z_S is a standard normal deviate and a test of the hypothesis $H_0 : \theta_1 = \ldots = \theta_k = 0$ is made by comparing Z_S to a table of values from the standard normal distribution.

For instance, Z_S is significant at the $\alpha = 0.05$ level if it is larger than $Z_\alpha = 1.645$. Rosenthal then asked, if an observed value of Z_S is already larger than Z_α, how many studies with z_i-values averaging zero (i.e., with $p_i = 0.5$) would need to be added to reduce the value of Z_S to Z_α. He asked what value of N would satisfy the formula

$$\frac{\sum_{i=1}^{k} z_i}{\sqrt{k + N}} < Z_\alpha,$$

where Z_α is the α-level upper-tail critical value of the normal distribution. The formula translates to several that can be solved for N. One formula based on the Stouffer test value Z_S is

$$N > k \left[\frac{Z_S}{Z_\alpha}\right]^2 - k. \tag{7.1}$$

Substituting $Z_S = \sum z_i / \sqrt{k}$ into (7.1) leads to an alternate form involving the sum of z-values $\sum z_i$, specifically

$$N > \left[\frac{\sum z_i}{Z_\alpha}\right]^2 - k. \tag{7.2}$$

Arithmetically, adding N studies with an average $z_i = 0$ is equivalent to adding N studies *all* with $z_i = 0$. However, those two sets of studies could look very different in reality (e.g., a set of effects with average $z_i = 0$ could also be highly variable) and it is quite unlikely that in practice one would find a set of studies with identical null results.

The logic of the failsafe N depends on the idea that Stouffer's test increases when individual studies produce large z_i-values (and decreases when negative z_i-values are added). Large z_i-values result because of two influences. One is the sample size in the ith study. Large samples are more likely than small ones to produce significant results (i.e., small p_is). Small p_i-values produce large z_i-values, regardless of the size of the true effect (θ_i in my notation). All other things being equal, the larger the number of subjects in the studies in a review, the larger the failsafe N is likely to be (because Stouffer's test would likely be larger). However, a second factor also leads to large z_i-values and large Stouffer test values. That is, small p-values and large z_i-values also occur when the true effect in study i is large. So the second and more interesting force that leads to large z_i-values is the true size of the effect. Because both large n_i and large observed effects lead to large Stouffer tests (and impact the value of the failsafe N) there is no clear relationship between the number of participants in the studies in a review, or the number of studies in the review, and the size of the failsafe N.

Rosenthal argued that when the failsafe N is large relative to the number of observed studies k, the results of the meta-analysis can be considered robust to publication bias, because it would be very unlikely to find a large number of additional unpublished results. Rosenthal did not provide statistical criteria for deciding whether the failsafe N is big enough, but suggested an *ad hoc* rule of thumb that raises concern if the failsafe N is less than $5k + 10$. More recently Mullen *et al.* (2001) proposed to monitor this rule of thumb by computing the failsafe ratio $N/(5k + 10)$. If this ratio exceeds 1, they argue, the 'weight of evidence does appear to be sufficiently tolerant for future results' (Mullen *et al.*, 2001, p. 1454).

EXAMPLES

In his seminal 1979 paper Rosenthal drew on his own research on the self-fulfilling prophecy for an example, and computed a file drawer $N = 3263$ for his initial review (Rosenthal, 1969) of $k = 94$ experiments. An updated review of the same literature (Rosenthal, 1976), including $k = 311$ studies, led to a failsafe N of $49\,457$. These impressively large numbers likely had the effect of encouraging other meta-analysts to compute the failsafe index.

It is easy to find similarly large numbers in the literature. Gershoff (2002) reported failsafe Ns for 11 sets of studies of the effects of corporal punishment on such outcomes as aggression, mental health, and criminal behavior. Her failsafe Ns ranged from a low of 538 for a set of 8 studies on corporal punishment and mental health to a high of 201 197 for 27 studies of corporal punishment and aggression. Gershoff's results make clear that the failsafe N does not depend directly on either the numbers of studies or the numbers of participants in the studies reviewed. Gershoff reported $N = 22\,419$ for 13 studies of parent–child relationship quality with 2216 participants, but a lower $N = 538$ for the 8 mental-health studies with 5851 participants, more than double the number of subjects in the 13 studies (that produced a larger failsafe N). Clearly having studies with more participants does not guarantee a larger failsafe N.

Computational example: Expectancy data

An example using Raudenbush's (1984) teacher expectancy data set illustrates the computation of the failsafe N. Table 7.1 shows results from the 19 randomized studies, with one-tailed probability values (p_i) ranging from 0.0158 to 0.9255 (where p-values between 0.5 and 1.0 reflect findings in the reverse direction). As described elsewhere, in these studies students were randomly assigned to either a control or 'expectancy induction' group. Teachers were told to expect superior performance by the students in the expectancy condition. The comparison of interest is the intelligence test performance of students in the two conditions, so θ_i represents a mean difference or effect size.

The p used in the computation is an upper-tail p from the test of the mean difference – Table 7.1 also shows the effect sizes (d_i) and the t-test statistics (t_i) for each study. Each t is a test of the null hypothesis that the true difference between experiment and control is less than or equal to zero. The Stouffer sum of Zs test is significant for these data, with a value of $Z_S = 10.64/\sqrt{19} = 2.44(p = 0.007)$. Applying (7.1), we obtain

$$N > k\left[\frac{Z_S}{Z_\alpha}\right]^2 - k = 19\left[\frac{2.44}{1.645}\right]^2 - 19 = 22.80,$$

that is, adding just 23 additional studies averaging $p = 0.5$ could bring the Stouffer test statistic to a non-significant value of 1.645. Note: Values of test statistics may differ slightly from those shown if computations are made to a greater degree of accuracy than used here. Here $z(p_i)$ and $-2\log(p_i)$ values were rounded to 2 decimal places before sums were computed.

For the expectancy example the failsafe N of 23 is less than the criterion of $5 \times 19 + 10 = 105$ studies, and in fact is just above the number of studies in the review, so we would be quite concerned about the potential impact of publication bias on these results.

Table 7.1   Example of failsafe N computation for teacher expectancy data.

Citation	Number of observations in						
	experimental group	control group	d_i	t_i	Upper tail p_i	$z(p_i)$	$-2\ln(p_i)$
Carter, 1971	22	22	0.54	1.79	0.0403	1.75	6.43
Claiborn, 1969	26	99	−0.32	−1.45	0.9255	−1.44	0.15
Conn, 1968	60	198	0.12	0.81	0.2081	0.81	3.14
Evans and Rosenthal, 1969	129	348	−0.06	−0.58	0.7196	−0.58	0.66
Fielder *et al.*, 1971	110	636	−0.02	−0.19	0.5768	−0.19	1.10
Fine, 1972	80	79	−0.18	−1.13	0.8709	−1.13	0.28
Fleming and Anttonen, 1971	233	224	0.07	0.75	0.2274	0.75	2.96
Flowers, 1966	43	38	0.18	0.81	0.2106	0.80	3.12
Ginsburg, 1971	65	67	−0.07	−0.40	0.6559	−0.40	0.84
Grieger, 1971	72	72	−0.06	−0.36	0.6403	−0.36	0.89
Henrikson, 1971	19	32	0.23	0.79	0.2155	0.79	3.07
Jose and Cody, 1971	72	72	−0.14	−0.84	0.7988	−0.84	0.45
Keschock, 1971	24	24	−0.02	−0.07	0.5275	−0.07	1.28
Kester, 1969	75	74	0.27	1.65	0.0508	1.64	5.96
Maxwell, 1971	32	32	0.80	3.20	0.0011	3.07	13.65
Pellegrini, 1972 (Study 1)	11	22	1.18	3.20	0.0016	2.95	12.87
Pellegrini, 1972 (Study 2)	11	22	0.26	0.70	0.2433	0.70	2.83
Rosenthal and Jacobson, 1968	65	255	0.30	2.16	0.0158	2.15	8.30
Rosenthal *et al.*, 1974	77	339	0.03	0.24	0.4061	0.24	1.80
Sum						10.64	69.78

Computational example: Passive smoking data

An example using the passive smoking data set (Hackshaw *et al.*, 1997) illustrates the computation of a much larger failsafe N. Table 7.2 shows the 37 studies with their probability values. The relative risks (the focus of the analysis) represent cancer risk for women non-smokers whose husbands smoked. Higher relative risks (RR_i values) indicate studies showing greater risks of cancer. The null model for the Stouffer test is that none of the populations studied had a relative risk above 1 ($H_0 : \theta_i \leq 1$ for $i = 1$ to k). The significance values used here and shown in Table 7.2 come from a z test computed by dividing the log-transformed relative risk by its standard error.

Table 7.2 Example of failsafe N computation for 37 passive smoking studies.

Citation	RR_i	$\ln(RR_i)$	$SE[\ln(RR_i)]$	p_i	$z(p_i)$	$-2\log(p_i)$
Chan et al., 1982	0.75	−0.288	0.282	0.8460	−1.02	0.33
Correa, 1983	2.07	0.727	0.477	0.0635	1.53	5.51
Trichopoulos, 1983	2.13	0.756	0.298	0.0056	2.54	10.37
Buffler, 1984	0.80	−0.223	0.439	0.6944	−0.51	0.73
Kabat, 1984	0.79	−0.236	0.582	0.6572	−0.40	0.84
Lam, 1985	2.01	0.698	0.313	0.0129	2.23	8.70
Garfinkel et al., 1985	1.23	0.207	0.213	0.1660	0.97	3.59
Wu et al., 1985	1.20	0.182	0.481	0.3524	0.38	2.09
Akiba et al., 1986	1.52	0.419	0.282	0.0689	1.48	5.35
Lee et al., 1986	1.03	0.030	0.466	0.4747	0.06	1.49
Koo et al., 1987	1.55	0.438	0.277	0.0571	1.58	5.73
Pershagen et al., 1987	1.03	0.030	0.267	0.4560	0.11	1.57
Humble et al., 1987	2.34	0.850	0.541	0.0580	1.57	5.69
Lam et al., 1987	1.65	0.501	0.180	0.0027	2.78	11.82
Gao et al., 1987	1.19	0.174	0.190	0.1805	0.91	3.42
Brownson et al., 1987	1.52	0.419	0.696	0.2736	0.60	2.59
Geng et al., 1988	2.16	0.770	0.352	0.0143	2.19	8.49
Shimizu et al., 1988	1.08	0.077	0.267	0.3864	0.29	1.90
Inoue et al., 1988	2.55	0.936	0.631	0.0690	1.48	5.35
Kalandidi et al., 1990	1.62	0.482	0.299	0.0535	1.61	5.85
Sobue, 1990	1.06	0.058	0.184	0.3755	0.32	1.96
Wu-Williams et al., 1990	0.79	−0.236	0.127	0.9683	−1.86	0.06
Liu et al., 1991	0.74	−0.301	0.425	0.7609	−0.71	0.55
Jockel, 1991	2.27	0.820	0.563	0.0727	1.46	5.24
Brownson et al., 1992	0.97	−0.031	0.112	0.6072	−0.27	1.00
Stockwell et al., 1992	1.60	0.470	0.337	0.0817	1.39	5.01
Du et al., 1993	1.19	0.174	0.299	0.2803	0.58	2.54
Liu et al., 1993	1.66	0.507	0.419	0.1135	1.21	4.35
Fontham et al., 1994	1.26	0.231	0.100	0.0105	2.31	9.11
Kabat et al., 1995	1.10	0.095	0.294	0.3727	0.32	1.97
Zaridze et al., 1995	1.66	0.507	0.200	0.0056	2.54	10.38
Sun et al., 1996	1.16	0.148	0.191	0.2183	0.78	3.04
Wang et al., 1996	1.11	0.104	0.258	0.3428	0.40	2.14
Garfinkel, 1982	1.18	0.166	0.137	0.1135	1.21	4.35
Hirayama, 1984	1.45	0.372	0.182	0.0205	2.04	7.78
Butler, 1988	2.02	0.703	0.735	0.1694	0.96	3.55
Cardenas et al., 1997	1.20	0.182	0.150	0.1115	1.22	4.39
Sum					34.28	158.83

The Stouffer test value for these studies is 5.64 ($p \leq 0.0001$). Applying (7.1), we compute

$$N > k\left[\frac{Z_S}{Z_\alpha}\right]^2 - k = 37\left[\frac{5.64}{1.645}\right]^2 - 37 = 397.94$$

or 398 studies. This failsafe *N* is much larger than the number of studies (37) and also passes Rosenthal's criterion for robustness of $5k + 10 = 5 \times 37 + 10 = 195$ studies. According to this criterion, the passive smoking studies are not likely to be subject to modification due to publication bias.

Computational example: Interview data

From the interview data set analyzed by McDaniel *et al.* (1994) we obtain 160 correlations between job interviews and job performance. Individual study results are not tabled here in the interest of space. For these results *p*-values are obtained by first computing a *t* test of the hypothesis that $\rho_i = 0$ in each study versus the alternative that ρ_i is positive. The *t* test for the *i*th study result is

$$t_i = \frac{r_i}{\sqrt{(1 - r_i^2)/(n_i - 2)}}$$

(see Howell, 2002, p. 274). The upper-tail *p* for this test is then obtained and the associated *z*-value is computed for the Stouffer test. The Stouffer test value for the 160 studies is 29.48, which is quite large and highly significant. This leads to a very large failsafe *N*:

$$N > k \left[\frac{Z_S}{Z_\alpha}\right]^2 - k = 160 \left[\frac{29.48}{1.645}\right]^2 - 160 = 51\,225.80.$$

According to this computation over 50 000 studies averaging no result would be needed to reduce the Stouffer test for the interview data to non-significance. Clearly there is little chance that this number of additional studies would be located, so we conclude that the relationship between interview scores and job performance is not susceptible to publication bias.

ASSUMPTIONS OF THE FAILSAFE *N*

Several assumptions underlie the computation of the failsafe *N*. The first is that the unpublished or omitted studies, on average, show a null result. Several researchers have questioned whether it is reasonable to make such an assumption when other (located) studies have found large and significant results. Begg and Berlin (1988), for instance, argued that the choice of zero for the 'added' average result is likely to be biased. It is instructive to ask what would occur if the added studies had, on average, results that were the *opposite* of the studies included in the synthesis. It would then take relatively fewer studies to reduce Z_S to non-significance. In fact adding exactly the same number of studies (*k*) with *z*-values opposite in sign to those observed would lead to a much-reduced Stouffer value of $Z_S = 0$ (with $p = 0.5$). It would take even fewer added results that opposed those observed to bring Z_S to just below Z_α.

Sutton *et al.* (2000) also pointed out that the failsafe *N* does not directly incorporate sample size information, and they argued this is a weakness. In fact, the effect of adding *N* studies (averaging zero effect) would be the same whether those studies all had samples of 5 subjects or 5000 subjects. This occurs because the added studies are constrained to have null effects, so their sample sizes would not impact their hypothetical z_i-values. However, since the *observed* Z_S and its component z_i-values depend on both the observed effect sizes and their sample sizes, information on both of these factors indirectly impacts the value of N. Indeed, when any study's null hypothesis is false, its z_i-value has a distribution that depends on both the sample size and the true value of θ_i.

Finally, several writers have noted other statistical issues. Orwin (1983) pointed out the absence of a statistical model for both versions of the failsafe N. As such, the failsafe numbers serve only as heuristic devices and issues such as sampling variation and the distribution of the failsafe N statistic are ignored. Iyengar and Greenhouse (1988) also noted that the failsafe N computation ignores heterogeneity in the obtained results, which is clearly a concern of most meta-analysts. While the failsafe approach has intuitive appeal, on technical grounds there is little to support the widespread use of the number and much to raise concern about its application.

VARIATIONS ON THE FAILSAFE N

A variety of modifications of Rosenthal's procedure have been developed, in part to address some of the concerns just mentioned. These include alternative failsafe computations (Iyengar and Greenhouse, 1988; Orwin, 1983), as well as methods for estimating the numbers of unpublished articles that may exist (e.g., Gleser and Olkin, 1996). As might be expected, different failsafe formulas may lead to very different failsafe numbers (Pham *et al.*, 2001). Only a few methods are examined here to illustrate that point.

Gleser and Olkin's estimates of the numbers of unpublished studies

Gleser and Olkin (1996) took a statistically grounded approach that falls somewhere between the quick failsafe methods discussed here and other more sophisticated selection modeling approaches (see Chapter 9). They provide two methods for estimating the number of unpublished studies. Both methods are based on analyses of the observed probabilities from the studies in a review, and both assume that the null hypothesis (of no effects in any population studied) is correct. Because their second approach is fairly complex it is not discussed here. Gleser and Olkin note, however, for both of their proposed estimation techniques, that 'because the null hypothesis is assumed true, the present approach is inappropriate for studying the effects of publication bias and selection on effect sizes' (1996, p. 2503). Thus when strong evidence suggests that the null model is false, their methods must be very cautiously applied.

Their first approach assumes that the k observed probabilities are the smallest from among a potential set of $N + k$ results. The largest observed value $p_{(k)}$ is then used to obtain an unbiased estimate of N, via

$$\hat{N} = \frac{k(1 - p_{(k)}) - 1}{p_{(k)}}. \tag{7.3}$$

So for example, if the largest p in a set of $k = 10$ values is $p_{(10)} = 0.05$, then

$$\hat{N} = \frac{k(1 - 0.05) - 1}{0.05} = 19k - 20 = 170,$$

suggesting that 170 additional unpublished studies may exist. When the largest p is quite large, the estimate $\hat{N}$ is likely to be small. For both the expectancy data and the passive smoking data, the largest ps were above 0.90, a range not even tabled by and Gleser and Olkin. Applying (7.3) to these data sets produces estimates of

$$\hat{N} = \frac{k(1 - p_{(k)}) - 1}{p_{(k)}} = \frac{19(1 - 0.9255) - 1}{0.9255} = 0.45$$

for the expectancy studies and

$$\hat{N} = \frac{k(1 - p_{(k)}) - 1}{p_{(k)}} = \frac{37(1 - .9683) - 1}{.9683} = 0.18$$

for the passive smoking data. Essentially these estimates suggest that no unpublished studies exist. For the interview data the largest p-value is $p_{(160)} = 0.743$, which leads to $\hat{N} = 54$.

Orwin's N_{es}

Rosenthal's approach was based on finding a significant p-value summary. However, while they are simple and widely applicable, p-value summaries are not as informative as effect size analyses (Becker, 1987; Darlington and Hayes, 2000). Orwin (1983) therefore extended the idea of the failsafe N to the (standardized) effect size metric, asking how many effect sizes averaging a particular value $\bar{d}_{FS}$ would be needed to reduce an observed mean effect size $\bar{d}_O$ to a particular criterion level (d_C). Orwin's approach leads to

$$N_{es} = \frac{k(\bar{d}_O - d_C)}{d_C - \bar{d}_{FS}}. \tag{7.4}$$

Orwin's approach differs from Rosenthal's, in that it does not require that the effect size $\bar{d}_O$ be reduced to non-significance, but rather to an arbitrary level d_C set by the reviewer. For example, a reviewer who had found a mean effect of $\bar{d}_O = 0.25$ standard deviations in a review might ask how many additional studies would be needed to reduce the effect to $d_C = 0.10$ – a tenth of a standard deviation in size. Although Orwin proposed N_{es} for standardized mean differences, the formula can also be applied to other effect indices such as correlations, risk ratios, and the like.

Sometimes Orwin's N_{es} can produce values quite different from Rosenthal's failsafe N. Orwin's failsafe N_{es} can be computed for the three example data sets, using as $\bar{d}_O$ the observed mean effect in each data set. For the $k = 19$ expectancy studies a mean effect size under the fixed-effects model is $\bar{d}_O = 0.06$ standard deviations. We might want to determine how many studies averaging zero effect

($\bar{d}_{FS} = 0$) would need to be added to reduce the mean to $d_C = 0.05$. Substituting these values into (7.4), we compute

$$N_{es} = \frac{k(\bar{d}_O - d_C)}{d_C - \bar{d}_{FS}} = \frac{19\,(0.06 - 0.05)}{0.05 - 0} = 3.8,$$

that is, that only 4 studies need be added. To reduce the mean effect to $d_C = 0.01$ would require the addition of many more studies averaging zero: specifically, $N_{es} = 19(0.06 - 0.01)/(0.01 - 0) = 95$ additional studies would be needed.

For the $k = 37$ passive smoking studies Hackshaw *et al.* (1997) reported an average relative risk for women non-smokers of 1.24. How many studies with a relative risk of 1.0 would it take to reduce the combined value to a relative risk of 1.05? Since we are working with relative risks the computations must be carried out on a log scale, with the observed log risk ratio being 0.185, the effect in the missing studies being 0.0, and the criterion effect being 0.049. Orwin's formula suggests that

$$N_{es} = \frac{k(\overline{\log RR}_O - \log RR_C)}{\log RR_C - \overline{\log RR}_{FS}} = \frac{37\,(0.185 - 0.049)}{0.049 - 0.0} = 103$$

studies with log risk ratios of zero (risk ratios of 1.0) would be needed to reduce the average log risk ratio to 0.049 (risk ratio of 1.05). This is a much smaller number than the nearly 400 studies suggested by Rosenthal's failsafe N.

The final example uses the average result for the $k = 160$ interview studies, which is a mean correlation of 0.21. Our example sets a target of $r_C = 0.15$ and asks how many studies averaging a zero correlation ($\bar{r}_{FS} = 0$) would be required to reach this target. Again using (7.4), Orwin's approach suggests

$$N_{es} = \frac{k(\bar{r}_O - r_C)}{(r_C - \bar{r}_{FS})} = \frac{160\,(0.21 - 0.15)}{0.15 - 0} = 64.$$

So adding only 64 more studies showing no correlation would reduce the mean correlation to 0.15. To reduce the mean correlation further, to 0.05, requires many more studies. In fact, $N_{es} = 512$ studies would be required to achieve this much smaller criterion correlation value. However, this is still a much smaller number of additional studies than the original failsafe N of over 50 000. (This computation is appropriate when the researcher is combining correlations directly. In the event that the mean correlation will be represented in the Fisher z scale, then the Orwin computations should also use the Fisher z transformation.)

A Fisher failsafe N

It is also instructive to consider versions of the failsafe N that can be derived from other tests of combined significance, such as Fisher's (1932) test based on the sum of logs of observed p values. Fisher's test is

$$\sum_{i=1}^{k} -2\ln(p_i),$$

which is distributed as chi-square with $2k$ degrees of freedom under the null hypothesis. For the expectancy data, the Fisher test value is shown in the last row of Table 7.1, and its value is 69.78. When compared to the chi-square distribution with $df = 38$, the test is significant with $p = 0.0013$. Similarly, for the passive smoking and interview data, the Fisher test value is highly significant. For the passive smoking studies it is 158.83 ($df = 74$, $p < 0.0001$) and for the interview data the Fisher test is 2331.58 ($df = 320$, $p < 0.0001$).

We can ask how many studies with p_i-values equal to 0.5 must be added to those observed to bring Fisher's test to non-significance. This requires a bit more computation than Rosenthal's original failsafe N, because the addition of each $p_i = 0.5$ result also changes the degrees of freedom of the test. However, this is easily done in any spreadsheet or even with a hand calculator by adding a term (or several terms) at a time and checking the value of Fisher's test and its probability. Adding a study with $p_i = 0.5$ is equivalent to adding a term to the Fisher test statistic equal to $-2 \ln(0.5) = 1.386$. Computationally this process assumes that each added study has $p = 0.5$, or that the average of the added $-2 \log(p_i)$ values is 1.386. Because the log transformation is not linear, this is not directly equivalent to assuming that the average p for the hypothetical additional studies is equal to 0.5.

Returning to the expectancy data set used above, adding one study with $-2 \log(p_i) = 1.386$ to the set of 19 studies produces a new Fisher value of 71.14 with $p = 0.0018$ ($df = 40$). The p value associated with Fisher's test goes above 0.05 for the 19 expectancy studies when we have added $N_F = 18$ more studies with $p = 0.5$. The augmented Fisher test statistic is 94.70 ($69.75 + 18 \times 1.386$), with $df = 74$ and $p = 0.053$. Recall that the original failsafe N for the expectancy data was 23 additional studies. This example shows that two fairly similar failsafe Ns can be associated with the same data.

Next consider the passive smoking data. Adding one hypothetical study with $p = 0.5$ produces a Fisher test value of 160.22 ($df = 76$, $p = 0.0004$). The significance level of the Fisher test surpasses 0.05 after the addition of 79 effects with $p = 0.5$, when the test equals 268.32 ($158.83 + 79 \times 1.386$). The probability value of the Fisher test is then $p = 0.051$ ($df = 232$). The failsafe value obtained using Fisher's test – $N_F = 79$ – is quite a bit smaller than Rosenthal's failsafe N of 397. Indeed, adding roughly twice the number of studies that were observed, but all with $p = 0.5$, produces a non-significant Fisher combined significance test. This would lead us to be much more concerned about the possible effect of publication bias than Rosenthal's computation.

The interview data show a large Fisher failsafe N, with over 3200 additional studies needed to bring Fisher's summary to non-significance. But even at $N_F = 3227$ this count of potential additional studies is still much lower than Rosenthal's failsafe N of 51 226. It is likely these differences in failsafe Ns relate to the power of the two combined significance summaries from which they are derived. Becker (1985) showed that Fisher's summary is very powerful to detect single populations that deviate from the null model. Thus one might expect that in general it would require more studies with $p = 0.5$ to bring Fisher's test to a non-significant level when one or even a small number of the study results are extreme.

Table 7.3   Examples of failsafe N computations for all data sets.

	Data set		
	Expectancy	Passive smoking	Interviews
Number of studies (k)	19	37	160
Rosenthal's failsafe N	23	398	51 226
Fisher failsafe N_F	18	79	3 227
Gleser and Olkin $\hat{N}$	0	0	54
Orwin N_{es}[a]	4 or 95*	103	64 or 512*
Average observed effect	0.06	1.24	0.21
Orwin criterion (d_C)	0.05 or 0.01	1.05	0.15 or 0.05

[a] Value is dependent on the effect size reduction specified in the formulae

SUMMARY OF THE EXAMPLES

At this point it is useful to examine the full set of results of the example computations from the three failsafe N formulas. Table 7.3 shows the values for the three data sets, with two example values shown for the Orwin N_{es}. For each data set, the three failsafe formulas give values that span considerable ranges. For the expectancy data, the largest failsafe $N_{es} = 95$ (Orwin's value for reducing the mean to 0.01 standard deviations) is more than 5 times the size of the smallest other value (from the Fisher test) $N_F = 18$. The largest value for the passive smoking data, Rosenthal's $N = 398$, is roughly four times Orwin's $N_{es} = 103$. And for the interview data the largest value, again Rosenthal's $N = 51 226$, is more than 100 times one of the smaller values, $N_{es} = 512$ (Orwin's value for reducing the mean correlation to 0.05). The estimated numbers of unpublished studies from the Gleser and Olkin approach are always quite low compared to the failsafe values.

The incredible range of values shown here reveals one of the greatest weaknesses of the failsafe N computations – it is difficult to interpret the values without a statistical criterion. Also there is virtually no consistency across the different failsafe computations, leading to the possibility of a researcher choosing the failsafe N that gives results favorable to some particular desired outcome. For both of these reasons extreme caution is needed should a researcher decide to apply any of the available failsafe N methods.

APPLICATIONS OF THE FAILSAFE N

Though it is conceptually attractive and relatively simple to compute, Rosenthal's failsafe N is also, unfortunately, prone to misinterpretation and misuse. Early explications of its problems were given by Iyengar and Greenhouse (1988) and Carson *et al.* (1990). However, problems with its application persist. Three examples will suffice to show the kinds of misinterpretations that may be found in the literature.

Some researchers have taken the failsafe N to represent the number of unpublished studies that ought to be added to a set of collected published documents.

Smith *et al.* (1994) computed the failsafe N for a set of 60 studies (with over 300 effects) on social support and health outcomes to be 6. They wrote 'six studies of comparable sample sized were deemed necessary to change the combined results of those published' (Smith *et al.*, 1994, p. 355). They then included six dissertations and one additional unpublished manuscript and added those sources to their review to 'offset publication bias'. The original intent in developing the failsafe N was clearly not to indicate a specific number of unpublished studies that should be added to a review. When unpublished studies are included in a synthesis (which is recommended), an exhaustive search should be done and all available unpublished sources should be included.

A different error in interpretation was made by Kavale and Forness (1996) in a synthesis of studies on the relationship between learning disabilities and social skills. Their review of 152 studies produced a failsafe N of 11 and Orwin's (1983) failsafe effect size N_{es} equal to 78. They concluded 'the obtained database ($n = 152$) was sufficient for ruling out the "file drawer problem" as a rival hypothesis' (1996, p. 229). However, this is exactly the opposite of the conclusion Rosenthal would have drawn, as the two failsafe numbers are both much lower than the actual number of studies in the review. This review's conclusions are highly susceptible to modification because of publication bias.

Finally, reviewers often simply ignore Rosenthal's criterion and make vague conclusions based on computed failsafe Ns. Astin *et al.* (2000) applied the failsafe N in their synthesis of randomized studies of 'distant healing', which includes such activities as prayer, spiritual healing, Therapeutic Touch, and the like. The computation for 16 trials on 2139 patients led to $N = 63$. Applying Rosenthal's criterion, we see that an N larger than $5 \times 16 + 10 = 90$ studies would be desired to assure resilience against publication bias. In spite of their N not reaching that criterion, Astin *et al.* (2000, p. 907) wrote that their failsafe value suggested 'that the significant findings are less likely to be the result of a "file-drawer effect" (that is, the selective reporting and publishing of only positive results)'.

CONCLUSIONS

While the failsafe N has been recognized as an ingenious invention, it has several drawbacks. No statistical model underlies Rosenthal's formula and there is no clear-cut and justifiable statistical criterion for what constitutes a 'large' failsafe N. Also, some of the assumptions underlying the computation are problematic. The formula itself does not directly acknowledge the average size of, or the variation in, effects that have been observed. Also the only way to address the role of categorical moderator variables is to compute failsafe numbers for all subsets of the data defined by the moderator of interest. There is no way to incorporate continuous moderators into failsafe analyses. While Orwin's version of the failsafe N incorporates information about the magnitude of observed effects, it does not address variation in effects. Overall, failsafe procedures do not assist the researcher in determining what they seek to know: the magnitude of the population effect. Rather, the procedures provide some quantification to what is already known, that results of meta-analyses based on small numbers of studies yield conclusions that can be easily altered with the addition of a few more studies.

More critically, different versions of the failsafe *N* produce very different results. Finally, the failsafe *N* is often misinterpreted, and even when it is not used incorrectly, the interpretations that have been made are often vague. Given the other approaches that now exist for dealing with publication bias, the failsafe *N* should be abandoned in favor of other, more informative analyses such as those described elsewhere in this book.

ACKNOWLEDGEMENT

I would like to thank Hannah Rothstein and Michael McDaniel for helpful input on this chapter.

REFERENCES

Astin, J.A., Harkness, E. and Ernst, E. (2000). The efficacy of 'distant healing': A systematic review of randomized trials. *Annals of Internal Medicine*, **132**, 903–910.
Becker, B.J. (1985). Tests of combined significance: Hypotheses and power considerations. Unpublished doctoral dissertation, University of Chicago.
Becker, B.J. (1987). Applying tests of combined significance in meta-analysis. *Psychological Bulletin*, **102**, 164–171.
Becker, B.J. and Morton, S.C. (2002). Publication bias: Methods and practice in evidence-based medicine and social science. Paper presented to Statistical Challenges for Meta-Analysis of Medical and Health-Policy Data Symposium held at the Mathematical Sciences Research Institute, Berkeley, California.
Begg, C.B. and Berlin, J.A. (1988). Publication bias: A problem in interpreting medical data (with discussion). *Journal of the Royal Statistical Society, Series A*, **151**, 419–463.
Carson, K.P., Schriesheim, C.A. and Kinicki, A.J. (1990). The usefulness of the 'fail-safe' statistic in meta-analysis. *Educational and Psychological Measurement*, **50**, 233–243.
Darlington, R.B. and Hayes, A.F. (2000). Combining independent *p* values: Extensions of the Stouffer and binomial methods. *Psychological Methods*, **5**, 496–515.
Fisher, R.A. (1932). *Statistical Methods for Research Workers* (4th edn). London: Oliver & Boyd.
Gershoff, E.T. (2002). Corporal punishment by parents and associated child behaviors and experiences: A meta-analytic and theoretical review. *Psychological Bulletin*, **128**, 539–579.
Gleser, L.J. and Olkin, I. (1996). Models for estimating the number of unpublished studies. *Statistics in Medicine*, **15**, 2493–2507.
Hackshaw, A.K., Law, M.R. and Wald, N.J. (1997). The accumulated evidence on lung cancer and environmental tobacco smoke. *British Medical Journal*, **315**, 980–988.
Howell, D.C. (2002). *Statisical Methods for Psychology* (5th edn). Pacific Grove, CA: Duxbury.
Iyengar, S. and Greenhouse, J.B. (1988). Selection models and the file drawer problem. *Statistical Science*, **3**, 109–135.
Kavale, K.A. and Forness, S.R. (1996). Social skill deficits and learning disabilities: A meta-analysis. *Journal of Learning Disabilities*, **29**, 226–237.
McDaniel, M.A., Whetzel, D., Schmidt, F.L. and Maurer, S. (1994). The validity of the employment interview: A comprehensive review and meta-analysis. *Journal of Applied Psychology*, **79**, 599–616.

Mullen, B., Muellerleile, P. and Bryant, B. (2001). Cumulative meta-analysis: A consideration of indicators of sufficiency and stability. *Personality and Social Psychology Bulletin*, **27**, 1450–1462.

Orwin, R.G. (1983). A fail-safe *N* for effect size in meta-analysis. *Journal of Educational Statistics*, **8**, 157–159.

Pham, B., Platt, R., McAuley, L., Klassen, T.P. and Moher, D. (2001). Is there a 'best' way to detect and minimize publication bias? An empirical evaluation. *Evaluation and the Health Professions*, **24**, 109–125.

Raudenbush, S.W. (1984). Magnitude of teacher expectancy effects on pupil IQ as a function of the credibility of expectancy induction: A synthesis of findings from 18 experiments. *Journal of Educational Psychology*, **76**, 85–97.

Rosenthal, R. (1969). Interpersonal expectations. In R. Rosenthal and R.L. Rosnow (eds), *Artifact in Behavioral Research*. New York: Academic Press.

Rosenthal, R. (1976). *Experimenter Effects in Behavioral Research* (enlarged edn). New York: Irvington.

Rosenthal, R. (1979). The 'file drawer problem' and tolerance for null results. *Psychological Bulletin*, **86**, 638–641.

Smith, C.E., Fernengel, K., Holcraft, C., Gerald, K. and Marien, H. (1994). Meta-analysis of the associations between social support and health outcomes. *Annals of Behavioral Medicine*, **16**, 352–362.

Stouffer, S.A., Suchman, E.A., DeVinney, L.C., Star, S.A. and Williams, R.M., Jr. (1949). *The American Soldier: Adjustment during Army Life* (Vol. 1). Princeton, NJ: Princeton University Press.

Sutton, A.J., Song, F., Gilbody, S.M. and Abrams, K.R. (2000). Modeling publication bias in meta-analysis: A review. *Statistical Methods in Medical Research*, **9**, 421–445.

CHAPTER 8

The Trim and Fill Method

Sue Duval

Division of Epidemiology and Community Health, School of Public Health, University of Minnesota, USA

KEY POINTS

- The 'trim and fill' method adjusts a meta-analysis for the impact of missing studies. This should be seen as a sensitivity analysis of the potential effect that missing studies may have had on the observed result.
- The method relies on scrutiny of one side of a funnel plot for asymmetry, assumed due to publication bias. It appears to give results that match the subjective visual assessment of a funnel plot.
- The 'trim and fill' method has been shown to give comparable results to more complex methods.
- Computational ease makes the technique accessible to all meta-analysts.

INTRODUCTION

In this chapter we present a new technique, the 'trim and fill' method developed in Duval and Tweedie (2000a, 2000b) for estimating and adjusting for the number and outcomes of missing studies in a meta-analysis. It has the advantage of being computationally simple, and in practical situations seems to perform as well as or better than other methods.

Publication Bias in Meta-Analysis – Prevention, Assessment and Adjustments   Edited by H.R. Rothstein, A.J. Sutton and M. Borenstein   © 2005 John Wiley & Sons, Ltd

HISTORY AND DEVELOPMENT

By the end of the 1990s, several methods for investigating publication bias had emerged. Several authors have developed methods of considerable complexity (Iyengar and Greenhouse, 1988; Hedges, 1992; Dear and Begg, 1992; Givens *et al.*, 1997; see Chapter 9, this volume) or of excessive simplicity (Rosenthal, 1979; Sugita *et al.*, 1992; see Chapter 7, this volume). There are also a number of relatively simple quantitative methods for detecting publication bias, including the rank correlation test of Begg and Mazumdar (1994), the regression-based test of Egger *et al.* (1997; see Chapter 6, this volume) and the methods described by Gleser and Olkin (1996).

Dear and Dobson (1997), commenting on the more complex and highly computer-intensive methods of Dear and Begg (1992), Givens *et al.* (1997) and Hedges (1992), noted 'previous methods have not been much used ...[and] ...the value of any new statistical methodology depends, in part, on the extent to which it is adopted'; they also noted that 'the culture of meta-analysis has traditionally favoured very simple methods'. DuMouchel and Harris (1997), also commenting on these methods, stated that 'attempts to assess publication bias beyond simple graphs like the funnel plot seem to involve a *tour de force* of modelling, and as such are bound to run up against resistance from those who are not statistical modelling wonks'.

These comments provided the impetus for the development of the trim and fill method. The idea was to develop a method that addressed these issues. The first of these was to make use of the subjective visual impression of the funnel plot, already widely in use as a graphical means of assessing publication bias, in a more objective fashion.

The choice of axes for the funnel plot can change its appearance considerably with respect to asymmetry (Vandenbroucke, 1988; Tang and Liu, 2000; Sterne and Egger, 2001; see also Chapter 5, this volume). Trim and fill is independent of the method of display of the precision measure and can help to resolve this issue. The method, however, is dependent on the metric employed for the effect size itself, and may give different conclusions based on this choice. It does appear to match the subjective impression of bias seen in funnel plot displays, and moreover gives results consistent with those from other more complex methods.

The second aim was to provide a more accessible method to the wider meta-analytic audience. It is possible to perform trim and fill using readily available software (e.g., Comprehensive Meta Analysis, Stata) and several groups have written their own programs (see Chapter 11). This computational ease is likely to increase its use, giving meta-analysts a simple tool to, at least, address the issue of publication bias that manifests itself as funnel plot asymmetry.

The trim and fill algorithm can be easily described. Simply put, we trim off the 'asymmetric' right-hand side of a funnel plot (see Chapter 5), assumed to be affected by publication bias, after estimating how many studies are in this outlying part; use the symmetric remainder to estimate the true 'centre' of the funnel; and then replace the trimmed studies and their missing 'counterparts' around the centre. The final estimate of the true overall effect, and also its variance, are then based on the 'filled' funnel plot.

MODELLING PUBLICATION BIAS

The funnel plot

Perhaps the most common method used to detect publication bias in a meta-analysis is the funnel plot (see Chapter 5). This is a graph of the effect found in each study against the size of the study, measured in some consistent way. In what follows, we make the additional assumption that publication bias is the underlying reason for the observed funnel plot asymmetry, and use the two terms interchangeably. Other reasons why the funnel plot may exhibit this behaviour have been proposed (Egger *et al.*, 1997; Sterne *et al.*, 2000; see also Chapters 5 and 6, this volume). The trim and fill method will produce positive results for any such funnel plot asymmetry and this should be recognized in any analysis using the method (see Chapter 5).

Figure 8.1 depicts such a funnel plot, using data from 11 studies of the effect of using gangliosides in reducing case fatality and disability in acute stroke (Candelise and Ciccone, 1998). We analyse this example in more depth below. Each of the studies supplies an estimate T_i of the effect in question in the ith study, and an estimate of the variance v_i within that study. Typically in the epidemiological literature, these T_i represent log relative risks or log odds ratios. Note that we have reversed the direction of the effect size given by the Cochrane Collaboration, so that positive log odds ratios as shown here are protective. This is consistent with the assumption of the trim and fill method, which relies on scrutiny of the left-hand side of the funnel plot for missing studies. This makes the method one-sided, in contrast to several other methods (see Chapter 10). The rationale behind this can be easily seen by inspection of the key assumption of the trim and fill method, which is formalized below.

The funnel plot here uses $1/\sqrt{v_i}$ (the reciprocal of the standard error) to depict the study size, so that the most precise estimates (e.g., those from the largest studies) are at the top of the funnel, and those from less precise, or smaller, studies are at the

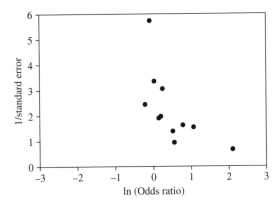

Figure 8.1 Funnel plot of the effect of gangliosides in acute ischaemic stroke. The horizontal axis depicts log odds ratios, and the vertical axis gives reciprocals of standard errors, for each study in the meta-analysis.

base of the funnel. In theory, this plot should depict a 'funnel' shape centred around the true overall mean that we are trying to estimate. The fact that there may be publication (or asymmetry) bias in this data set might be subjectively inferred since what should be a funnel shape is asymmetric: there is an appearance of missing studies in the bottom left-hand corner, and the assumption is that, whether because of editorial policy or author inaction or other reasons, these papers (which show, say, no significance, or perhaps the reverse effect (namely an overall effect less than zero) from that envisaged when carrying out the studies (namely an overall effect greater than zero)) are the ones that might not be published. This particular data set provides an extreme example of what is typically seen in the literature and is a good illustration of the trim and fill method.

The visual impression of bias in a funnel plot, such as Figure 8.1, is purely subjective. The 'trim and fill' method we now discuss provides a more objective approach based on the rationale behind the interpretation of a funnel plot; the concept underlying it is simple, and it is computationally relatively straightforward to use, as we show below.

Assumptions and the 'Suppressed Bernoulli Model'

In the standard structure for a meta-analysis (in the absence of publication bias) we assume we have n individual studies, all of which are addressing the same problem; and that there is some global 'effect size' θ which is relevant to the overall problem, and which each study attempts to measure. For each $i = 1, \ldots, n$, study i produces an *effect size* T_i which estimates θ, and an *estimated within-study variance* v_i.

The method could equally apply to effect size measures such as log relative risks, risk differences or log mortality ratios in clinical or epidemiological trials. In social science applications, an appropriately transformed correlation measure and the standardized mean difference are typical effect sizes that could be analysed using trim and fill.

In Duval and Tweedie (2000a, 2000b), we modify this standard model to account for publication bias. We assume that in addition to the n observed studies, there are k_0 relevant studies which are not observed due to publication bias. The value of k_0, and the effect sizes which might have been found from these k_0 studies, are unknown and must be estimated; and uncertainty about these estimates must be reflected in the final meta-analysis inference. The key assumption behind our non-parametric method is that *the suppression has taken place in such a way that it is the k_0 values of the T_i with the most extreme leftmost values that have been suppressed.*

The trim and fill method makes the assumption that it is the most 'negative' or 'undesirable' studies which are missing. A simple 'flipping' or reflection of the funnel about the y-axis (achieved in practice by multiplying all effect sizes by -1) to ensure that the method always scrutinizes the left-hand side of the funnel plot can be performed in those meta-analyses where the outcome measure is presented in the opposite direction. This might be expected to lead to a truncated funnel plot as in Figure 8.1, for example.

ADJUSTING FOR PUBLICATION BIAS

Some researchers do not agree with the idea of adjusting the results of a meta-analysis for publication bias, and are against the idea of imputing 'fictional' studies into a meta-analysis (Begg, 1997). We certainly would not rely on results of imputed studies in forming a final conclusion. In contrast, we stress that we believe any adjustment method should be used primarily as a form of sensitivity analysis, to assess the potential impact of missing studies on the meta-analysis, rather than as a means of adjusting results *per se*.

Estimation of the number of missing studies

The key to this method lies in estimating the number of missing studies. We now describe the non-parametric approach which we shall use. This is illustrated in Figure 8.2 where we have estimated, using trim and fill, that the number of missing studies is 5; replaced 5 symmetrically as indicated by the open circles, in the manner described below; and recovered a visually symmetric funnel plot. Rather more importantly, after 'filling' the funnel plot we obtain an overall estimate of the odds ratio of 1.01 with 95 % confidence interval (0.82, 1.26), which is considerably reduced from that estimated from the original data (1.11 with 95 % confidence interval (0.88,1.39)). This is shown at the bottom of Figure 8.2.

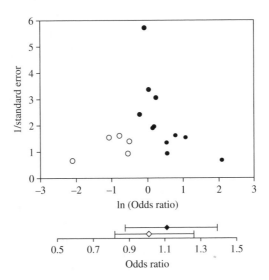

Figure 8.2 A filled funnel plot, as in Figure 8.1, but with the open circles denoting the imputed missing studies. Note that the 'funnel' is now much more visually symmetric. The bottom panel gives the odds ratios and their 95 % confidence intervals before and after allowing for funnel plot asymmetry, in the original scale.

Estimators assuming the global effect size is known

In order to form our estimators we need to use only the ranks of the absolute values of the observed effect sizes, and the signs of those effect sizes around θ. In describing the method, we first assume that θ is known. An iterative method is described later and used in practice, since typically θ is unknown.

We write X_i for the observed values of $T_i - \theta$, and denote the ranks of the absolute values $|X_i|$ as $rank_i$: these ranks run from 1 to n.

We let $\gamma^* \geq 0$ denote the length of the rightmost run of ranks associated with positive values of the observed X_i; and we also use S_{rank} which is the sum of the $rank_i$ *for the positive X_i only*. This is just the Wilcoxon statistic for this set of data.

In order to illustrate these calculations, a worked example is given in Table 8.1 for the data in Figure 8.1.

Based on these quantities, we define several estimators of k_0, with the simplest and most useful given by

$$R_0 = \gamma^* - 1,$$

$$L_0 = [4S_{rank} - n(n+1)]/[2n-1].$$

In using either of these estimators, we will round to the nearest non-negative integer to obtain R_0^+ and L_0^+ respectively, since in practice we need to trim whole studies.

Choice of estimator

It turns out that both of these estimators have good properties. Both have low bias, and as n gets larger the estimator R_0 becomes preferable to L_0 in terms of having a relatively smaller variance. Initial simulations also show that L_0 is more robust than R_0 against certain data configurations that might occur under some circumstances. For example, if there is just one very negative value of X_i followed by a missing collection of studies (which violates the exact assumption that we have made, but not the spirit of publication bias), then R_0 must be zero, although L_0 may be non-zero. The technical details are beyond the scope of the book, but the interested reader may find more details of the properties of the trim and fill estimators in Duval and Tweedie (2000a, 2000b).

In the original articles describing the method, another estimator, Q_0, was presented. L_0 is a simple linearization of Q_0; however, Q_0 does not possess the more desirable properties of L_0 and R_0. For that reason, we restrict our discussions to the more useful estimators, L_0 and R_0, although the interested reader may find the derivation and properties of Q_0 in the original papers (Duval and Tweedie, 2000a, 2000b). Some software packages may allow the user to calculate all three estimators, but we caution against use of Q_0 for the reasons presented in the original articles.

In practice, we would advise using both L_0 and R_0 before making a judgement on the real number that might be suppressed: if both agree, then conclusions are obvious, and if there is disagreement one should perhaps use the resulting values as a basis for a sensitivity analysis.

Trimming and filling funnel plots

The estimators of k_0 we have used above depend on knowing the value of θ, because they rely on knowledge of whether a given observation is to the left or the right of θ. This is clearly not information we would have in practice, and assuming that $\theta = 0$ will lead to an obvious bias if in fact $\theta > 0$.

In Duval and Tweedie (2000a, 2000b), we formally describe how to carry out an iterative algorithm using the estimators above; an informal description of the main ideas is as follows. The iteration is simple in concept. We first estimate θ using a standard fixed- or random-effects model. Using this value, we use one of the estimators of k_0 (say L_0) to decide how many unmatched values there might be around the initial estimate of θ. We trim off this many values, and this leaves a more symmetric funnel plot. A cautionary note should be given here. The original articles (Duval and Tweedie 2000a, 2000b) used random-effects models to trim studies, often preferred because they take into account between-study heterogeneity. However, since they give studies with smaller precision more weight relative to the fixed-effect model, they can be more affected by publication bias (Cooper and Hedges, 1994). Research is under way to investigate the best combination of fixed- and random-effects models for trimming and filling, respectively.

On this 'trimmed' set of data we then re-estimate θ using a fixed or random effects model, typically getting a value to the left of our previous value due to the studies we have trimmed. Using this new central value, we re-estimate the number of missing studies, and then repeat the trimming process. We have found in practice that this stabilizes on real examples after only two or three iterations.

When we have a final estimate $\bar{T}$ and a final estimate k_0^* we 'fill' the funnel with the imputed missing studies. We do this simply by taking the rightmost k_0^* studies, symmetrically reflecting their values of T_i around the estimate $\bar{T}$, and using their values of v_i for the imputed new studies. As a final step we then re-estimate θ using the observed and imputed studies, and also use the observed and imputed studies to estimate a standard error for the effect size, corresponding to that we would have seen if all these studies had been observed.

FORMAL TESTS FOR PUBLICATION BIAS

Some preliminary work using the distributional properties of the estimators of k_0 to formulate tests for the existence of publication bias was described in Duval and Tweedie (2000b). We do not reproduce those results here since the main application of the trim and fill method is to provide a means of adjusting for the possible effects of missing studies, rather than statistical testing *per se*. Nonetheless, the tests based on the trim and fill estimators appear to be quite powerful if there are more than five or six missing studies in a meta-analysis (Duval and Tweedie, 2000b; see also Chapter 10, this volume).

APPLYING 'TRIM AND FILL' IN PRACTICE

Fixed versus random effects

There is controversy regarding the use of fixed-effects and random-effects models in meta-analysis, irrespective of the question of studying publication bias. Random-effects models are more affected by funnel plot asymmetry than fixed-effects models, due to the less precise studies receiving more weight in the former case. Some authors (see Chapter 10) give more credence to the fixed-effects 'version' of trim and fill, now implemented as an option in Stata's 'metatrim' macro, for example. The original articles used random-effects models for the iterative portion of the algorithm, in part due to the recommendations of a number of reports at that time (NRC, 1992: Sutton *et al.*, 1998). As the method has evolved and received considerable attention, the more conservative approach using fixed effects appears to be more tenable, and should encourage users to implement both methods in any assessment.

Sensitivity analysis

How much do we need to be concerned about publication bias? It is clear from simulations that we might wish to change the inferences made if we had the full picture. In the real examples we have examined here and elsewhere, we have found the same thing: in those data sets where we assess considerable publication bias, the filled funnel plot may lead to conclusions which are quantitatively different from those in the original data.

The main goal of this method should be seen as providing methods for sensitivity analyses rather than actually finding the values of missing studies. We are not interested in the exact imputed values. We are, however, interested in how much the value of θ might change if there are missing studies.

Comparison with other methods

The empirical assessment of publication bias is addressed in some detail in Chapter 10. Here we briefly present the results of several assessments which have specifically compared the trim and fill method to other methods (see also Chapter 10). Sutton *et al.* (2000) compared the results of three methods for the detection of publication bias (Begg rank correlation, Egger regression, and trim and fill) to test for the existence of missing studies in 48 meta-analyses from the Cochrane Database of Systematic Reviews (see Chapter 10). Pham *et al.* (2001) applied several methods to a set of 14 meta-analyses including both published and unpublished randomized trials, and concluded that the trim and fill method 'provides a reasonable approximation of the number of missing studies'. Gilbody *et al.* (2000) assessed potential publication bias in a meta-analysis of risperidone studies in psychiatry. They concluded that no single test is effective, but rather a set of tests and analyses should be used. Four methods, including the trim and fill method, were demonstrated in Soeken and Sripusanapan (2003) using a contrived data set. Publication bias in ecology and evolution was investigated by Jennions and Møller (2002) using trim and fill. Their recommendation in this area was that 'future literature reviews assess the robustness of their main conclusions by correcting for potential publication bias using the 'trim and fill' method.' How well trim and fill works largely depends on the somewhat untestable assumption that the most extreme results in one direction are those which have been suppressed. Since we cannot test this assumption, we recommend the method as a tool for sensitivity analysis.

LIMITATIONS AND FUTURE RESEARCH

Impact of assumptions

The assumption that the missing studies are the most negative studies may be questionable in some situations. Light and Pillemer (1984) describe a situation in which the shape of the funnel plot can be distorted in a different manner. If studies with significant results are more likely to be published when the true treatment effect is zero, then results of small studies can only be statistically significant when they are away from zero, either positive or negative at either end of the funnel. This would result in a hollow 'tunnel' inside the funnel rather than a gap in one corner. This possibility is not accounted for using the trim and fill method. Interestingly, this is not a common finding in the meta-analyses we have studied using this method.

Other study-related factors may distort the appearance of the funnel plot. Petticrew *et al.* (1999) demonstrated that when estimates from observational studies investigating the association of hostility and coronary heart disease were adjusted by a quality score, a very asymmetric funnel plot became much more symmetric. Theoretically, any factor related to both outcome and study size has the potential to distort the appearance of a funnel plot, and clearly in such cases misleading results could be obtained using a method related to the funnel plot to address publication bias.

Other suppression mechanisms

The common picture of a truncated funnel plot, as in Figure 8.1, suggests that we are concerned with papers not appearing in the literature because of their Y-values. We might assume that there is a threshold for this effect, and that suppression takes place by leaving out all studies with T_i below that (unknown) threshold. This is no doubt simplistic, but does reflect the type of truncation that is picked up by 'eyeballing' a funnel plot.

A different, commonly used, scenario (Dear and Begg, 1992; Hedges, 1992; Givens *et al.*, 1997) is that suppression of a study depends on its p-value against the null hypothesis $\theta = 0$. Taking the one-sided view of this, we would therefore say that there is greater probability of suppression as the test variable $Z_i = T_i/\sqrt{v_i}$ (which is standard normal under the null hypothesis) becomes smaller (or more negative). Again we might take a threshold approach, and model this by saying that suppression takes place by leaving out all studies with Z_i below some (unknown) threshold, and including all above: this is exactly the assumption of Gleser and Olkin (1996).

Of course, in many cases, the two scenarios are the same: the most negative values of T_i have the largest one-sided p-values. This does, however, depend on the structure of the set of v_i. Now both the T_i under the Y-value scenario and the Z_i under the p-value scenario, provided the null hypothesis of no effect is true, meet the Bernoulli assumptions: these do not require that the X_i are identically distributed, but merely assume that each X_i is from a (possibly different) symmetric distribution around θ. However, if the null hypothesis is false then the Z_i under the p-value scenario do not meet this assumption.

A number of authors (cf. DuMouchel and Harris, 1997) have pointed out that the p-value scenario is rather simplistic, since it fails to acknowledge the role of other criteria, such as size of study, in the decision as to whether a study is published. Misakian and Bero (1998) found that p-values may be the critical determinant in delaying publication, although study size does appear to have some effect.

In some ways the Y-value scenario allows for many of these issues. It assumes that (one-sided) non-significance is, in the main, likely to lead to suppression; but if such a poor p-value is from a large but only slightly 'negative' study, as may well happen, then this is not likely to be the type of study that is actually omitted. In this way the Y-value scenario seems the more appropriate.

Work on the robustness of the trim and fill method to other suppression mechanisms is in progress. We would expect that the results would be rather more conservative under a p-value suppression mechanism, and some preliminary results support this. In any meta-analysis subjected to the trim and fill method, one could certainly look at which studies would satisfy both the p-value and the Y-value criteria so as to qualitatively assess robustness to this assumption.

Simulation studies

Here we briefly summarize results from several simulation studies which have been reported since the introduction of the trim and fill method. Much of the initial simulation work appears as part of Duval's PhD thesis (Duval, 1999), available from the author on request.

Sterne (2000) reported on the basis of simulation studies that the trim and fill method detects 'missing' studies in a substantial proportion of meta-analyses, even in the absence of bias. He warns that uncritical application of the method could mean adding and adjusting for non-existent studies in response to funnel plot asymmetry arising from random variation. While this is certainly of concern and true of any statistical method, no method should ever be applied in an uncritical fashion. The emphasis when applying the trim and fill method should be on the interpretation of the meta-analysis with and without consideration of publication bias, and as such the method provides a tool for the meta-analyst to do this. More results on 'false positive' rates for trim and fill appear in Duval and Tweedie (2000b).

A report by Terrin *et al.* (2003) showed that when studies in a meta-analysis are heterogeneous, trim and fill may inappropriately adjust for publication bias where none exists. This is hardly surprising, given that trim and fill will detect and adjust for funnel plot asymmetry, independent of the reasons for such asymmetry. Heterogeneity should always be explored using appropriate methods, and the results of trim and fill should always be tempered with the realization of alternative explanations for the finding of missing studies.

EXAMPLES

A worked example – gangliosides and stroke

We now work through the data in the Cochrane Database meta-analysis of gangliosides in acute ischaemic stroke, indicating how we achieve the final outcome in Figure 8.2.

In Table 8.1 we give the working for this example (with values given to two decimal places). The following gives a step-by-step approach to implementing the trim and fill method using the data in Table 8.1.

1. Order the observed data, here log odds ratios, according to their magnitude. The first two columns represent the original log odds ratios and their standard errors in ascending order.
2. Obtain an overall estimate of the effect using these data. Here this leads to an initial meta-analysed estimate (using a random-effects model) of $\theta_1 = 0.105$, which we assume is affected by publication bias.

Table 8.1 The calculations in trimming and filling: ganglioside example.

In OR	SE	Centred $\text{lnOR}-\theta_1$	Signed ranks	Centred $\text{lnOR}-\theta_2$	Signed ranks	Centred $\text{lnOR}-\theta_3$	Signed ranks	In OR	SE
		Original data		Omit 4 rightmost		Omit 5 rightmost		Filled data	
−0.20	0.41	−0.31	−6	−0.23	−5	−0.22	−5	−0.20	0.41
−0.07	0.18	−0.17	−5	−0.10	−2	−0.08	−2	−0.07	0.18
0.04	0.30	−0.06	−2	0.01	1	0.03	1	0.04	0.30
0.16	0.53	0.06	1	0.14	3	0.15	3	0.16	0.53
0.21	0.51	0.11	3	0.18	4	0.20	4	0.21	0.51
0.27	0.33	0.16	4	0.24	6	0.25	6	0.27	0.33
0.53	0.74	0.42	7	0.50	7	0.51	7	0.53	0.74
0.56	1.08	0.46	8	0.54	8	0.55	8	0.56	1.08
0.80	0.62	0.69	9	0.77	9	0.78	9	0.80	0.62
1.08	0.66	0.97	10	1.05	10	1.06	10	1.08	0.66
2.11	1.55	2.01	11	2.09	11	2.10	11	2.11	1.55
								−0.50	0.74
								−0.54	1.08
								−0.77	0.62
								−1.05	0.66
								−2.09	1.55

3. Centre the original data around this initial meta-analysed mean obtained from step 2. This leads to the values in column 3.
4. Rank the absolute values of the centred values and then give these ranks the sign (+ or −) associated with the centred value. This gives the signed ranks in column 4.
5. Sum the positive ranks only to obtain an initial value for S_{rank}. In these data, $S_{rank} = 1 + 3 + 4 + 7 + 8 + 9 + 10 + 11 = 53$.
6. Insert this value into the equation for L_0 (say). This gives an estimate of 3.81 missing studies at the first iteration; this is then rounded to give an estimate of 4. It is also easy to see that the run of positive ranks is $\gamma^* = 5$, which gives $R_0 = 4$ also. (This equality between R_0 and L_0 does not always happen!)
7. Trim off the rightmost (largest) 4 values from the original data set on the basis of the obtained value of L_0, and re-estimate the overall mean from the remaining 7 $(11 - 4)$ studies. Here we find $\theta_2 = 0.028$.
8. Centre the original data around this new value. We find that the ranks (given in column 5) now change because of the new centring, giving the values in column 6.
9. Recalculate S_{rank} based on these newly obtained ranks. We find now that $S_{rank} = 1 + 3 + 4 + 6 + 7 + 8 + 9 + 10 + 11 = 59$, $L_0 = 4.95$, which is rounded to an estimate of 5; we find at this iteration that $R_0 = 5$ also.
10. Iterating again, trim off the rightmost 5 values (based on L_0^+) from the original data set. Re-estimate the overall mean from the remaining 6 $(11 - 5)$ studies to find $\theta_3 = 0.014$.
11. Iterate until the ranks stabilize, that is, the estimate of S_{rank} [or L_0^+] does not change. Further iteration for the data centred around θ_3 does not change the ranks, as shown in columns 7 and 8, and the algorithm terminates.
12. Finally, 'fill' the original data with the symmetric 5 (the converged value of L_0^+) data points and their standard errors, as in columns 9 and 10 of Table 1. This leads to the final estimate of $\bar{T}$ with 95 % confidence interval as illustrated in Figure 8.2.

Common data sets

In the examples to follow, we analyse the data sets described in Appendix A for funnel plot asymmetry using trim and fill. We present the overall results adjusted according to L_0^+ only, and we show the adjustment due to imputing studies according to the value of L_0^+ in the figures. The original data are shown as solid symbols and the 'filled' data as open symbols. All calculations used in the 'trimming' part of the algorithm were performed using the random-effects model. For simplicity, calculations on the observed data and on the 'filled' data are given for the random-effects model only.

In those examples where categorical moderators were examined, the trim and fill algorithm was applied separately in each subgroup defined by stratification of the data according to levels of the moderator. This will not necessarily produce results comparable to those methods which perform adjustment via meta-regression models (see Chapter 6). We also stress that the original application of trim and fill was to a single funnel plot, hence the present application to stratified funnel plots is not yet validated and should therefore be considered exploratory.

Passive smoking example

Here we investigate 37 studies which assess the risk of lung cancer in non-smoking women exposed to spousal smoking, using data reported by Hackshaw *et al.* (1997). The data are presented as risk ratios and associated confidence intervals in the table, although all calculations were done on log-transformed values. Log risk ratios are shown in the funnel plot depiction of these data. Table 8.2 and Figure 8.3 show the way in which trim and fill behaves on this example.

After filling seven studies using L_0^+, the adjusted risk ratio is attenuated a little (going from 1.24 to 1.19) but still implicates passive smoking as a risk factor for lung cancer. The impact of the change in the risk ratio due to adjustment is small; using the adjusted compared to the unadjusted estimate would change the posterior

Table 8.2 Analysis of the passive smoking data.

Model	Number of studies	Value of k_0	θ	95 % CI
Observed	37	0	1.24	(1.13,1.36)
'Filled'	44	7	1.19	(1.08,1.31)

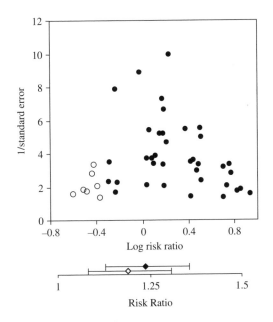

Figure 8.3 A filled funnel plot of the passive smoking data, with the open circles denoting the imputed missing studies. The bottom panel gives the risk ratios and their 95 % confidence intervals before and after allowing for funnel plot asymmetry, in the original scale.

probability of lung cancer by a factor of only 0.96. This suggests that although L_0^+ indicates that perhaps seven studies are missing, the practical impact is negligible.

Validity of the employment interview example

Correlations between employment interview scores and job performance in 160 studies were analysed in detail in McDaniel *et al.* (1994). Data were originally given as correlation coefficients and have been transformed to Fisher z-values for analysis purposes.

Data were subgrouped according to prior hypotheses given by the original authors and individually subjected to trim and fill analyses. The indentation in Table 8.3 indicates the hierarchical structure of the data after such stratification, and results before and after adjustment are given on the original correlation scale. A funnel plot, stratified on whether the interview was structured or unstructured, is given in Chapter 5.

The studies which reported structured interview data (106 studies) had substantially higher correlations than those which reported unstructured interviews (39 studies; $r = 0.265$ versus $r = 0.187$). Assessment of funnel plot asymmetry in these two strata indicates asymmetry in those studies that used structured interviews, and not in those that used unstructured interviews (18 missing studies versus 0 missing studies). The adjusted correlation for the structured subgroup was attenuated to $r = 0.205$ after imputing 18 studies, much closer to the unadjusted value in the unstructured subgroup.

Further investigation of the structured interview group was conducted using a variable collected to reflect whether the interview content was job-related (job),

Table 8.3 Analysis of validity of employment interview data.

Model	Number of studies	Value of k_0	θ	95 % CI
Structured				
Observed	106	0	0.265	(0.223,0.308)
'Filled'	124	18	0.205	(0.156,0.253)
Job				
Observed	89	0	0.264	(0.216,0.310)
'Filled'	90	1	0.260	(0.212,0.307)
Situational				
Observed	16	0	0.283	(0.209,0.353)
'Filled'	22	6	0.242	(0.171,0.310)
Admin				
Observed	8	0	0.254	(0.125,0.376)
'Filled'	11	3	0.206	(0.086,0.321)
Research				
Observed	8	0	0.304	(0.219,0.384)
'Filled'	11	3	0.278	(0.191,0.366)
Unstructured				
Observed	39	0	0.187	(0.138,0.235)
'Filled'	39	0	0.187	(0.138,0.235)

psychological or job-related-situational (situational). Only one study reported data on psychological content via a structured interview and was not analysed further. Modest funnel plot asymmetry was noted in the 'situational' studies (16 studies; 6 missing studies), in contrast to very little or no asymmetry in the studies reporting job-related interview data (89 studies; 1 missing study).

Teacher expectancy example

This example uses data from 19 randomized studies examining the hypothesis that teachers' expectations influence students' IQ intelligence test scores. Raudenbush (1984) reported 19 studies where students were randomly assigned either to an experimental 'expectancy induction group' or to a control group. Teachers of students in the experimental group were led to believe that their students were 'likely to experience substantial intellectual growth'. The data are in the form of standardized mean differences (d_i) and their standard errors. Raudenbush (1984) also hypothesized that the magnitude of teacher expectancy effects might be related to the length of time that the teacher and student were in contact prior to the expectancy induction. Teachers who had had more contact with students were expected to be less vulnerable to the expectancy induction. Amount of prior contact was categorized into two groups: contact for one week or less (the low-contact group), and contact for more than one week (the high-contact group).

In a meta-analysis ignoring funnel plot asymmetry, the random-effects overall estimate was $\theta = 0.084$ with 95 % CI $(-0.023, 0.191)$ (Table 8.4). Filling with the three missing studies indicated by L_0^+, we obtain an overall estimate of $\theta = 0.027$ with 95 % CI $(-0.098, 0.152)$. The overall effect is considerably reduced from that estimated from the original data. After allowing for even this small amount of asymmetry, the estimate of θ is reduced by two-thirds. This finding warrants further investigation, as we now proceed to show.

We examine two subgroups of the data, according to a variable dichotomized to examine possible effects due to length of contact between teacher and pupil prior to the experiment. We find that application of trim and fill to each of the

Table 8.4 Analysis of teacher expectancy data.

Model	Number of studies	Value of k_0	θ	95 % CI
Overall				
Observed	19	0	0.084	$(-0.023, 0.191)$
'Filled'	22	3	0.027	$(-0.098, 0.152)$
Low-contact				
Observed	8	0	0.373	$(0.170, 0.576)$
'Filled'	8	0	0.373	$(0.170, 0.576)$
High-contact				
Observed	11	0	-0.021	$(-0.102, 0.059)$
'Filled'	11	0	-0.021	$(-0.102, 0.059)$

subgroups defined by amount of prior contact reveals no missing studies in either subgroup. This analysis seems to indicate that the apparent heterogeneity in teacher expectancy effects found in the overall analysis may indeed be accounted for by the amount of prior contact. Careful inspection of the data reveals essentially two superimposed funnel plots with minimal overlap (see Chapter 5). Studies where teachers and pupils had little contact produced rather larger effects than those studies where the contact had been longer – in this latter case, effects were close to zero.

These data are a good example of the care needed in interpretation of findings of simple pooled analyses in which there is considerable heterogeneity and no investigation of covariates as an explanation for such variability has been carried out. The importance of investigating systematic effects as an explanation of both heterogeneity and funnel plot asymmetry before considering publication bias is recommended.

CONCLUSIONS

The trim and fill method uses simple symmetry assumptions and an iterative approach to estimate the number of missing studies. By removing studies which have no symmetric counterparts, we find an estimate of the overall mean θ which takes asymmetry bias into account. This estimate will have the wrong variance, being based on a smaller sample than is estimated (or indeed known) to exist. However, it can be expected (if the symmetry assumptions are reasonable) to provide the 'correct' mean, and we can re-estimate the variance by imputing the values of the missing studies around that mean.

In the examples we have illustrated in detail, the trim and fill method matches the subjective impression of bias given by the funnel plot. In other examples we have examined, it also appears to give results consistent with more complex methods. Results are also consistent with those from methods which only consider the existence of bias: when the Begg and Mazumdar (1994) or Egger *et al.* (1997) tests indicate existence of publication bias, our approach also indicates missing studies; when no bias is assessed by these methods, we estimate a small number of missing studies, if any.

It is clear from simulations that we might wish to change the inferences made if we had the full picture. In the real examples we have examined here and elsewhere, we have found the same thing: in those data sets where we assess considerable asymmetry bias, the filled funnel plot may lead to conclusions which are quantitatively different from those in the original data.

Nonetheless, the main goal of this work should be seen as providing methods for sensitivity analyses rather than actually finding the values of missing studies. We are not interested in the exact imputed values. We are, however, interested in how much the value of θ might change if there are missing studies: and from that perspective, the trim and fill approach does seem to give good indications of which meta-analyses do not suffer from bias, and which need to be evaluated much more carefully.

REFERENCES

Begg, C.B. (1997). Comment on Givens, G.H., Smith, D.D. & Tweedie, R.L. (1997). Publication bias in meta-analysis: A Bayesian data-augmentation approach to account for issues exemplified in the passive smoking debate (with discussion). *Statistical Science*, **12**, 241–244.

Begg, C.B. and Mazumdar, M. (1994). Operating characteristics of a rank correlation test for publication bias. *Biometrics*, **50**, 1088–1101.

Candelise L. and Ciccone A. (1998). Gangliosides for acute ischaemic stroke (Cochrane Review). In *The Cochrane Library*, Issue 1, 1998. Chichester: John Wiley & Sons, Ltd.

Borenstein, M., Hedges, L., Higgins, J. & Rothstein, H. (2005) *Comprehensive Meta Analysis, Version 2*. Englewood, NJ: Biostat.

Cooper, H. and Hedges, L.V. (eds) (1994). *The Handbook of Research Synthesis*. New York: Russell Sage Foundation.

Dear, K. and Begg, C. (1992). An approach for assessing publication bias prior to performing a meta-analysis. *Statistical Science*, **7**, 237–245.

Dear, K. and Dobson, A. (1997). Comment on Givens, G.H., Smith, D.D. & Tweedie, R.L. (1997), Publication bias in meta-analysis: A Bayesian data-augmentation approach to account for issues exemplified in the passive smoking debate (with discussion). *Statistical Science*, **12**, 245–246.

DuMouchel, W. and Harris, J. (1997). Comment on Givens, G.H., Smith, D.D. & Tweedie, R.L. (1997), Publication bias in meta-analysis: A Bayesian data-augmentation approach to account for issues exemplified in the passive smoking debate (with discussion). *Statistical Science*, **12**, 244–245.

Duval, S.J. (1999). Effects of publication bias in meta-analysis. PhD thesis, Department of Preventive Medicine and Biometrics, University of Colorado Health Sciences Center, Denver, CO.

Duval, S.J. and Tweedie, R.L. (2000a). A non-parametric 'trim and fill' method of accounting for publication bias in meta-analysis. *Journal of the American Statistical Association*, **95**, 89–98.

Duval, S.J. and Tweedie, R. L. (2000b). Trim and fill: A simple funnel-plot-based method of testing and adjusting for publication bias in meta-analysis. *Biometrics*, **56**, 455–463.

Egger, M., Davey, S.G., Schneider, M. and Minder, C. (1997). Bias in meta-analysis detected by a simple, graphical test. *British Medical Journal*, **315**, 629–634.

Gilbody, S.M., Song, F., Eastwood, A.J. and Sutton, A. (2000). The causes, consequences and detection of publication bias in psychiatry. *Acta Psychiatrica Scandinavica*, **102**, 241–249.

Givens, G.H., Smith, D.D. and Tweedie, R.L. (1997). Publication bias in meta-analysis: A Bayesian data-augmentation approach to account for issues exemplified in the passive smoking debate (with discussion). *Statistical Science*, **12**, 221–250.

Gleser, L.J. and Olkin, I. (1996). Models for estimating the number of unpublished studies. *Statistics in Medicine*, **15**, 2493–2507.

Hackshaw, A.K., Law, M.R. and Wald, N.J. (1997). The accumulated evidence on lung cancer and environmental tobacco smoke. *British Medical Journal*, **35**, 980–988.

Hedges, L. (1992). Modeling publication selection effects in meta-analysis. *Statistical Science*, **7**, 227–236.

Iyengar, S. and Greenhouse, J.B. (1988). Selection models and the file drawer problem. *Statistical Science*, **3**, 109–135.

Jennions, M.D. & Møller, A.P. (2002). Publication bias in ecology and evolution: An empirical assessment using the 'trim and fill' method. *Biological Reviews*, **77**, 211–222.

Light, R. and Pillemer, D. (1984). *Summing Up: the Science of Reviewing Research*. Cambridge, MA: Harvard University Press.

McDaniel, M.A., Whetzel, D., Schmidt, F.L. and Maurer, S. (1994). The validity of the employment interview: A comprehensive review and meta-analysis. *Journal of Applied Psychology*, **79**, 599–616.

Misakian, A.L. and Bero, L.A. (1998). Publication bias and research on passive smoking: Comparison of published and unpublished studies. *Journal of the American Medical Association*, **280**, 250–253.

NRC Committee on Applied and Theoretical Statistics (1992). *Combining Information: Statistical Issues and Opportunities for Research.* National Academy Press, Washington, DC.

Petticrew, M., Gilbody, S. & Sheldon, T. A. (1999). Relation between hostility and coronary heart disease. Evidence does not support link. *British Medical Journal*, **319**, 917.

Pham, B., Platt, R., McAuley, L., Klassen, T.P. and Moher, D. (2001). Is there a 'best' way to detect and minimize publication bias? *Evaluation & The Health Professions*, **24**, 109–125.

Raudenbush, S.W. (1984). Magnitude of teacher expectancy effects on pupil IQ as a function of the credibility of expectancy induction. *Journal of Educational Psychology*, **76**, 85–97.

Rosenthal, R. (1979). The 'file-drawer problem' and tolerance for null results. *Psychological Bulletin*, **86**, 638–641.

Soeken, K.L. and Sripusanapan, A. (2003). Assessing publication bias in meta-analysis. *Nursing Research*, **52**, 57–60.

StataCorp. 2003. *Stata Statistical Software: Release 8.0.* College Station, TX: StataCorp LP.

Sterne, J.A.C. (2000). High false positive rate for trim and fill method. Electronic response to Sutton *et al.* (2000). http://www.bmj.com/cgi/eletters/320/7249/1574#E1.

Sterne, J.A.C. and Egger, M. (2001). Funnel plots for detecting bias in meta-analysis: Guidelines on choice of axis. *Journal of Clinical Epidemiology*, **54**, 1046–1055.

Sterne, J.A.C., Gavaghan, D. and Egger, M. (2000). Publication and related bias in meta-analysis: Power of statistical tests and prevalence in the literature. *Journal of Clinical Epidemiology*, **53**, 1119–1129.

Sugita, M., Kanamori, M., Izuno, T. and Miyakawa, M. (1992). Estimating a summarized odds ration whilst eliminating publication bias in meta-analysis. *Japanese Journal of Clinical Oncology*, **22**, 354–358.

Sutton, A.J., Abrams, K.R., Jones, D.R., Sheldon, T.A. and Song, F. (1998). Systematic reviews of trials and other studies.*Health Technology Assessment*, **2**, 1–276.

Sutton, A.J., Duval, S.J., Tweedie, R.L., Abrams, K.R. and Jones, D.R. (2000). Empirical assessment of effect of publication bias on meta-analyses. *British Medical Journal*, **320**, 1574–1577.

Tang, J.-L. and Liu, J.L.Y. (2000). Misleading funnel plot for detection of bias in meta-analysis. *Journal of Clinical Epidemiology, **53**, 477–484.

Terrin, N., Schmid, C.H., Lau, J. & Olkin, I. (2003). Adjusting for publication bias in the presence of heterogeneity. *Statistics in Medicine*, **22**, 2113–2126.

Vandenbroucke, J.P. (1988). Passive smoking and lung cancer: A publication bias? *British Medical Journal*, **296**, 391–392.

CHAPTER 9

Selection Method Approaches

Larry V. Hedges
University of Chicago, USA

Jack Vevea
University of California, Santa Cruz, USA

KEY POINTS

- Observed effect sizes in a meta-analysis depend on an effect size model and a selection model.
- The selection model describes the mechanism by which effect estimates are selected to be observed.
- The effect size model describes what the distribution of effect sizes would be if there were no publication selection.
- If the selection model is known, it is relatively easy to 'correct for' the effects of selection.
- Assuming various selection models provides means for sensitivity analyses.
- If the selection model is unknown, it may still be possible to correct for selection by estimating the selection model from the observed effect size data.
- Selection model methods are ideally suited to analysis of studies with heterogeneous effects.

Publication Bias in Meta-Analysis – Prevention, Assessment and Adjustments   Edited by H.R. Rothstein, A.J. Sutton and M. Borenstein   © 2005 John Wiley & Sons, Ltd

INTRODUCTION

There are various approaches that the researcher confronting the problem of publication bias may take. Other chapters discuss graphical methods for detecting bias, tools for assessing how severe bias would need to be in order to overturn one's conclusion, and models that attempt to correct for the presence of publication bias. The present chapter describes a specific class of corrective model: models that correct for bias using a weight function to represent the process of selection.

Correcting for publication bias requires some model for the sampling behavior of the observed effect sizes that incorporates the selection process. In constructing such a model it is useful to distinguish two aspects of that model: the effect size model and the selection model. The effect size model specifies what the distribution of the effect size estimates would be (at least in form) if there were no selection. The selection model specifies how this effect size distribution is modified by the selection process. If the selection model were known, it would be relatively straightforward to obtain an estimate of the unselected distribution of effect size estimates by 'inverting' the selection process.

Selection models involve one or more parameters that govern the selection process. Unfortunately these parameters are typically unknown. One strategy for using explicit selection models is to estimate from observed effect size data the parameters of the selection model along with the parameters of the effect size model. However, the parameters of the selection model may be difficult or impossible to estimate from a collection of observed effect size estimates, particularly if the number of studies is small. The difficulty in estimating selection models implies that precise correction of effect sizes for publication bias is also difficult. An alternative to estimating the parameters of the selection model is to assume specific values of the selection model parameters to carry out sensitivity analyses. This permits, for example, the evaluation of bounds for the impact of selection on effect size analyses.

EFFECT SIZE MODELS FOR SELECTION BIAS

Selection models can be applied with any effect size model, and effect size models can take many different forms. To simplify the discussion, we concentrate the exposition on the general mixed-effects model and note that many other effect size models (such as fixed-effects models) are special cases of the model we consider. We outline the model with covariates (between-study predictors of effect size) and treat the case of estimating a single common or mean effect size as a special case of the models without covariates.

The general mixed-effects model

When selection is absent, it is appropriate to represent the distribution of sample effects using the usual mixed-effects general model. Suppose that we have a series of n studies with effect size estimates $T_1, \ldots, T_n$ of effect parameters $\theta_1, \ldots, \theta_n$. The ith study provides the estimate T_i of parameter θ_i such that

$$T_i \sim N(\theta_i, \sigma_i^2),$$

where σ_i^2 depends primarily on sample size, and hence is approximately known, and θ_i is an unknown parameter. Suppose that the effect parameter of the ith study θ_i depends on a linear combination of q known predictor variables:

$$\theta_i = \beta_0 + \beta_1 x_{i1} + \cdots + \beta_q x_{iq} + \xi_i, \tag{9.1}$$

where ξ_i is a random effect, usually assumed to have a normal distribution with zero mean and unknown variance τ^2 (the residual between-studies variance component), so that

$$\xi_i \sim N(0, \tau^2).$$

Denote the vector of predictor variables from the ith study as $\mathbf{x}_i = (x_{i1}, \ldots, x_{iq})'$, and the vector of regression coefficients as $\boldsymbol{\beta} = (\beta_1, \ldots, \beta_q)'$, where q denotes the number of predictors.

The observed statistic T_i from study i is used to test the null hypothesis $H_0 : \theta_i = 0$ by means of the test statistic

$$Z_i = T_i / \sigma_i.$$

The one-tailed p-value associated with that test is

$$p_i = 1 - \Phi(Z_i) = \Phi(-Z_i) \tag{9.2}$$

(when the positive tail is the one of interest), where $\Phi(z)$ denotes the standard normal cumulative distribution function. (Note that in the primary study, this test was most likely conducted using an exact distribution such as t or F; nevertheless, the p-value based on the standard normal distribution provides a good approximation.)

Special cases
The simple random-effects model (the mixed model with no covariates) corresponds to defining $q = 0$ so there are no covariates. In this case $\boldsymbol{\beta} = (\beta_0)$ and β_0 is the mean of the random effects. Also τ^2, the residual between-studies variance component, is just the entire between-studies variance component.

The general fixed-effects model with covariates corresponds to constraining the residual between-studies variance component to be 0. The simple fixed-effects model (the fixed-effects model with no covariates) corresponds to defining $q = 0$ so there are no covariates while constraining the residual between-studies variance component to be 0. In this case $\boldsymbol{\beta} = (\beta_0)$ and β_0 is the common effect size.

SPECIFYING SELECTION MODELS

Iyengar and Greenhouse (1988) suggested the use of weighted distributions (Rao, 1965, 1985) to describe selection models in meta-analysis. The selection models that we examine here can be formulated in terms of weighted distributions. In these models, a non-negative weight function $w(T)$ is used to model the selection process by giving the likelihood that an effect estimate T will be observed if it occurs. To make this idea precise we need to distinguish effect size estimates that may occur before selection from those that are observed after selection.

Let T^* be the random variable representing the effect estimate before selection, and let the density function of T^* be $f(t|\theta)$, which depends on parameter θ. Denote the weight function by $w(t|\omega)$, which depends on parameter ω. Then the weighted density of the observed effect estimate T is given by

$$g(t|\theta, \omega) = \frac{w(t|\omega)\, f(t|\theta)}{\int w(t|\omega) f(t|\theta)\, dt}. \tag{9.3}$$

Whenever the weight function is not a constant, the sampling distribution of the observed effect size T differs from that of the unselected effect size T^*, and this difference is a way of describing publication bias. Inference for the parameters θ that describe the *unselected* distribution of effect size estimates is carried out using the weighted distribution $g(t|\theta, \omega)$.

To understand why selection makes a difference, it is helpful to write $T^* = \theta + \varepsilon^*$, and then let $\bar{\varepsilon}^* = \varepsilon^*/\sigma$ be the standardized sampling error. Using the fact that T^* is normally distributed about θ with variance σ^2 (and hence $\bar{\varepsilon}^*$ is normally distributed about 0), express the expected value of T using (9.3) as

$$E[T] = \int tg(t|\theta, \omega)dt = \theta + \frac{\int \bar{\varepsilon} w(\theta + \bar{\varepsilon}\sigma|\omega)\phi(\bar{\varepsilon})d\bar{\varepsilon}}{\sigma \int w(\theta + \bar{\varepsilon}\sigma|\omega)\phi(\bar{\varepsilon})d\bar{\varepsilon}}. \tag{9.4}$$

If larger values of T^* are more likely to be observed (given that they occur) than smaller values, $w(t|\omega)$ is a monotonic increasing function of t, and the second term (the bias) is always positive. Moreover, since σ is usually a decreasing function of the within-study sample size n, it follows that the second term of (9.4), the bias due to selection, tends to zero as n tends to infinity. It can also be shown that if $w(t|\omega)$ is a monotonic increasing function of t, then the distribution of T (the observed effect size estimates) is stochastically larger than the distribution of T^* (the unselected effect size estimates); see Iyengar and Zhao (1994).

Classes of selection models

Two somewhat different classes of explicit selection models have been proposed for meta-analysis. In one class, the weight is typically presumed to depend on the effect size estimate only through the p-value associated with the study – or equivalently through the ratio T/σ (Hedges 1984, 1992; Iyengar and Greenhouse, 1988). A second class of weight functions depends on the effect size estimate and its standard error separately (Copas, 1999; Copas and Shi, 2000, 2001). In both cases other parameters determine the probability of selection of given T/σ (in the first class of models) or T and σ (in the second class of models). In either case, one can attempt to estimate the selection parameters, or one can assume some or all of these parameters are known for the purposes of analysis or sensitivity analysis. Either of these selection processes produces non-ignorable missing data, which leads to biases in estimation (Little and Rubin, 2002).

Weight functions that depend only on the *p*-value

Weight functions that depend only on the *p*-value from a test of the statistical significance of the effect size have the longest history in meta-analysis. They are based on the conceptual model that decisions about the conclusiveness of research results are often based on statistical significance. Let T_i be the observed effect size estimate from study i and let p_i be the one-tailed *p*-value given in (9.2).

Lane and Dunlap (1978) and Hedges (1984) considered the extreme case where *only* studies with statistically significant effect sizes (i.e. $p < 0.05$ two-tailed, or $p > 0.975$ or $p < 0.025$ one-tailed) are observed; this corresponds to a weight of zero if the study is non-significant, and one otherwise. Iyengar and Greenhouse (1988) utilized two weight functions with known functional form: one in which the weight increased from zero to one as the two-tailed *p*-value decreased from 1.0 to 0.05, and one in which the weight was constant but non-zero when $1 > p > 0.05$. Both of these weight functions assigned weight of unity to values in which $0 < p < 0.05$.

More recent work has focused on the problem of estimating the form of the weight function from the effect size data. Dear and Begg (1992) estimate the weight function as a step function assuming that both relative weights and location of the steps (point discontinuities) are unknown. Several studies (Hedges, 1992; Vevea *et al.*, 1993; Vevea and Hedges, 1995) used a similar procedure except that the location of the steps was assumed to be known a priori. With a sufficient number of steps, a weight function with any functional form can be reasonably well approximated.

In this chapter, we follow the approach described by Hedges (1992) in which the weights depend on effect size only through the one-tailed *p*-values, which preserve information about the direction of the effect. This is an important feature because it seems plausible that direction may sometimes influence selection decisions. Moreover, two-tailed selection is a special case of one-tailed selection in which the weight function is symmetric about $p = 0.5$. The use of one-tailed *p*-values in the model in no way implies an assumption that the original tests of the effects were conducted as one-tailed tests.

Weight functions that depend on *T* and *σ*

A different suggestion for a weight function was made in a series of papers by Copas (1999; Copas and Li, 1997; Copas and Shi, 2000). The basic idea of the selection model is that the probability of selection depends on both sample size (more precisely, standard error of the effect size estimate) and the size of the estimated effect, being an increasing function of each. These ideas are operationalized for a selection model by hypothesizing a selection variable

$$z = a + b/\sigma + \delta,$$

which is defined in terms of two parameters a and b and a random quantity δ (which is assumed to have a standard normal distribution). The selection variable z determines the probability of selection via

$$w(z) = \bar{\Phi}(z). \tag{9.5}$$

The parameter a determines the minimal probability of selection and the parameter b determines how fast the probability of selection increases as the standard error σ decreases. The expected value of the selection variable z is given by

$$E\{z\} = a + b/\sigma.$$

The selection variable z may be correlated with T^*, the effect size estimates before selection. If the correlation ρ between z and T^* is non-zero, then the expected value of T (observed after selection) is not the same as the expected value of T^* (before selection). In other words, selection introduces bias in the effect size estimates.

The probability of selection of study i is given by

$$w(T_i, \sigma_i) = \Phi\left(\frac{a + b/\sigma_i + \rho_i(T - \mathbf{x}_i\boldsymbol{\beta})/\sqrt{\tau^2 + \sigma_i^2}}{\sqrt{1 - \rho_i^2}}\right), \tag{9.6}$$

where $\boldsymbol{\beta}$ is a vector of unknown regression coefficients, $\mathbf{x}_i$ is a vector of known covariate values for study i, τ^2 is the variance of the effect size parameter distribution in the unselected population of studies and ρ_i is the correlation between z_i and T_i^* which depends on a parameter ρ of the selection model and σ_i and τ via

$$\rho_i = \frac{\sigma_i\rho}{\sqrt{\tau^2 + \sigma_i^2}}.$$

It is important to recognize that this selection model cannot, in general, be estimated from effect size data. Consider the case where $b = 0$; then the selection model says that the probability of selection for all studies is $\Phi(a)$, but this cannot be estimated from the data unless the number of studies in the unselected population is known. In other cases, (e.g., when b and ρ are small) the data contain very little information about the parameters a and b.

Note also that this weight function depends on $\boldsymbol{\beta}$ and τ^2, and thus on the distribution of effect size parameters in the unselected population. The fact that the weight function depends on the parameters of the effect size model complicates estimation of this model, leading to problems of non-robustness observed in econometric selection models that are similar in form (Little, 1985).

Copas and his colleagues (Copas, 1999; Copas and Shi, 2000, 2001) have proposed evaluating the goodness of fit of their weight function selection model by testing whether, after taking selection into account, the standard error σ_i still predicts effect size. Their test is equivalent to adding σ_i as the $(q + 1)$th predictor in the effect size model and testing the statistical significance of this predictor via the usual likelihood ratio test.

METHODS THAT ASSUME A KNOWN SELECTION MODEL

Methods that assume a particular extreme selection model can be useful because they provide a bound for the likely effects of selection on as few as one observed estimate. They also have the virtue that tables can be prepared for their implementation without the aid of specialized software. Essentially these methods permit the researcher to 'correct' each observed estimate for selection.

Two-tailed selection of only significant results

Given any particular weight function, it is possible to obtain the maximum likelihood estimator of θ given the observed estimate T under that selection model by maximizing the likelihood under the weighted distribution model. The problem is equivalent to obtaining the maximum likelihood estimate of the normal mean under an unusual truncation model. The maximum likelihood estimates cannot usually be obtained in closed form, but it is possible to tabulate them as a function of θ and σ. This approach was used by Hedges (1984) and Hedges and Olkin (1985) for the standardized mean difference (except that they used its exact distribution).

Maximum likelihood estimation of effect size under the extreme model of two-tailed selection can be obtained by numerically discovering the value of θ that maximizes the likelihood under the selection model. For large values of T, the maximum is almost identical to T, but as the value of T decreases, the maximum becomes smaller than T. As T approaches the minimum observable (statistically significant) value of T, the value of $\hat{\theta}$ decreases to a smaller, but still positive, value. One way to describe the relation between the maximum likelihood estimator and the observed effect size is that the former is the latter 'corrected by shrinking toward zero' by an amount that depends on the size of T in relation to the minimum significant effect size.

It can be shown that the proportionate amount of this correction depends only on the size of the test statistic $z = T/\theta$. Thus tabulating the proportionate correction factor $(\hat{\theta}/T)$ as a function of z will permit the computation of $\hat{\theta}$ for any effect size regardless of θ. Table 9.1 presents such a tabulation. Only values of z greater than 1.96 are tabulated because these are the only values possible given the extreme selection model. Note that for effect size estimates that are just barely significant ($z = 1.96$), the proportionate correction factor is 0.247, so the maximum likelihood estimate is only a quarter of the observed effect size. When the observed effect size reaches a higher level of statistical significance, the proportionate correction factor increases rapidly. When $z = 2.6 (p < 0.005)$ it is greater than 0.5, when $z = 3.0$ the proportionate correction factor is nearly 0.85, and it becomes essentially 1.0 for $z > 4$.

To use Table 9.1, enter the table with the test statistic value $z = T/\theta$ and obtain the corresponding correction factor. Multiply this factor by the observed effect size T to obtain the maximum likelihood estimate of θ given extreme two-tailed selection. Only positive values are tabled because the problem is symmetric about $T = 0$. For negative T-values, compute the estimate as the negative of the estimate that corresponds to $|T|$.

This method illustrates the idea that if a literature has been affected by publication bias, that effect is present in every statistical result we encounter in that literature. The method provides a means of adjusting each of the individual effect size estimates directly. While a collection of estimates adjusted in this way could be combined into a combined estimate, it would be more satisfactory to use the methods discussed later in this chapter to obtain a combined estimate from a collection of studies.

Table 9.1 Ratio $\hat{\theta}/T^*$ of the observed effect size T^* to the maximum likelihood estimate $\hat{\theta}$ of effect size under extreme two-tailed selection as a function of $Z^* = T^*/\sigma$.

Z^*	$\hat{\theta}/T^*$	Z^*	$\hat{\theta}/T^*$	Z^*	$\hat{\theta}/T^*$	Z^*	$\hat{\theta}/T^*$
1.96	0.247	2.55	0.510	3.15	0.889	3.75	0.973
2.00	0.252	2.60	0.565	3.20	0.901	3.80	0.976
2.05	0.259	2.65	0.620	3.25	0.912	3.85	0.979
2.10	0.268	2.70	0.666	3.30	0.922	3.90	0.982
2.15	0.278	2.75	0.704	3.35	0.931	3.95	0.984
2.20	0.290	2.80	0.740	3.40	0.938	4.00	0.986
2.25	0.304	2.85	0.771	3.45	0.945	4.10	0.989
2.30	0.321	2.90	0.797	3.50	0.951	4.20	0.992
2.35	0.342	2.95	0.820	3.55	0.957	4.30	0.994
2.40	0.370	3.00	0.841	3.60	0.962	4.50	0.996
2.45	0.408	3.05	0.859	3.65	0.966	4.70	0.998
2.50	0.455	3.10	0.875	3.70	0.970	4.90	0.999
						5.20	1.000

Maximum likelihood estimation under one-tailed selection

Values of the observed effect size that are just large enough to be observed (just large enough to be significant) provide little information about the magnitude of the true effect size under the extreme one-tailed selection model. This corresponds to the fact that when the observed effect size is small, the likelihood as a function of θ is very flat for small positive and negative values of θ. When the observed effect size is too small (less than about 2.1σ), the likelihood actually increases as θ decreases and is unbounded as a function of θ. Under these conditions, maximum likelihood estimation of effect size is not possible. Because of these problems with maximum likelihood estimation from a single effect size under one-tailed selection, we do not pursue it further here.

METHODS FOR WEIGHT FUNCTIONS THAT DEPEND ONLY ON THE p-VALUE

If selection is a function of the one-tailed p-value, the weight function $w(p_i)$ determines the likelihood that data from an individual study with p-value p_i are observed. Suppose that the weight function has k intervals on which it is constant. Denote the left and right endpoints of the jth interval by a_{j-1} and a_j respectively, and set $a_0 = 0$ and $a_k = 1$. If a study's p-value falls within the ith such interval, we denote its weight by ω_i. We expect that the weight functions, as functions of p, will be the same for all studies, so that

$$w(p_i) = \begin{cases} \omega_1, & \text{if } 0 < p_i \leq a_1, \\ \omega_j, & \text{if } a_{j-1} < p_i \leq a_j, \\ \omega_k, & \text{if } a_{k-1} < p_i \leq 1. \end{cases} \tag{9.7}$$

We can also define the weight function to be a function of study's effect size T_i and its conditional variance σ_i^2. The subscript i is needed on the weight function because the p-value depends on both T_i and σ_i. Thus

$$w(T_i, \sigma_i) = \begin{cases} \omega_1, & \text{if } -\sigma_i \Phi^{-1}(a_1) < T_i \leq \infty, \\ \omega_j, & \text{if } -\sigma_i \Phi^{-1}(a_j) < T_i \leq -\sigma_i \Phi^{-1}(a_{j-1}), \\ \omega_k, & \text{if } -\infty < T_i \leq -\sigma_i \Phi^{-1}(a_{k-1}). \end{cases} \tag{9.8}$$

Since we have no information beyond the observed T_i-values, and, in particular, since we lack knowledge of how many studies existed before selection occurred, the weights must be relative rather than absolute. Accordingly, it is convenient to constrain the value of one of the weights, and let the others represent the chance of studies within an interval being observed, relative to the chance of a study from the constrained interval being observed. Our convention is to set $\omega_1 = 1.0$. Hence, if the weight for the third p-value interval were 0.5, it would indicate that studies from the third interval are only half as likely to be observed as studies from the first interval. Note that it is often desirable to rescale the weights (and recompute their standard errors) after estimation so that the largest rescaled weight is unity. The standard errors of the rescaled weights can be computed using the asymptotic approximation for the variance of the ratio of two random variables

$$\text{var}\left(\frac{\hat{\omega}_i}{\hat{\omega}_{\max}}\right) \approx \left(\frac{\hat{\omega}_i^2}{\hat{\omega}_{\max}^2}\right)^2 \left[\frac{\text{var}(\hat{\omega}_i)}{\hat{\omega}_i^2} + \frac{\text{var}(\hat{\omega}_{\max})}{\hat{\omega}_{\max}^2} + \frac{2\text{cov}(\hat{\omega}_i, \hat{\omega}\max)}{\hat{\omega}_i \hat{\omega}_{\max}}\right].$$

The combined probability model[*]

The weighted probability density function of T_i given the weight function $w(T_i, \sigma_i)$ and the parameters $\boldsymbol{\beta}$ and $\boldsymbol{\omega} = (\omega_1, \ldots, \omega_k)'$ is

$$f(T_i | \boldsymbol{\beta}, \boldsymbol{\omega}, \sigma_\theta) = \frac{w(T_i, \sigma_i)\phi((T_i - \mathbf{x}_i'\boldsymbol{\beta})/\eta_i)}{\sigma_i A_i(\boldsymbol{\beta}, \boldsymbol{\omega}, \sigma_\theta)}, \tag{9.9}$$

where

$$\eta_i^2 = \sigma_i^2 + \sigma_\theta^2,$$

$$A_i(\boldsymbol{\beta}, \boldsymbol{\omega}, \sigma_\theta) = \int_{-\infty}^{\infty} \eta_i^{-1} w(T_i, \sigma_i)\phi\left(\frac{T_i - \mathbf{x}_i'\boldsymbol{\beta}}{\eta_i}\right) dT_i,$$

and $\phi(z)$ denotes the standard normal probability density function. Since A_i can be expressed as the sum of normal integrals over the intervals where $w(T_i, \sigma_i)$ is constant,

$$A_i(\boldsymbol{\beta}, \omega, \sigma_\theta) = \sum_{j=1}^{k} \omega_j B_{ij}(\boldsymbol{\beta}, \sigma_\theta), \tag{9.10}$$

where $B_{ij}(\boldsymbol{\beta})$ is the probability of a normally distributed random variable with mean $\mathbf{x}_i'\boldsymbol{\beta}$ and variance σ_i^2 being assigned a weight value of ω_j. That is,

$$B_{i1} = 1 - \Phi((b_{i1} - \mathbf{x}_i'\boldsymbol{\beta})/\eta_i),$$

$$B_{ij} = \Phi((b_{i,j-1} - \mathbf{x}_i'\boldsymbol{\beta})/\eta_i) - \Phi((b_{ij} - \mathbf{x}_i'\boldsymbol{\beta})/\eta_i), \qquad 1 < j < k,$$

$$B_{ik} = \Phi((b_{i,k-1} - \mathbf{x}_i'\boldsymbol{\beta})/\eta_i),$$

where the b_{ij} are the left endpoints of the intervals assigned weight ω_j in the ith study, that is,

$$b_{ij} = -\sigma_i \Phi^{-1}(a_j).$$

The joint likelihood for the data $\mathbf{T} = (T_1, \ldots, T_n)'$ is given by

$$\ell(\boldsymbol{\beta}, \boldsymbol{\omega}|T, \sigma_\theta) = \prod_{i=1}^{n} \frac{w(T_i, \sigma_i)\phi((T_i - \mathbf{x}_i'\boldsymbol{\beta})/\eta_i)}{\eta_i A_i(\boldsymbol{\beta}, \boldsymbol{\omega}, \sigma_\theta)},$$

and the log likelihood is

$$L = \log(\ell) \propto \sum_{i=1}^{n} \log w_i(T_i, \boldsymbol{\omega}) - \frac{1}{2} \sum_{i=1}^{n} \left(\frac{T_i - \mathbf{x}_{i'}\boldsymbol{\beta}}{\eta_i} \right)^2 - \sum_{i=1}^{n} \log \eta_i$$
$$- \sum_{i=1}^{n} \log \left[\sum_{j=1}^{k} \omega_j B_{ij}(\boldsymbol{\beta}, \sigma_\theta) \right]. \tag{9.11}$$

The log likelihood may be altered to reflect a fixed-effects approach by setting the variance component to zero and omitting the variance component from the vector of parameters to be estimated.

Estimation*

Estimation of the model parameters can be accomplished numerically through the standard multivariate Newton–Raphson method. The Newton–Raphson method works by calculating a direction from the vector of first derivatives (the gradient) of the function to be optimized (the log likelihood, in this case). Provisional parameter estimates are adjusted in that direction by an amount that is determined by the rate of change of the gradient. In pure Newton–Raphson, this rate of change is determined from the matrix of second derivatives of the function (the Hessian matrix). After such a step, the gradient is recalculated, and another step is taken. The process repeats until subsequent steps are small and the gradient is near zero. Various quasi-Newton methods employ a similar principal but approximate the gradient or Hessian by numerical methods.

Specialized software is needed to implement these methods. The first derivatives of the log likelihood are given below to facilitate preparation of estimation programs using equation solvers such as those found in *Numerical Recipes*, IMSL, or NAG software libraries. The estimation proceeds smoothly when the number of studies is large enough (e.g., there are several effect size estimates per weight function interval) and reasonable starting values (e.g., the usual estimates of the mean and between-studies variance component in random-effects models) are chosen. However, it is important to realize that even with rather large data sets, the selection model is estimated only imprecisely, and consequently sharp conclusions about the form of the selection model are rarely possible. Nevertheless, simulation results have shown that even when the selection model is poorly estimated, the associated adjustment to the effect estimate can be quite accurate (Hedges and Vevea, 1996).

The first derivatives are presented here as likelihood equations; that is, they are set equal to zero. The solution of these likelihood equations (by Newton–Raphson or by other optimization methods) provides maximum likelihood estimates of the model's parameters. The likelihood equations for the unconstrained weights $(\omega_2, \ldots, \omega_k)'$ are

$$\frac{\partial L}{\partial \omega_j} = \frac{c(j)}{\omega_j} - \sum_{i=1}^{n} \frac{B_{ij}}{\sum_{m=1}^{k} \omega_m B_{im}} = 0,$$

where $c(j)$ is the count of studies assigned the weight ω_j. The equations for the linear predictors are

$$\frac{\partial L}{\partial \beta_j} = \sum_{i=1}^{n} X_{ij}\left(\frac{Y_i - \Delta_i}{\eta_i^2}\right) - \sum_{i=1}^{n} \frac{\sum_{m=1}^{k} \omega_m \partial B_{im}/\partial \beta_j}{\sum_{m=1}^{k} \omega_m B_{im}} = 0,$$

where

$$\frac{\partial B_{im}}{\partial \beta_j} = \begin{cases} \dfrac{X_{ij}}{\eta_i} \phi\left(\dfrac{b_{i1} - \Delta_i}{\eta_i}\right), & \text{if } m = 1; \\[2ex] \dfrac{X_{ij}}{\eta_i}\left(\phi\left(\dfrac{b_{im} - \Delta_i}{\eta_i}\right) - \phi\left(\dfrac{b_{i,m-1} - \Delta_i}{\eta_i}\right)\right), & \text{if } 1 < m < k; \\[2ex] -\dfrac{X_{ij}}{\eta_i} \phi\left(\dfrac{b_{i,k-1} - \Delta_i}{\eta_i}\right), & \text{if } m = k. \end{cases}$$

The equation for the variance component is

$$\frac{\partial L}{\partial \sigma^2} = \sum_{i=1}^{n} \frac{(Y_i - \Delta_i)^2}{2\eta_i^4} - \sum_{i=1}^{n} \frac{1}{2\eta_i^2} - \sum_{i=1}^{n} \frac{\sum_{m=1}^{k} \omega_m \partial B_{im}/\partial \sigma^2}{\sum_{m=1}^{k} \omega_m B_{im}} = 0,$$

where

$$\frac{\partial B_{im}}{\partial \sigma^2} = \begin{cases} \dfrac{b_{i1} - \Delta_i}{2\eta_i^3} \phi\left(\dfrac{b_{i1} - \Delta_i}{\eta_i}\right), & \text{if } m = 1, \\[2ex] \dfrac{b_{im} - \Delta_i}{2\eta_i^3} \phi\left(\dfrac{b_{im} - \Delta_i}{\eta_i}\right) - \dfrac{b_{i,m-1} - \Delta_i}{2\eta_i^3} \phi\left(\dfrac{b_{i,m-1} - \Delta_i}{\eta_i}\right), & \text{if } 1 < m < k, \\[2ex] \dfrac{b_{i,k-1} - \Delta_i}{2\eta_i^3} \phi\left(\dfrac{b_{i,k-1} - \Delta_i}{\eta_i}\right), & \text{if } m = k. \end{cases}$$

If estimates for a fixed effects model are desired, the variance component is fixed to zero, and likelihood equations involving the variance component are not used.

Standard errors (or, if needed, estimates of covariances between parameter estimates) may be obtained from the negative square root of the relevant element of the inverted information matrix. The information matrix is the expectation of the Hessian matrix that is used to determine the distance of a step in the Newton–Raphson iterations. In practice, it is convenient and effective for these weight function models to use the *observed* information matrix, which is just the Hessian itself when optimization has converged – see also Thisted (1988) for a discussion of this approach. We do not present the second derivatives here; interested readers will find them available elsewhere (Vevea and Hedges, 1995).

A useful characteristic of maximum likelihood estimation is that a test for the significance of constrained parameters is available, facilitating a general way of examining whether simpler models nested within more complex models are adequate. The test is accomplished by considering negative two times the difference in log likelihoods between the two models. Under the null hypothesis that the simpler model is adequate, this statistic is distributed as chi-square with degrees of freedom equal to the difference in number of free parameters. The test is called a likelihood-ratio chi-square test, and is often denoted G^2. In the present context, one such nested model is the one that constrains the weight function to be constant. Hence, the likelihood-ratio test provides a means of assessing the statistical significance of publication bias.

Bayesian approaches to selection models

The selection model can also be approached from a Bayesian perspective. Givens *et al.* (1997) and Silliman (1997) describe the use of data augmentation to obtain Bayes estimates of the joint posterior distribution of the selection model parameters and the effect size model parameters. One particularly interesting aspect of this approach is that, in principle, it provides direct estimates of the number of unobserved studies and the distribution of their effect sizes. In practice such estimates are very imprecise. Larose and Dey (1995) also examined a similar approach to this problem.

A priori specification of weight functions

It is possible to employ an approach similar to the weighted model just described when there are insufficient data to estimate the weights adequately. The method (see Vevea and Woods, in press) is similar in spirit to the models that assume an extreme selection function described earlier in this chapter. However, instead of specifying a single, simple extreme selection function, one specifies fixed values for weights for multiple *p*-value intervals, and estimates the model for effects conditional on the weight function, using the method of maximum likelihood. The likelihood and likelihood equations are similar to those of the full model in which weights are estimated; however, the weights are not estimated, and derivatives involved in estimation of the weights are not used.

By applying a variety of weight function specifications representing different forms and different severity of selection, it is possible to conduct a sensitivity analysis using the approach. One may find that the estimated effect size model is not much affected regardless of what form of selection is specified. In that case, it is difficult to credit the proposition that publication bias represents a reasonable explanation for the magnitude of effect one observes. On the other hand, it may be that the estimated effect size model can be altered at will by changing the specified weight function. Under that circumstance, publication bias may represent a plausible counter-explanation for a finding that an effect is present.

METHODS FOR WEIGHT FUNCTIONS THAT DEPEND ON BOTH T AND σ

Now consider the weight function described by Copas and his colleagues, discussed earlier in this chapter, which depends on both T and σ. The weight function is given in (9.6).

The weighted probability model[*]

The weighted probability density function of T_i given the weight function $w(T_i, \sigma_i)$ (see (9.6)) and the parameters $\boldsymbol{\beta}$ and τ^2 of the effect size model and a, b, and ρ of the selection model is

$$f(T_i | \boldsymbol{\beta}, \tau, a, b, \rho) = \frac{w(T_i, \sigma_i)\phi\left((T_i - \mathbf{x}_i\boldsymbol{\beta})/\eta_i\right)}{\eta_i \Phi(a + b/\sigma_i)};$$

here

$$\eta_i^2 = s_i^2 + \tau^2,$$

where

$$s_i = \frac{\sigma_i}{\sqrt{1 - M(a + b/\sigma_i)[a + b/\sigma_i + M(a + b/\sigma_i)]\rho^2}}$$

and $M(x) = \Phi(-x)/\phi(x)$ is Mills' ratio.

The likelihood of the joint data $\boldsymbol{T} = (T_1, \ldots, T_k)'$ is given by

$$\ell(\boldsymbol{\beta}, \tau, a, b, \rho | \boldsymbol{T}) = \prod_{i=1}^{k} \frac{w(T_i, s_i)\phi\left((T_i - \mathbf{x}_i\boldsymbol{\beta})\eta_i\right)}{\eta_i \Phi(a + b/s_i)}$$

and the log likelihood is

$$L(\boldsymbol{\beta}, \tau, a, b, \rho | \boldsymbol{T}) = \sum_{i=1}^{k} \log[w(T_i, s_i)] - \log(\eta_i) - (T_i - \mathbf{x}_i\boldsymbol{\beta})^2/2\eta_i^2 \\ - \log[\Phi(a + b/s_i)]. \tag{9.12}$$

Estimation[*]

It is not always possible to estimate all of the parameters of the selection model and effect size model simultaneously. Even when it is possible, the likelihood function is quite flat, suggesting that there is very little information about the parameters a

and b in the data. Consequently, Copas (1999) and Copas and Shi (2000, 2001) have suggested fixing values of the selection parameters a and b, then estimating ρ and the parameters of the effect size model ($\boldsymbol{\beta}$ and τ) conditional on those values of a and b to provide a sensitivity analysis. By estimating ρ, $\boldsymbol{\beta}$, and τ for many combinations of a and b, one can understand the degree to which selection may have occurred and its possible influence on the parameters of the effect size model.

The likelihood equations do not lend themselves to closed-form solutions, but estimation can be accomplished by numerically solving the likelihood equations

$$\frac{\partial L}{\partial \rho} = -\sum_{i=1}^{k} \frac{\sigma_i T_i - \sigma_i \mathbf{x}_i \boldsymbol{\beta} + a\rho\sigma_i \eta_i + b\rho\eta_i}{\sigma_i \eta_i (\rho^2 - 1)\sqrt{1-\rho^2}}$$
$$\times M\left(\frac{a + b/\sigma + \rho(T - \mathbf{x}_i\boldsymbol{\beta})/\sqrt{\tau^2 + \sigma^2}}{\sqrt{1-\rho^2}}\right) = 0,$$

$$\frac{\partial L}{\partial \beta_j} = \sum_{i=1}^{k} \frac{x_{ij}(T_i - \mathbf{x}_i\boldsymbol{\beta})}{\eta_i^2} - \frac{\rho x_{ij}}{\eta_i\sqrt{1-\rho^2}}M\left(\frac{a + b/\sigma + \rho(T - \mathbf{x}_i\boldsymbol{\beta})/\sqrt{\tau^2 + \sigma^2}}{\sqrt{1-\rho^2}}\right) = 0,$$

where β_j is the jth element of $\boldsymbol{\beta}$ and x_{ij} is the value of the jth covariate for the ith study, and

$$\frac{\partial L}{\partial \tau} = -\sum_{i=1}^{k} \frac{(T_i - \mathbf{x}_i\boldsymbol{\beta})^2 - 1}{\eta_i^2} + \frac{\rho\tau(T - \mathbf{x}_i\boldsymbol{\beta})}{\eta^3\sqrt{1-\rho^2}}$$
$$\times M\left(\frac{a + b/\sigma + \rho(T - \mathbf{x}_i\boldsymbol{\beta})/\sqrt{\tau^2 + \sigma^2}}{\sqrt{1-\rho^2}}\right) = 0,$$

where $M(x)$ is Mills' ratio.

As in the case of the weight functions that depend only on p, the likelihood equations are solved using the Newton–Raphson algorithm or other numerical methods.

APPLYING WEIGHT FUNCTION MODELS IN PRACTICAL META-ANALYSIS

Considerations for step function methods

Several things should be considered in the practical use of these methods. Perhaps the most important consideration is when to use the methods. They may perform poorly if the number of studies is small for two reasons. The first is that there is relatively little information in the data about the selection model, and small sample sizes mean there is very little information with which to estimate the parameters of the selection model. Small sample sizes also limit the number of steps that can be used in the weight function (which may limit the accuracy with which the weight function can approximate the selection function).

Choice of the step boundaries

The choice of boundaries or cutpoints for the weight function is not entirely *ad hoc*. Selection is a psychosocial phenomenon, in that results are selected, at least in part, based on interpretations of their conclusiveness. Experimental research on the interpretation of statistical results by psychologists (Rosenthal and Gaito, 1963, 1964; Nelson *et al.*, 1986) has shown that people tend to perceive a result to be much more conclusive if its *p*-value is less than one of the conventional levels of significance such as 0.05 or 0.01. Thus the selection of a discontinuous weight function seems quite reasonable.

It also seems reasonable, then, to fix the endpoints of *p*-value intervals a priori by using conventional critical alpha levels and other psychologically important points rather than to estimate them from the data. When one-tailed *p*-values are used to determine the weight function, one psychologically important point occurs at $p = 0.5$, where the effect magnitude is zero. For analysis of questions for which a large number of studies exist, it may further be desirable to set other cutpoints at *p*-values that produce approximately equal numbers of studies within the intervals. If we define ten intervals, for example, we might set cutpoints at 0.001, 0.005, 0.01, 0.025, 0.05, 0.10, 0.20, 0.30, 0.50, and 1.00. If we wish to investigate the possibility that significant negative outcomes are also favored, the second tail of the *p*-value distribution can be investigated by setting cutpoints at values such as 0.95, 0.975, or 0.995. In practice, it is rare to find data sets in which sufficient studies are present at both extremes to allow estimation of such weights; however, two-tailed specifications can be useful in sensitivity analysis.

Robustness of nonparametric weight function methods

Methods that correct for selection work roughly by comparing the *observed* distribution of effect size estimates to what would be expected given the effect size model and adjusting the selection model to obtain correspondence. Thus the *assumption* about the form of the distribution of the (unselected) effect size estimates embodied in the effect size model provides the information necessary to estimate the selection model. If the assumption is grossly wrong, the estimated selection model will be wrong and any corrections based on it must also be in error, perhaps grossly so.

An extensive series of simulation studies by Hedges and Vevea (1996) sheds some light on the robustness of estimates from the selection model to misspecification of the distribution of the random effects. In their study τ^2 was about the half as large as the average of the σ_i^2. They examined $k = 25$, 50, and 100, and several weight functions that resulted in selection of from 50 % and 85 % of the estimates that were generated. They examined performance for several non-normal distributions that constituted violations of the assumptions of the model, including heavy-tailed distributions (Student's *t* with 3 degrees of freedom), skewed distributions (chi-squared with 5 degrees of freedom), and bimodal distributions (mixtures of two normals with means 4 standard deviations apart).

It is somewhat surprising that they found the procedure to be rather robust. Specifically, they found that estimates of the mean from the selection model had bias only 5–15 % as large as that of the mean of the observed effect sizes. They also found that whenever there was non-null selection, the estimates of the mean effect size based on the selection model had smaller mean squared error than the unadjusted estimates. Of course, there is a penalty in variance for estimating the selection model, that is, the estimates derived from the selection model had larger variance. When there was real selection, the reduction in bias more than compensated for the increase in variance, but when there was no selection, the mean of the observed (unadjusted) estimates was superior (as expected).

Such robustness might seem surprising, but it is less so when the operation of the selection model is considered in detail. It is useful to consider the simplest case, in which there are no study-level covariates and all of the studies have the same mean. The selection model assumes that the unselected distribution of effect size estimates is a mixture of normal distributions with variances given by $\sigma_1^2, \ldots, \sigma_k^2$, with a normal mixing distribution with variance τ^2. Then the distribution of the unselected effect size estimates is a normal mixture of normal distributions, which is itself normal. The assumptions are wrong if the random effects (the ξ_i) have a non-normal distribution. If the variance of the random effects is small in comparison to the sampling error variances, then the mixture will be very similar to that of a normal mixture of normals, even if the ξ_i have a non-normal distribution. Therefore we would expect some degree of robustness as long as the variance τ^2 of the random effects is not too large in comparison with the sampling error variances $\sigma_1^2, \ldots, \sigma_I^2$.

Practical considerations in using weight functions that depend on both T and σ

These methods also perform poorly when the number of studies is small, and again lack of information in the data to estimate the selection model is a key reason. This model is even more difficult to estimate from the data than the previous selection model (and cannot be estimated at all for certain parameter values). However, the primary use of this method is for sensitivity analyses rather than strict estimation. Evaluating the model for a variety of values of the a and b parameters is quite useful in determining whether the results of analyses change dramatically as a function of selection under the model.

HOW WELL DO WEIGHT FUNCTION MODELS WORK?

What can go wrong in estimating selection models?

Perhaps the biggest threat to the validity of inferences in either of these two methods is model misspecification. If either the effect size model or the selection model is incorrect, the inferences that result may be far from correct. Unexplained heterogeneity in the effect size model is a particular concern, because it suggests that there may be unmeasured covariates that explain the heterogeneity. If these unmeasured covariates are correlated with both σ and T, they may induce the appearance of publication selection (e.g., asymmetry in the funnel plot and a relation between T and σ) when none exists *or* mask publication bias that might exist.

When the number of studies is small, two problems arise. There may be numerical problems in obtaining stable estimates of the model parameters. More importantly, the standard errors of estimates will be large, perhaps so large as to make any specific inferences impossible or meaningless. For example, a result might be highly statistically significant in a conventional analysis and the estimates after correction for selection might be essentially identical. However, the corrected estimate might be far from statistically significant because the standard error was so much larger.

It is not always clear how to interpret such results. Indeed, it is difficult even to state how large the number of studies must be in order to avoid the problems; this can depend to a great degree on the idiosyncrasies of the particular data set. However, it may be possible to combat these problems to a degree by reducing the number of parameters to be estimated. This can be accomplished by specifying a simpler weight function where there are fewer cutpoints defining a smaller number of p-value ranges where the probability of selection is equal. While this eases the burden of estimating the weight function, results may very to a considerable degree based on the choice of interval boundaries. Under such circumstances, it is better to regard the model as a tool for sensitivity analysis, and to consider variation in results with different cutpoint specifications. Moreover, it is important in such cases to compare conclusions from the weight function model with results from other approaches, such as the trim-and-fill methods and the model with a priori specification of weights.

What can weight function models do well?

In view of potential problems like those just described, the question of when these models are useful arises: Why go to the trouble of implementing something that is complex and may not even work? The answer is that weight function models excel in two situations for which alternative approaches (e.g. trim and fill – see Chapter 8) may create a spurious impression of publication bias. A recent simulation study (Terrin *et al.*, 2003) shows that correctly specified weight function models are more successful at avoiding such spurious findings.

The first situation in which spurious findings of publication bias may occur is the case where subpopulations exist that have systematically different sample sizes and systematically different effect sizes from the rest of the studies. Under such circumstances, the funnel plot may exhibit a classic pattern of asymmetry that has nothing to do with publication bias. If the factors that distinguish these subpopulations are specified in the effect size model, the weight function methods will tend not to mistake the funnel plot asymmetry for publication bias.

The second situation in which weight function models excel is when there is a legitimate inverse association between effect size and sample size that arises because the designers of the primary studies incorporated power analysis during the planning stages. When sound power analysis is conducted prior to a decision about sample size, a quite natural tendency for studies with large effects to have small sample sizes arises, because the designers of the investigations, expecting large effects, concluded that larger samples were not needed. The phenomenon can produce funnel plot asymmetry that is visually indistinguishable from publication bias. Yet, if the factors that led the original investigators to expect large effects are identified in the effect size model, the weight function model will not mistake the cause of the asymmetry.

EXAMPLES

The effects of environmental tobacco smoke

Our first example uses the data from 37 studies of the effects of environmental tobacco smoke reported by Hackshaw *et al.* (1997). Each study provided an estimate of the odds ratio comparing the likelihood of lung cancer for non-smoking spouses of smokers relative to the likelihood of lung cancer in non-smoking spouses of non-smokers. The effect size model was the simple random effects model with no covariates,

$$T_i \sim N(\beta_0, \sigma_i^2 + \tau^2).$$

The model was estimated for log-transformed odds ratios. (The log transformation is commonly employed to simplify the sampling variance for each effect size.)

Selection depending only on *p*-values, with weight function estimated
In our application of the non-parametric weight function method, the selection model used a step function with cutpoints at $p = 0.05$, 0.10, and 0.50. Thus the weight function took values

$$w(p) = \begin{cases} \omega_1 \equiv 1, & \text{if } 0 \leq p < 0.05, \\ \omega_2, & \text{if } 0.05 \leq p < 0.10, \\ \omega_3, & \text{if } 0.10 \leq p < 0.50, \\ \omega_4, & \text{if } 0.50 \leq p < 1.00. \end{cases}$$

The cutpoints were selected to yield a rather even distribution of observed effects in each interval. In this case there were 7 effects in the first interval, 8 in the second, 16 in the third, and 6 in the last interval.

The maximum likelihood estimates of β_0 and τ^2 (with standard errors in parentheses) assuming no selection (i.e. constraining $\omega_2 = \omega_3 = \omega_4 = 1$) are given in the first column of Table 9.2. The pooled log odds ratio is about one quarter, and the variance component is quite small.

Table 9.2 Effects of environmental tobacco smoke: Results of the selection model depending only on *p* and using the estimated weight function.

Parameter	Unadjusted estimate (S.E.)	Adjusted estimate (S.E.)
Pooled log odds ratio	0.22	0.13
	(0.050)	(0.129)
Variance component	0.019	0.031
	(0.015)	(0.027)

The maximum likelihood estimates of the parameters of the selection model (and their standard errors) were

$$\omega_2 = 2.48(1.70),$$
$$\omega_3 = 1.01(0.85),$$
$$\omega_4 = 0.42(0.49).$$

Alternatively, one can express the estimated weights as proportions of the largest weight. In that case, the first weight is fixed at $1/2.48 = 0.40$. The other weights (and their standard errors) become

$$\omega_2 = 1.00(0.69),$$
$$\omega_3 = 0.39(0.35),$$
$$\omega_4 = 0.17(0.09).$$

Viewing the weights this way can aid interpretation. These weights are inherently relative to one another, thus it is natural to use weights relative to the studies most likely to be observed. Figure 9.1 depicts the estimated selection function graphically.

The likelihood ratio test statistic for publication bias (the test of the hypothesis that $\omega_2 = \omega_3 = \omega_4 = 1$) was $G^2 = 7.02$ with 3 degrees of freedom and was not statistically significant ($p = 0.071$). Note that all of the estimated weights are within about a standard error of 1, but the weight for negative findings is more than one standard error smaller than 1, suggesting that negative results (reduced log odds ratio) are less likely to be observed than positive findings. Thus the selection model provides some reason to believe that publication selection exists in these data.

The estimate of the pooled log odds ratio adjusted for publication selection, and the adjusted variance component, are presented in the second column of Table 9.2. Note that the pooled effect has been attenuated by about 41 %. Comparing these

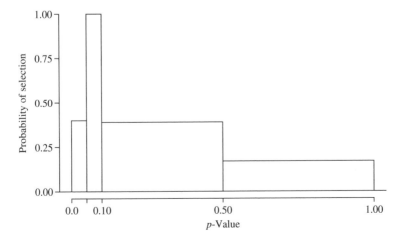

Figure 9.1   Estimated weight function for effects of environmental tobacco smoke.

adjusted values with the unadjusted values suggests that publication selection *could* have inflated the estimate of the pooled log odds ratio.

It is worth noting that the estimate of the weight function has a large standard error, even though it was specified in such a way that only three weights were estimated. A necessary consequence is that the standard error of the adjusted odds ratio estimate is large. Indeed, the unadjusted estimate falls within the bounds of a $100(1 - \alpha)\%$ confidence interval for the adjusted odds ratio at any commonly used value of α. Hence, this is a case in which it may be appropriate to investigate further using different a priori specifications of the weight function.

Selection depending only on *p*-values, with weight function specified a priori

Table 9.3 defines four different a priori specifications for the weight function (denoted 'weak one-tailed', 'strong one-tailed', 'weak two-tailed' and 'strong two-tailed'). We applied each of these weight functions to the analysis, and considered the impact of the hypothetical selection patterns on the effect model. Table 9.4 presents adjusted estimates, along with the unadjusted maximum likelihood estimates discussed under 'Selection depending only on *p* values, with weight function estimated') for comparison. Note that standard errors for these adjusted estimates are not provided. That is because the estimates are based on entirely hypothetical selection functions, and conventional standard errors would be correct conditional on the proposition that these selection functions are precisely known. Hence it would be seriously misleading to interpret the standard errors. The sensitivity analysis under the most extreme selection condition resulted in an estimate attenuated

Table 9.3   Specification of weights for a priori weight functions.

p-value interval	Probability of observing effect			
	Weak one-tailed selection	Strong one-tailed selection	Weak two-tailed selection	Strong two-tailed selection
0.000–0.005	1.00	1.00	1.00	1.00
0.005–0.010	0.99	0.99	0.99	0.99
0.010–0.050	0.95	0.90	0.95	0.90
0.050–0.100	0.90	0.75	0.90	0.75
0.100–0.250	0.80	0.60	0.80	0.60
0.250–0.350	0.75	0.55	0.75	0.50
0.350–0.500	0.65	0.50	0.60	0.25
0.500–0.650	0.60	0.45	0.60	0.25
0.650–0.750	0.55	0.40	0.75	0.50
0.750–0.900	0.50	0.35	0.80	0.60
0.900–0.950	0.50	0.30	0.90	0.75
0.950–0.990	0.50	0.25	0.95	0.90
0.990–0.995	0.50	0.20	0.99	0.99
0.995–1.000	0.50	0.10	1.00	1.00

Table 9.4 Effects of environmental tobacco smoke: Results of the selection model depending only on p and using a priori weight functions.

Parameter	Unadjusted estimate	Estimate adjusted for			
		Weak one-tailed selection	Strong one-tailed selection	Weak two-tailed selection	Strong two-Tailed selection
Pooled log odds ratio	0.22	0.16	0.12	0.18	0.14
Variance component	0.020	0.019	0.019	0.018	0.018

by approximately 45%; moreover, the estimate under both the strong one-tailed and the strong two-tailed condition reached a range that could be considered to be of minimal clinical interest. Hence, this exploratory approach supports the idea that publication bias may partly account for the unadjusted outcome.

Selection depending on both T and σ

We also applied the selection model depending on both T and σ. In our application of this model, we computed estimates for the five (a, b) combinations suggested by Copas and Shi (2001): $a = 0.99$, $b = 0.80$; $a = 0.80$, $b = 0.50$; $a = 0.60$, $b = 0.30$; $a = 0.40$, $b = 0.10$; and $a = 0.20$, $b = 0.01$. The maximum likelihood estimates of τ^2, ρ, and β (and its standard error) are given in Table 9.5. Examining this table, we see that for some values of a and b (e.g., $a = 0.40$, $b = 0.10$ or $a = 0.20$, $b = 0.01$) the estimates of β are considerably smaller than the value of 0.25 estimated assuming no selection. It is interesting that the value of β estimated for $a = 0.4$, $b = 0.1$ is almost identical to that estimated using the other selection model.

Validity of the employment interview

Our second example uses data from 106 studies of the validity of the employment interview reported by McDaniel *et al.* (1994). The result from each study was expressed as a correlation coefficient, and all analyses were carried out in the metric of Fisher z-transforms of the correlations. A subset of the data set consisting of

Table 9.5 Environmental tobacco smoke: Results of the selection model depending on both T and σ.

a	b	β_0	S.E.	τ^2	ρ	Likelihood-ratio test	p
0.99	0.80	0.254	0.058	0.033	0.56	1.917	0.166
0.80	0.50	0.250	0.058	0.032	0.99	1.834	0.176
0.60	0.30	0.229	0.058	0.032	0.57	1.901	0.168
0.40	0.10	0.174	0.069	0.031	0.78	0.354	0.552
0.20	0.01	0.127	0.098	0.030	0.07	0.143	0.661

correlations for interviews that were structured (as opposed to unstructured) exhibits apparently severe funnel plot asymmetry. The data may be further divided according to whether the content of the interview was job-related, psychological, or situational, and whether the interview was conducted for research or for administrative reasons. The possibility exists that the observed funnel plot asymmetry may be accounted for, in part, by an explanatory model for variation in effects as a function of those variables. This suggests the use of a mixed-effects model with three covariates, dummy-coded for whether the interview was job-related (x_{i1}) or situational (x_{i2}), with psychological content as the reference category, and for administrative purpose (x_{i3}), vs. research (reference category). (In principle, interactions could be estimated; however, the sparseness of the implied cells made it impossible to implement such a model.) Thus the effect size model was

$$\theta_i = \beta_0 + \beta_1 x_{i1} + \beta_2 x_{i2} + \beta_3 x_{i3} + \varepsilon_i, \qquad \varepsilon_i \sim N(\beta_0, \sigma_i^2 + \tau^2).$$

The sample size for the analysis was 103, after listwise deletion of three cases with missing data.

Selection depending only on p-values, with weight function estimated
The selection model used a step function with cutpoints at $p = 0.001$, 0.01, 0.05, 0.20, and 0.50. Thus the weight function took values

$$w(p) = \begin{cases} \omega_1 \equiv 1, & \text{if } 0 \leq p < 0.001, \\ \omega_2, & \text{if } 0.001 \leq p < 0.01, \\ \omega_3, & \text{if } 0.01 \leq p < 0.05, \\ \omega_4, & \text{if } 0.05 \leq p < 0.20, \\ \omega_5, & \text{if } 0.20 \leq p < 0.50, \\ \omega_6, & \text{if } 0.50 \leq p < 1.00. \end{cases}$$

(Note that this larger data set permitted the estimation of a more detailed series of steps than in the previous example.)

The cutpoints were selected to yield a rather even distribution of observed effects in each interval. In this case there were 29 effects in the first interval, 15 in the second, 19 in the third, 15 in the fourth, 12 in the fifth, and 13 in the last interval.

The mixed-effects model encountered problems with convergence when estimated by the selection modeling software; hence, it was necessary to employ a fixed-effects analysis. The data set without a variance component is heterogeneous; the Q test for heterogeneity has the value 460.26 on 98 degrees of freedom ($p < 0.001$), and the maximum likelihood estimate of the variance component (achieved with other software) is 0.026. However, the mixed-effects parameter estimates do not vary greatly from the fixed-effects estimates. Hence, the fixed-effects selection analysis is not likely to mislead.

The maximum likelihood estimates of β (with standard errors in parentheses) assuming no selection (that is constraining $\omega_2 = \omega_3 = \omega_4 = \omega_5 = \omega_6 = 1$) are

Table 9.6   Validity of the employment interview: Results of the selection model depending only on p and using the estimated weight function.

Parameter	Unadjusted estimate (S.E.)	Adjusted estimate (S.E.)
Intercept	0.187	0.227
	(0.104)	(0.094)
Slope for job-related content	0.123	0.091
	(.102)	(0.091)
Slope for situational content	0.134	0.115
	(.108)	(0.096)
Slope for administrative purpose	−0.097	−0.096
	(0.021)	(0.020)

presented in Table 9.6. The maximum likelihood estimates of the parameters of the selection model (and their standard errors) were

$$\omega_2 = 0.86(0.34),$$

$$\omega_3 = 1.12(0.48),$$

$$\omega_4 = 0.93(0.46),$$

$$\omega_5 = 1.35(0.76),$$

$$\omega_6 = 3.66(2.25).$$

These weights, rescaled relative to the interval with the highest weight, become

$$\omega_2 = 0.23(0.04),$$

$$\omega_3 = 0.31(0.09),$$

$$\omega_4 = 0.25(0.07),$$

$$\omega_5 = 0.37(0.17),$$

$$\omega_6 = 1.00(0.61),$$

with the weight for the first interval (0.000 to 0.001) fixed at 0.27. The estimated weight function is represented graphically in Figure 9.2.

The likelihood ratio test statistic for publication bias (the test of the hypothesis that $\omega_2 = \omega_3 = \omega_4 = \omega_5 = \omega_6 = 1$) was $G^2 = 11.18$ with 5 degrees of freedom and was statistically significant ($p = 0.048$). However, the estimated weights reverse the usual pattern: studies with p-values in the *least* significant interval are observed more frequently than we would expect. Thus the selection model, to the degree that it adjusts the estimated parameters of the explanatory model, is likely to *inflate* them rather than attenuate them. On first consideration, this is counterintuitive. It is conceivable, though, that we are seeing a tendency for the literature to be sympathetic to the idea that employment interviews are invalid.

The estimates of the parameters of the model for pooled effects adjusted for publication selection are presented in the second column of Table 9.6. Comparing

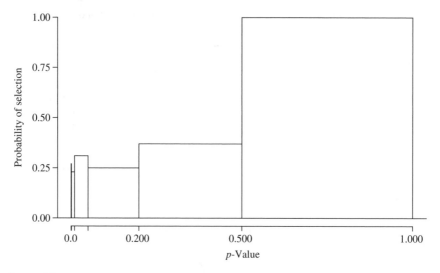

Figure 9.2   Estimated weight function for validity of the employment interview.

these adjusted values with the unadjusted values suggests that the impact of selection is minimal in this data set. The model intercept is slightly inflated, but increments associated with job-related interviews (β_1) and situational interviews (β_2) move in a compensatory fashion, so that predicted values for several of the design cells would be essentially unchanged.

Selection depending only on p-values, with weight function specified a priori
We performed a sensitivity analysis, applying the four hypothetical weight functions described in Table 9.3; the results are presented in Table 9.7. The reference values in the first column differ slightly from those we have just seen, because in the

Table 9.7   Validity of the employment interview: Results of the selection model depending only on p and using a priori weight functions.

Parameter	Unadjusted estimate	Estimate adjusted for			
		Weak one-tailed selection	Strong one-tailed selection	Weak two-tailed selection	Strong two-Tailed selection
Intercept	0.13	0.10	0.06	0.12	0.12
Slope for job-related content	0.14	0.14	0.16	0.14	0.13
Slope for situational content	0.19	0.19	0.21	0.18	0.17
Slope for administrative purpose	−0.04	−0.04	−0.04	−0.04	−0.04
Variance component	0.026	0.026	0.028	0.026	0.026

context of the sensitivity analysis we were able to implement the mixed-effects model. Only under the 'strong one-tailed' scenario is there any appreciable change in the parameter estimates. That is the scenario that is least compatible with the results of the analysis in which the weight function was estimated, rather than fixed. This analysis, then, taken in the larger context, suggests that publication bias is not a plausible explanation for the presence of validity effects.

Selection depending on both T and σ

We also applied the selection model depending on both T and σ to the validity of the employment interview data. We computed estimates for the five (a, b) combinations suggested by Copas and Shi (2001): $a = 0.99$, $b = 0.80$; $a = 0.80$, $b = 0.50$; $a = 0.60$, $b = 0.30$; $a = 0.40$, $b = 0.10$; and $a = 0.20$, $b = 0.01$. The maximum likelihood estimates of τ^2, ρ, β_0, β_1, β_2, and β_3 (and their standard errors) are given in Table 9.8. Examining this table, we see that the estimates of β vary somewhat from the values estimated assuming no selection. In general, it seems that the values do not differ from the values estimated using no selection as much as those estimated using the other selection model.

Teacher expectancy effects

Our third example uses data from the 18 studies of teacher expectancy effects of Raudenbush (1984). The results of each study were expressed as a standardized mean difference. The effect size model was a fixed effects model with one covariate, dummy-coded for the amount of prior contact between children and their teacher when the expectancy was induced ($x_{i1} = 1$ if there was more than one week of prior contact). Thus the effect size model was

$$\theta_i = \beta_0 + \beta_1 x_{i1} + \varepsilon_i, \qquad \varepsilon_i \sim N(\beta_0, \sigma_i^2)$$

Selection depending only on p-values, with weight function estimated

We fitted a selection model used a step function with cutpoints at $p = 0.05$ and 0.30. Thus the weight function took values

$$w(p) = \begin{cases} \omega_1 \equiv 1, & \text{if } 0 \leq p < 0.05, \\ \omega_2, & \text{if } 0.05 \leq p < 0.30, \\ \omega_3, & \text{if } 0.30 \leq p < 1.00. \end{cases}$$

Table 9.8 Validity of the employment interview: Results of the selection model depending on both T and σ.

a	b	β_0	SE	β_1	SE	β_2	SE	β_3	SE	τ^2	ρ
0.99	0.80	0.133	0.231	0.174	0.253	0.217	0.257	−0.053	0.048	0.054	0.572
0.80	0.50	0.128	0.214	0.179	0.259	0.222	0.262	−0.053	0.048	0.054	1.0
0.60	0.30	0.115	0.196	0.186	0.267	0.233	0.268	−0.048	0.046	0.052	1.0
0.40	0.10	0.110	0.190	0.182	0.254	0.232	0.253	−0.040	0.045	0.048	1.0
0.20	0.01	0.103	0.189	0.166	0.236	0.214	0.237	−0.031	0.043	0.044	1.0

The cutpoints were selected to yield a rather even distribution of observed effects in each interval. In this case there were 4 effects in the first interval, 6 in the second, and 9 in the last interval.

This data set was extremely homogeneous, and we estimated a fixed-effects version of the selection model. The choice of a fixed-effects analysis was made not because of the homogeneity *per se*; indeed, such homogeneity could be an artifact of the very publication bias that the model is designed to detect. Moreover, it has been argued (Hedges and Vevea, 1998) that a finding of homogeneity is not necessarily a good reason to employ a fixed-effects model, as the practice may bias inferences about effect size estimates. However, when the observed effects are homogeneous, it can be difficult to estimate the variance component because of the border condition: variance components less than zero are impossible. In the present example, we were unable to estimate the mixed-effects model.

The maximum likelihood estimates of β (with standard errors in parentheses) assuming no selection (i.e., constraining $\omega_2 = \omega_3 = \omega_4 = 1$) are presented in the first column of Table 9.9. The maximum likelihood estimate of the parameters of the selection model (and their standard errors) were

$$\omega_2 = 0.89(0.80),$$

$$\omega_3 = 0.56(0.64).$$

The estimated weight function is represented graphically in Figure 9.3.

The likelihood ratio test statistic for publication bias (the test of the hypothesis that $\omega_2 = \omega_3 = 1$) was $G^2 = 0.43$ with 2 degrees of freedom and was not statistically significant ($p = 0.81$). Note that all of the estimated values of both weights are within one standard error of 1, meaning that there is relatively little selection. Thus the selection model provides no reason to believe that there is publication selection in these data.

The estimate of the model for the pooled standardized mean difference adjusted for publication selection appears in the second column of Table 9.9. Comparing these adjusted values with the unadjusted values suggests that publication selection had essentially no effect on these data. On the other hand, we must acknowledge that, because of the small data set, the weights are very poorly estimated. Moreover, the three-interval weight function, which is all we were able to estimate, might not be capable of representing any selection process adequately. Hence, it

Table 9.9 Teachers' expectations: Results of the selection model depending only on p and using the estimated weight function.

Parameter	Unadjusted Estimate (SE)	Adjusted Estimate (SE)
Intercept	0.349	0.321
	(0.079)	(0.111)
Slope for prior contact	−0.371	−0.363
	(0.089)	(0.109)

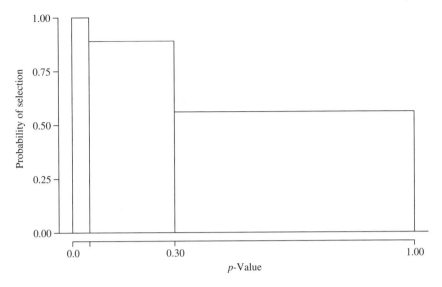

Figure 9.3 Estimated weight function for teachers' expectations.

may be unreasonable to expect the data to inform us about the selection process here

Selection depending only on *p*-values, with weight function specified a priori

It is especially appropriate in such a case to consider the question of whether a variety of putative weight functions *could* affect the outcome for this data set. Once again, we performed a sensitivity analysis to address that question.

We applied the same four hypothetical weight functions (see Table 9.3). The adjusted estimates are given in Table 9.10. Recall that the intercept here represents the predicted effect when the teachers have not known the children for long; hence, it represents the situation where theory would predict an effect. When the teachers are more acquainted with the children, the predicted effect is the sum of the intercept and the slope. Under all four selection scenarios, the conclusion is the same: the

Table 9.10 Teachers' expectations: Results of the selection model depending only on *p* and using a priori weight functions.

Parameter	Unadjusted estimate	Estimate adjusted for			
		Weak one-tailed selection	Strong one-tailed selection	Weak two-tailed selection	Strong two-Tailed selection
Intercept	0.35	0.32	0.30	0.33	0.29
Slope for prior contact	−0.37	−0.37	−0.36	−0.35	−0.31

Table 9.11   Teachers' expectations: Results of the selection model depending on both T and σ.

a	b	β_0	SE	β_1	SE	ρ	Likelihood-ratio test	p
0.99	0.80	0.349	0.079	-0.370	0.089	0.56	0.190	0.663
0.80	0.50	0.346	0.080	-0.368	0.090	0.99	0.183	0.669
0.60	0.30	0.338	0.083	-0.358	0.093	0.57	0.132	0.717
0.40	0.10	0.266	0.086	-0.322	0.092	0.78	3.017	0.082
0.20	0.01	0.242	0.108	-0.326	0.098	0.07	4.931	0.026

effect is about 0.3 where we would expect an effect, and about zero where we would not. Hence, we are in a strong position to argue that this analysis is relatively robust to the effects of publication bias.

Selection depending on both T and σ

We also applied the selection model depending on both T and σ to the data on teacher expectancies. We computed estimates for the five (a, b) combinations suggested by Copas and Shi (2001): $a = 0.99$, $b = 0.80$; $a = 0.80$, $b = 0.50$; $a = 0.60$, $b = 0.30$; $a = 0.40$, $b = 0.10$; and $a = 0.20$, $b = 0.01$. The maximum likelihood estimates of τ^2, ρ, β_1, and β_2 (and their standard errors) are given in Table 9.11. Examining this table, we see that the estimates of β_1 and β_2 are changed very little from those estimated assuming no selection, except for the values $a = 0.40$, $b = 0.10$ or $a = 0.20$, $b = 0.01$. However, these values are the ones where the likelihood ratio test for asymmetry of the residual funnel plot suggests that the selection model fits poorly. Thus we would conclude there is little evidence that publication bias substantially affected results of analyses of the teacher expectancy data.

CONCLUSIONS AND RECOMMENDATIONS

Weight function models provide a method to describe the publication selection model, determine its consequences for bias in estimation and adjust estimates to reduce bias. These methods provide considerable insight into the problem of publication bias and methods that might be used to adjust for its effects. Perhaps the most important insight that arises from these models is that the observed effect size data often have little information for estimating the selection process. This is particularly so when the number of studies is small. Thus any completely empirical method of adjustment for publication bias must be rather imprecise.

Weight function models are complex and involve a substantial amount of computation. Thus they are unlikely to be used routinely in meta-analysis. However, in cases where study effects are heterogeneous, simpler methods may give misleading results. When effects are heterogeneous across studies, weight function models are better suited than simpler methods to the task of examining publication bias. In such cases, particularly when the number of studies is not small, the benefits of weight function methods may outweigh their disadvantages of complexity.

REFERENCES

Copas, J.B. (1999). What works? Selectivity models and meta-analysis. *Journal of the Royal Statistical Association, Series A*, **162**, 95–109.

Copas, J.B. and Li, H.G. (1997). Inference for non-random samples (with discussion). *Journal of the Royal Statistical Society, Series B*, **59**, 55–95.

Copas, J.B. and Shi, J.Q. (2000). Reanalysis of epidemiological evidence on lung cancer and passive smoking. *British Medical Journal*, **320**, 417–418.

Copas, J.B. and Shi, J.Q. (2001). A sensitivity analysis for publication bias in systematic reviews. *Statistical Methods in Medical Research*, **10**, 251–265.

Dear, K.B.G. and Begg, C.B. (1992). An approach for assessing publication bias prior to performing a meta-analysis. *Statistical Science*, **7**, 237–245.

Givens, G.H., Smith, D.D. and Tweedie, R.L. (1997). Publication bias in meta-analysis: A Bayesian data-augmentation approach to account for issues exemplified in the passive smoking debate. *Statistical Science*, **12**, 221–250.

Hackshaw, A.K., Law, M.R. and Wald, N.J. (1997). The accumulated evidence on lung cancer and environmental tobacco smoke. *British Medical Journal*, **35**, 980–988.

Hedges, L.V. (1984). Estimation of effect size under nonrandom sampling: The effects of censoring studies yielding statistically insignificant mean differences. *Journal of Educational Statistics*, **9**, 61–85.

Hedges, L.V. (1992). Modeling publication selection effects in meta-analysis. *Statistical Science*, **7**, 246–255.

Hedges, L.V. and Olkin, I. (1985). *Statistical Methods for Meta-analysis*. New York: Academic Press.

Hedges, L.V. and Vevea, J.L. (1996). Estimating effect size under publication bias: Small sample properties and robustness of a random effects selection model. *Journal of Educational and Behavioral Statistics*, **21**, 299–332.

Hedges, L.V. and Vevea, J.L. (1998). Fixed and random-effects models in meta-analysis. *Psychological Methods*, **3**, 486–504.

Iyengar, S. and Greenhouse, J.B. (1988). Selection models and the file drawer problem. *Statistical Science* **3**, 109–135.

Iyengar, S. and Zhao, P.L. (1994). Maximum likelihood estimation for weighted distributions. *Statistics and Probability Letters*, **50**, 438–460.

Lane, D.M. and Dunlap, W.P. (1978). Estimating effect size: Bias resulting from significance criterion in editorial decisions. *British Journal of Mathematical and Statistical Psychology*, **31**, 107–112.

Larose, D.T. and Dey, D.K. (1995). *Grouped Random Effects Models for Bayesian Meta-analysis*. Storrs, CT: University of Connecticut, Department of Statistics.

Little, R.J.A. (1985). A note about models for selectivity bias. *Econometrica*, **53**, 1469–1474.

Little, R.J.A. & Rubin, D.B. (2002). *Statistical Analysis with Missing Data* (2nd edn). Hoboken, NJ: John Wiley & Sons, Inc.

McDaniel, M.A., Whetzel, D., Schmidt, F.L. and Maurer, S. (1994). The validity of the employment interview: A comprehensive review and meta-analysis. *Journal of Applied Psychology*, **79**, 599–616.

Nelson, N., Rosenthal, R. and Rosnow, R.L. (1986). Interpretation of significance levels by psychological researchers. *American Psychologist*, **41**, 1299–1301.

Rao, C.R. (1965). On discrete distributions arising out of methods of ascertainment. *Sankhyā, Series A*, **27**, 311–324.

Rao, C.R. (1985). Weighted distributions arising out of methods of ascertainment: What does a population represent? In A.C. Atkinson and S.E. Feinberg (eds), *A Celebration of Statistics: The ISI Centenary Volume* (pp. 543–569). New York: Springer.

Raudenbush, S.W. (1984). Magnitude of teacher expectancy effects on pupil IQ as a function of the credibility of expectancy induction. *Journal of Educational Psychology*, **76**, 85–97.

Rosenthal, R. and Gaito, J. (1963). The interpretation of levels of significance by psychological researchers. *Journal of Psychology*, **55**, 33–38.

Rosenthal, R. and Gaito, J. (1964). Further evidence for the cliff effect in the interpretation of levels of significance. *Psychological Reports*, **15**, 570.

Silliman, N.P. (1997). Hierarchical selection models with applications in meta-analysis. *Journal of the American Statistical Association*, **92**, 926–936.

Terrin, N., Schmid, C.H., Lau, J. and Olkin, I. (2003). Adjusting for publication bias in the presence of heterogeneity. *Statistics in Medicine*, **22**, 2113–2126.

Thisted, R.A. (1988). *Elements of Statistical Computing: Numerical Computation*. New York: Chapman & Hall.

Vevea, J.L. and Hedges, L.V. (1995). A general linear model for estimating effect size in the presence of publication bias. *Psychometrika*, **60**, 419–435.

Vevea, J.L. and Woods, C.W. (in press). Publication bias in research synthesis: Sensitivity analysis using a priori weight functions. Psychological Methods.

Vevea, J.L., Clements, N.C. and Hedges, L.V. (1993). Assessing the effects of selection bias on validity data for the General Aptitude Test Battery. *Journal of Applied Psychology*, **78**, 981–987.

CHAPTER 10

Evidence Concerning the Consequences of Publication and Related Biases

Alexander J. Sutton

Department of Health Sciences, University of Leicester, UK

KEY POINTS

- Due to its nature, direct evidence of publication bias is hard to find, but a number of isolated examples of negative studies being suppressed in the medical literature have been identified.
- Several global empirical assessments have been carried out which apply a number of statistical assessments relating to funnel plot asymmetry to collections of published meta-analyses.
- Despite inconsistencies in the design and analysis of these global assessments, their results are broadly consistent and suggest funnel asymmetry and hence that potentially publication bias is a problem in a proportion of published meta-analyses.
- A number of assessments looking at specific publication biases – time lag and language – have been carried out. These add support to the existence of publication and related bias.
- The majority of published meta-analyses do not consider the effect of publication bias on their results. However, researchers should always check for the presence of publication bias and perform a sensitivity analysis to assess the potential impact of missing studies.

Publication Bias in Meta-Analysis – Prevention, Assessment and Adjustments   Edited by H.R. Rothstein, A.J. Sutton and M. Borenstein   © 2005 John Wiley & Sons, Ltd

INTRODUCTION

In Chapter 2 a historical perspective on the problem of publication bias was presented in which evidence for the existence of publication and related biases was reviewed. This included (i) direct evidence, such as editorial policy; (ii) results of surveys of investigators, indicating that non-statistically significant results are less likely to be submitted for publication; (iii) follow-up of cohorts of studies registered at inception, suggesting non-statistically significant studies are less likely to be published, and, if they are, published more slowly than significant ones. Additionally, indirect evidence also exists which is persuasive. This includes quantification of an overrepresentation of significant results in published studies and negative correlations estimated between sample size and treatment effect in collections of published studies.

As described in previous chapters, if publication bias does exist in a published literature, then meta-analyses based on such data may be biased and misleading. For example, in a health-care context, this may mean that inferior or, in extreme cases, even harmful interventions may be wrongly recommended if policy is based on the biased evidence. This then leads to avoidable suffering of patients and wasted resources (Song *et al.*, 2000). Further, research on ineffective interventions may be needlessly and unknowingly replicated if initial research never reaches the public domain. Such wastefulness and even harm translates directly to other research domains.

In this chapter the empirical evidence of the *impact* and *consequences* of publication and related biases is examined. The direct evidence is reviewed in the next section, although this is brief since little direct evidence actually exists. The indirect evidence is reviewed later in this chapter. This summarizes, critiques and compares a number of empirical investigations in which several of the statistical methods for assessing publication (and related) bias, largely through funnel plot asymmetry (described in Part 2 of this book), have been applied to collections of published meta-analyses. Through such assessments, the impact of publication bias on the findings of research is considered.

DIRECT EVIDENCE CONCERNING THE CONSEQUENCES OF PUBLICATION AND RELATED BIASES

Song *et al.* (2000) review the evidence relating to direct consequences of publication bias in the medical literature from which this account borrows. Essentially, a small number of cases have been identified where studies, initially unpublished, showed treatments to be useless or possibly harmful, and have been unearthed at some later time. Inferences can then be made as to how clinical practice was likely to have changed if such studies had been reported earlier. For example, in a perinatal trial (Chalmers, 1990), routine hospitalization was associated with more unwanted outcomes in women with uncomplicated twin pregnancies, but this finding remained unpublished for seven years. Chalmers (1990, p. 1406) commented that, 'at the very least, the delay led to continued inappropriate deployment of limited resources; at worst it may have resulted in the continued use of a harmful policy'. In a similar instance, in 1980 a trial evaluated lorcainide against placebo in patients with myocardial infarction (Cowley *et al.*, 1993) and found evidence that it was, in fact, harmful (9/48 deaths compared to 1/47). The manufacturer stopped evaluation of locainide but did not publish the trial. Around a decade later excess deaths in two further trials (Cardiac Arrhythia Suppression Trial, 1989, 1992) evaluating related compounds encainide and flecainide were observed. These deaths might have been avoided or reduced if the findings of the earlier trial had been known.

There are, of course, probably many other instances where lack of reporting has caused wastage and harm, but the offending studies are never identified to be 'blamed'. For example, there has been a lot of controversy and discussion over the use of intravenous magnesium as a treatment for acute myocardial infarction. An initial meta-analysis (Yusuf *et al.*, 1993) suggested a modest benefit over standard practice, but a potentially harmful effect was seen in a subsequent very large trial (ISIS-4, 1995). Publication bias has been blamed for this discrepancy (Egger and Smith, 1995), although this has never been proven (i.e., no unpublished studies have subsequently come to light). Indeed, as in this case, often the most that can be done is to make indirect assessments of publication bias using meta-analyses; these are the topic of the next section.

INDIRECT EMPIRICAL EVIDENCE CONCERNING THE CONSEQUENCES OF PUBLICATION BIAS AS ASSESSED USING ASYMMETRY ASSESSMENTS OF META-ANALYSES

The statistical methods available to assess possible publication bias (described in previous chapters) have been applied to collections of published meta-analyses. The aim of such global assessments is to consider the potential impact publication bias may have on an area of science. To date, several such assessments have been carried out, primarily in the field of medicine (Egger *et al.*, 1997a; Sutton *et al.*, 2000; Song *et al.*, 2002; Sterne *et al.*, 2000), but also in biology (Jennions & Møller, 2002a). These are reviewed individually below, and then compared and contrasted in the discussion which follows.

Assessments of publication bias in medical randomized controlled trials

To the best of the author's knowledge, the first global assessment of bias in meta-analysis was assessed through examination of the asymmetry of the distribution of effect sizes by Egger *et al.* (1997a). Two collections of meta-analyses were identified for assessment, and a regression asymmetry test (see Chapter 6) was developed and applied to both collections.

The first collection was meta-analyses of randomized controlled trials (RCTs) identified through a hand-search of four of the leading general journals in medicine (*Annals of Internal Medicine, British Medical Journal, Journal of the American Medical Association* and *The Lancet*) between 1993 and 1996. The second collection was identified from the Cochrane Database of Systematic Reviews. For a meta-analysis to be included in either collection the outcome needed to be dichotomous and at least five primary studies had to be included. Thirty-seven meta-analyses from medical journals and 38 from Cochrane reviews were identified that satisfied these selection criteria. The mean of the intercept from the regression test was -1.00 (-1.50 to -0.49) and -0.24 (-0.65 to 0.17) for the journal and Cochrane meta-analyses, respectively (analysis carried out on the log odds ratio scale). This suggests, on average, that small studies estimate the treatment effect to be more beneficial. The intercept was significantly different (at a 10 % significance level) from zero in 5 (13 %) of the Cochrane meta-analyses and 14 (38 %) of the journal meta-analyses. Further details of this study and each of those described below are summarized in Table 10.1 for easy comparison.

Sterne *et al.* (2000) carried out a similar empirical assessment to Egger *et al.* (1997a), although they applied the rank correlation test (see Chapter 6) in addition to a regression test. They selected meta-analyses coming from the same four leading journals used by Egger *et al.* (1997a) and four specialist medical journals (*American Journal of Cardiology*, *Cancer*, *Circulation*, and *Obstetrics and Gynaecology*) both between 1993 and 1997. Inclusion criteria were the same as used by Egger *et al.* (1997a) described above. Seventy-eight meta-analyses were identified as eligible. Using the regression methods 21 (26.9 %) of these had an intercept significant at the 10 % level, and the corresponding figure for the rank correlation test was 10 (12.8 %). Hence, these findings broadly corroborate with those reported previously by Egger *et al.* (1997a).

Sutton *et al.* (2000) also examined the likely effect of publication bias on meta-analyses of RCTs. In this assessment 48 meta-analyses from the Cochrane Database of Systematic Reviews (1998, issue 3) were used. This assessment attempted to go beyond replicating those done previously by using the 'trim and fill' method, discussed in Chapter 8, to estimate the likely impact of publication bias on a collection of meta-analyses, rather than just identifying meta-analyses where asymmetry of the funnel was apparent. To be included in the assessment a meta-analysis had to include at least 10 trials (in this and the two assessments described above only one meta-analysis per systematic review was selected). Funnel plots for the 48 meta-analyses selected are provided in Figure 10.1. When a fixed-effect meta-analysis model was used, 26 out of 48 meta-analyses (54 %) were estimated to have at least one study missing. This figure decreased to 23 (48 %) when a random-effect model was used. For the meta-analyses considered, the number of missing trials was statistically significant at the 5 % level (i.e., asymmetry was greater than would be expected by chance 95 % of the time) if more than three were estimated missing. Eight meta-analyses reached this critical level under the random-effect model and 10 under the fixed-effect model. When the estimated 'missing' studies were imputed, inferences regarding the treatment effect would have changed using a 5 % significance level in four instances using a random-effect model and three with a fixed-effect model. For the random effect model, three meta-analyses (numbers 5, 10 and 13 in Figure 10.1) that originally suggested a significant treatment effect became non-significant, and one that was originally non-significant (30) changed to a significantly harmful treatment effect. The degree to which each odds ratio was reduced when 'trim and fill'

Table 10.1 Design characteristics and findings from indirect assessments of publication bias: Summary of results from the empirical assessments.

Investigation	Study designs of interest	Number of meta-analyses examined	Sources searched	Minimum number of studies in meta-analysis	Outcome measure	Statistical significance level	Begg's correlation-based asymmetry test		Egger's regression-based asymmetry test			Trim and fill			
							Proportion of meta-analysis statistically significant	Mean correlation coefficient (90/95% CI)	Weighted/ unweighted model	Proportion of meta-analysis statistically significant	Mean value of bias coefficient (90/95% CI)	Proportion with significant missing studies using fixed effects	Proportion with significant missing studies using random effects	Proportion of times inference changed using fixed effects	Proportion of times inference changed using random effects
Egger et al. (1997a)	Medical RCTs	37	Hand-search of 4 leading journals	5	Odds ratio	10%			Most extreme of both	38% (14/37)	−1.00 (−1.50 to −0.49)				
Egger et al. (1997a)	Medical RCTs	38	Cochrane Database of Systematic Reviews	5	Odds ratio	10%			Most extreme of both	13% (5/38)	−0.24 (−0.65 to 0.17)				
Sterne et al. (2000)	Medical RCTs	78	4 leading and 4 specialist journals	5	Odds ratio	10%	13% (10/78)	(not reported)	Unweighted	27% (21/78)	−0.54 (−0.65 to −0.42)				
Sutton et al. (2000)	Medical RCTs	48	Cochrane Database of Systematic Reviews	10	Odds ratio	5%	15% (7/48)	−0.41 (−0.74 to −0.09)	Unweighted	27% (13/48)	−0.35 (−0.67 to −0.08)	21% (10/48)	17% (8/48)	6% (3/48)	8% (4/48)
Song et al. (2002)	Medical diagnostic studies	28	DARE database	5	Diagnostic odds ratio	5%	21% (6/28)	−0.27 (−0.38 to −0.17)	Unweighted	43% (12/28)	−1.42 (−1.98 to −0.87)	39% (11/28)	25% (7/28)	14% (4/28)	7% (2/28)
Jennions and Moller (2002a)	Ecological and evolutionary	40	Hand-search of relevant journals + WebSpiris database	8	Pearson's r, Hedges' d, ln(response ratio) + other	5% (although probably 10% for trim and fill)	35% (14/40)	−0.20 (not reported)				50% (20/40) (2-sided inference)	38% (15/40) (2-sided inference)	8% (3/40) (2-sided inference)	20% (8/40) (2-sided inference)

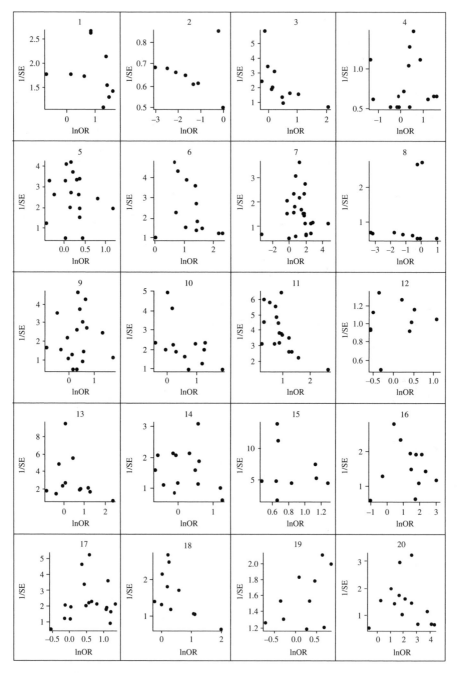

Figure 10.1 Funnel plots (log odds ratio vs. 1/standard error of the log odds ratio) of 48 meta-analyses identified in the Cochrane Database of Systematic Reviews by Sutton *et al.* (2000). Note that axis scales vary between plots.

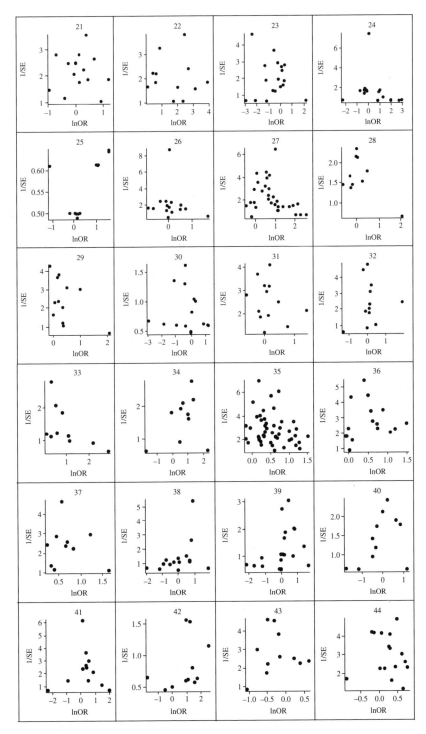

Figure 10.1   (Continued).

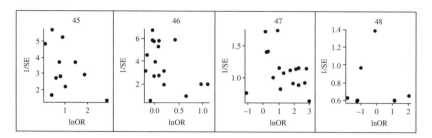

Figure 10.1 (Continued).

was applied was quite modest in many instances; however, in six studies with a positive treatment effect the reduction in the point estimate of the odds ratio was greater than 30 % (1, 3, 5, 10, 24 and 47), and in two meta-analyses with initially negative effects (30 and 48) the change in the odds ratio was even more striking.

Although not reported in the original paper due to paper size restrictions, Begg's and Egger's tests (Begg and Mazumdar, 1994; Egger *et al.*, 1997a) were also carried out on this collection of meta-analyses. Taking inferences at the 5 % statistical significance level, the Begg test was significant for 7 out of 48 meta-analyses (15 %) and the Egger test for 13 (27 %). Concordance between methods was not high, with only three meta-analyses showing significant asymmetry using all three statistical tests. However, this is not an entirely fair comparison, since the Begg and Egger tests are two sided and 'trim and fill' is one-sided, and two meta-analyses significant under the Begg and Egger tests were significant in the opposite direction from that expected (and investigated under 'trim and fill'). This issue is discussed further below. Only 16 (33 %) of the meta-analyses produced non-significant *p*-values for all three tests. This lack of agreement between methods has been noted elsewhere (Pham *et al.*, 2001).

On reflection, one of the most informative aspects of this assessment was the visual inspection of the funnel plots generated by the 48 meta-analyses (see Figure 10.1). I think the reader will agree that very few of these conform to the 'classic' funnel shape (with or without a missing portion) often discussed. This is partially due to the relatively small numbers of studies included in these meta-analyses (median 13, range 10 to 47). Recall that at least 10 studies had to be included for a meta-analysis to be selected and approximately 300 systematic reviews in the Cochrane Database did not meet this requirement. Hence, the typical size of meta-analyses done on RCTs does prohibit the asymmetry assessments that can be done.

Assessment of publication bias in medical studies evaluating diagnostic tests

Song *et al.* (2002) examine the asymmetry of funnel plots derived from meta-analyses of studies assessing the performance of diagnostic test accuracy in which a new technique was compared to the gold standard. Systematic reviews of diagnostic test accuracy were identified using the Database of Abstracts of Reviews of Effects (DARE) up to December 1999. A meta-analysis was included if it contained five or more primary studies and used a binary outcome measure. The diagnostic odds ratio was calculated and used as the outcome measure (Deeks, 2001). Funnel asymmetry was assessed using both the Begg and Egger tests, and 'trim and fill' in addition to informal examination of the funnel plots. Twenty-eight meta-analyses were identified for assessment from 20 systematic reviews. Of these 28 meta-analyses, 23 (82%) showed a positive correlation between test accuracy and variance (i.e., under conditions of no publication bias you would expect no association between performance of the test and the precision of the study, in the same way as you would expect a funnel plot to be symmetrical), and this relationship was statistically significant (5% level) in 6 (21%) cases. Similar findings were gained using Egger's regression asymmetry test where 25 (89%) meta-analyses had positive intercepts, 12 (43%) of which attained statistical significance (5% level). The 'trim and fill' analysis using a random-effect model suggested that 17 (61%) of the meta-analyses had some missing studies. This number reached statistical significance in 7 (25%) instances. The 'trim and fill' analysis suggested that the value of the diagnostic odds ratio has been overestimated by 21–30% in two (7%) meta-analyses, by 11–20% in seven (25%) meta-analyses, and by 5–10% in six (21%) meta-analyses. Figures were similar for fixed effect analyses, with 21 (75%) of the meta-analyses suggesting some studies were missing. These findings suggest the problems of publication bias are at least as bad for diagnostic studies as they are for RCTs.

Assessment of publication bias in studies in biological sciences

Jennions and Møller (2002a) make an assessment of funnel asymmetry in meta-analyses from the biological sciences in the field of ecology and evolution. To be included in their assessment a meta-analysis had to include a minimum of eight primary studies. Forty such meta-analyses were identified in the literature up until the end of 2000. Using mean effect size as outcome, funnel plots were examined and standardized effect sizes were used in the application of the rank correlation test and the 'trim and fill' method. Additionally, Rosenthal's failsafe N (see Chapter 7) was calculated for each meta-analysis. 'Trim and fill' estimated at least one missing study in 30 (75%) and 36 (90%) meta-analyses using random- and fixed-effect models, respectively. The number of meta-analyses in which the number of studies attained statistical significance was 15 (38%) for meta-analyses using a random-effect model and 20 (50%) using a fixed-effect model. For those studies using Pearson's r as the outcome measure, the mean percentage change was 57.4% and 36.5% ($n = 21$) for fixed- and random-effect models, respectively. The corresponding figures for Hedges' d are 30.3% and 17.0% ($n = 10$); and for the natural log of the response ratio 20.9% and 27.0% ($n = 7$). Inferences changed when comparing the 'trim and fill' adjusted results to the original analyses in 11 (28%) instances. Three of these were when a fixed-effect model was used and eight when a random-effect model was used. Curiously, inferences never changed at the

5 % significance level for any particular meta-analysis for both fixed- and random-effect analyses. These findings appear unusual but no explanation is given. (See below for a discussion regarding the significance level of these inferences.) The rank correlation test was also applied to these meta-analyses. The mean correlation was −0.20 (suggesting the more precise the study, the lower the effect size) and 14 (35 %) had a significant *p*-value at the 10 % level. Hence, there is some evidence that funnel asymmetry, and hence publication bias, is more of a problem here than in medical studies, but the points raised in the next section should be considered before drawing conclusions.

Critical appraisal and comparison of the empirical assessments

Obvious advantages of using formal statistical methods over subjective assessments of funnel plots, to assess publication bias in a meta-analysis, are the added objectivity and reproducibility. However, although the empirical assessments reviewed above used many of the same statistical techniques to assess the likely presence of publication bias through funnel asymmetry, implementation details vary between them that potentially could have important implications. These are discussed below with reference to Table 10.1.

The first factor to consider is the inclusion criteria for the selected meta-analyses. Since the assessment methods' power to detect asymmetry increases as the number of studies in a meta-analysis increases (power is also affected by the magnitude of the treatment effect and the distribution of study sizes) (Macaskill *et al.*, 2001), each meta-analysis within each assessment is uniquely powered. This means comparison of collections of meta-analyses of different sizes is potentially misleading. The inclusion criteria for the minimum number of studies within a meta-analysis required varied from five to ten between assessments which will have a large influence on the average power of meta-analyses within each assessment. Indeed, Sterne *et al.* (2000) comment that 50 % of the meta-analyses in their assessment had 10 or fewer studies – the minimum required in the assessment of Sutton *et al.* (2000).

Another important detail is that different investigations used different levels of statistical significance as cut-off values (5 % or 10 %), which should also be taken into account in any comparisons made. Hence, although *p*-values may provide a useful way of summarizing the assessment's findings, and have been reported here, methods of quantifying levels of asymmetry, such as the degree of correlation in Begg's test, the magnitude of the intercept in Egger's test and the magnitude of the change of the effect size using 'trim and fill', are more meaningful in general. This is particularly true for across-assessment comparisons. Hence, such quantification measures are reported, where available, in Table 10.1. Having said this, caution should be expressed in comparing such coefficients across different measures of effect; Jennions and Møller (2002a) use a range of outcome measures compared to odds ratios in the other investigations.

As outlined in previous chapters, all of these statistical methods perform differently with respect to their type I and type II error rates across different meta-analysis scenarios. This is why results can differ quite considerably between the statistical methods. Further close inspection also reveals that the studies did not implement some of the statistical methods in the same way. The regression test was implemented in a weighted and unweighted form by Egger *et al.* (1997a) and the more extreme result taken. Later, it was shown that there was no theoretical justification for the weighted version of the test (Deeks and Altman, 2001), and it is believed the remaining authors only considered the unweighted form. Thus, the estimation of bias by Egger *et al.* (1997a) is going to be exaggerated compared to the other implementations of the test.

For the 'trim and fill' method, all authors considered both fixed- and random-effect models and the same estimator (L_0 – see Chapter 8 for an explanation of the competing estimators) for the numbers of missing studies. As these assessments demonstrate, the difference between results of fixed- and random-effect analyses can be quite striking so it is advantageous that authors provide the results from both. This author is generally an advocate of random-effect models for meta-analysis, but since they are more affected by 'classic' funnel asymmetry than fixed-effect models, because less precise studies are given relatively more weight, I currently give more credence to the fixed-effect 'trim and fill' results until the performance of 'trim and fill' under both models is further assessed. It is very important not to overlook that 'trim and fill', as originally described (see Chapter 8 and references therein), is a one-sided (or one-tailed) assessment. That is, asymmetry is only looked for in the side of the funnel where a priori it is suspected to exist (usually the one corresponding to small or negative effects). This is in contrast to Begg's and Egger's tests which are both two-sided. It is interesting to note that in the assessments reviewed these tests attain statistical significance in the 'wrong' direction in some instances.

Jennions and Møller (2002a) would appear to apply 'trim and fill' to both sides of each funnel assessed, thus making their results more comparable with Begg's and Egger's tests but not with the other 'trim and fill' analyses. The performance of 'trim and fill' has not been assessed as a two-sided method and the procedure, as implemented, really tests at the ($5\% \times 2 =$) 10% level (i.e., 5% probability mass is being considered in both tails).

Finally, due to similar inclusion criteria, a proportion of the meta-analyses included by Egger *et al.* (1997a), Sutton *et al.* (2000) and Sterne *et al.* (2000) will be the same, making these assessments non-independent. (If they had been independent there would possibly have been a case for using meta-analysis to combine results across assessments.)

Despite all the above caveats and discrepancies, results across the assessments do appear reasonably consistent, with all of them finding considerable levels of funnel plot asymmetry in the meta-analyses they examine. Although funnel asymmetry cannot directly be interpreted as publication bias, it does raise fears that publication bias is a large enough problem to make the results from a proportion of meta-analyses (around 10% in the assessments discussed) misleading and exaggerate findings from a further sizeable proportion. Such a conclusion does make the assumption that the scale used in the analysis of each meta-analysis was the most the appropriate (Deeks and Altman, 2001). It should be realized that funnel plot asymmetry has been shown to be scale-dependent, and this is an issue that clearly requires further work (Tang and Liu, 2000; Sterne and Egger, 2001).

FURTHER INDIRECT EMPIRICAL ASSESSMENTS OF PUBLICATION AND RELATED BIASES USING META-ANALYSES

This section briefly reviews further indirect empirical investigations using collections of meta-analyses of publication and related biases. These do not use the statistical methods described in this part of the book, but do contribute to the totality of knowledge in the area.

Discordant large trials and meta-analyses

Not to investigate publication bias *per se*, but to assess the reliability of meta-analyses of RCTs more generally, Villar *et al.* (1995) estimated the agreement between the results of 30 meta-analyses with the largest trial removed compared with the results of the largest trial singularly. In terms of a funnel plot, this is equivalent to asking whether a meta-analysis of all studies but the largest agrees with the study at the top of the funnel. This investigation is of interest here as a likely cause of any discrepancy is likely to be funnel asymmetry and hence potentially publication bias, although there could be a number of other explanations, including differential design and conduct and recruited population issues for large trials compared to smaller ones. Twenty-four meta-analyses (80 %) estimated the treatment effect to be in the same direction (i.e., superior or inferior compared to the control group comparator) as the large trial. Hence, serious discrepancy was observed in 6 (20 %) cases.

In a similar investigation Egger *et al.* (1997a) identified meta-analyses for which a relevant large-scale RCT had been published afterwards. The concordance between the large trials and the previous meta-analyses was then assessed. The meta-analysis and the large trial were considered concordant when the estimate from the meta-analysis was in the same direction and within 30 % of the estimate from the large trial. Eight meta-analyses and corresponding large trial pairs were identified. Four of these pairs were considered concordant and four discordant. In all discordant pairs, the meta-analysis showed more beneficial effects than the subsequent large trial. Three out of four of the funnel plots from the discordant meta-analyses displayed asymmetry as assessed by the regression test, while the test was non-significant for all the concordant ones. Further assessment of these study and meta-analysis pairs was considered by identifying any further relevant (but smaller) trials that had been published after the meta-analysis. On doing this, the only inference that changed was that the test result for the fourth discordant pair also became statistically significant.

Assessment of time-lag biases

One form of publication bias is time-lag bias which is given specific treatment in Chapter 13. Evidence of time-lag bias exists from investigations following up studies registered with ethics committees (Stern and Simes, 1997) demonstrating that studies with significant results were more likely to be published quickly than those that were non-significant. In this study, the median time to publication of RCTs with significant results was 4.69 years, compared with 7.99 years for those with null results (other studies are reviewed in Chapter 13). One recent study (Jennions and Møller, 2002b) investigated time-lag bias by examining the relationship between time of publication and effect size in 44 published meta-analyses in ecology and evolutionary biology – 40 of these were the same as those used in the assessment using 'trim and fill' reviewed above (Jennions and Møller, 2002a). In this related assessment, the authors found evidence of a negative relationship between date of publication and effect size (correlation coefficient $= -0.133$ (95 % CI -0.189 to -0.062), $p < 0.01$). These findings suggest that, on average, reported study effects in a scientific area decrease over time providing evidence of time-lag bias. Such findings have not been established in other disciplines, including the social sciences.

Assessments of language biases

Language bias was defined in Chapter 2, and its presence and impact have been examined in a number of empirical assessments using a variety of study designs (Gregoire *et al.*, 1995; Egger *et al.*, 1997b; Moher *et al.*, 2000; Jüni *et al.*, 2002).

Gregoire *et al.* (1995) identified 28 meta-analyses in which an English-language restriction had been stated in the inclusion criteria. The investigators then attempted to identify non-English trials relevant to the 28 meta-analyses and included them in a reanalysis. Eleven articles were identified as having the potential to modify seven of the 28 identified meta-analyses. In one instance inference would have changed at the 5 % level if the non-English paper had been included. Surprisingly, however, this would have turned a non-significant effect significant. This is the opposite direction to that in which we perceive a language selection bias to be working as consensus thinking would suggest that researchers are more likely to publish non-significant studies in their native language, while being more motivated to publish significant results in prestigious English-language journals.

In a further study, Moher *et al.* (2000) identified 18 meta-analyses of RCTs in which non-English-language reported trials were included. An analysis comparing the results of these 18 meta-analyses against a reanalysis with the non-English-language reported trials removed found that treatment effects were, on average, only 2 % (relative odds ratio 0.98 (95 % CI 0.81 to 1.17)) lower in the analyses in which no non-English-language trials were included compared to those in which they were excluded.

Finally, Jüni *et al.* (2002) carried out a similar investigation to Moher *et al.* (2000) by identifying 50 meta-analyses of RCTs in which at least one non-English-language trial report was included. The effects in the English-language trials were compared with the non-English-language ones and it was found that, on average, treatment effect estimates were 16 % greater in non-English language trials (ratio of estimates 0.84 (95 % CI 0.74 to 0.97, $p = 0.01$)). When comparing the original meta-analyses to those with the non-English-language trials excluded, changes in the effect size ranged from a 42.0 % increase, indicating less benefit, to a 22.7 % decrease, indicating more benefit, but in 29 (58 %) of meta-analyses the effect size changed by less than 5 %.

DISCUSSION

Although direct evidence relating to the consequences of publication bias was briefly considered, this chapter has focused on assessment of the indirect evidence relating to publication and related biases through statistical assessment of the asymmetry of funnel plots from collections of meta-analyses. Six empirical assessments were reviewed which assessed levels of potential publication bias using statistical methodology. These were carried out predominantly in medicine. Although these assessments differed in quite subtle but potentially different ways, findings are broadly consistent.

As stressed throughout this part of the book, funnel asymmetry may be caused by factors other than publication and related biases, including: (i) true heterogeneity, implying the size of the effect does differ according to study size; and (ii) for artifactual reasons, such as the choice of effect measure (Deeks and Altman, 2001). Thus, statistical methods used in the assessments of funnel asymmetry are indirect and exploratory (Egger *et al.*, 1997a). Hence, any inferences regarding the levels and impact of potential publication bias can only be speculative. Additionally, there is a concern that in any specific case observed asymmetry detected by the statistical methods may be due solely to chance. This is true, but the play of chance should be greatly reduced when making global assessments on collections of meta-analyses rather than individual data sets.

At least for the assessments of the medical literature, high-quality sources were searched to identify the meta-analyses – that is, high-quality journals, or leading databases of systematic reviews. Little is known regarding the extent of publication bias of meta-analyses published in lesser or more specialized sources. Additionally, the majority of the medical assessments have been carried out on RCTs. Relating to the assessment of diagnostic studies, Song *et al.* (2002) observed that such studies can be easily conducted and abandoned as data are often collected as part of routine clinical care with the absence of a registration process for attempted evaluations. Similarly, many epidemiological meta-analyses are currently carried out assessing associations between potentially harmful exposures and diseases. No global empirical examination of such meta-analyses with regard to funnel asymmetry and publication bias has been published, although examination of individual meta-analyses such as those assessing the risk of lung cancer associated with environmental tobacco smoke has caused much attention (Copas and Shi, 2000; Givens *et al.*, 1997). Additionally, the problem of potentially extreme reporting biases in genetic association studies has recently been highlighted (Little *et al.*, 2002; Colhoun *et al.*, 2003). Concerns that publication and related biases may be greater in such literatures when compared to those with experimental designs have been raised (Sutton *et al.*, 2002). For instance, case–control studies often examine numerous exposures in the hope of finding potential associations with the disease of interest. There is a great temptation only to report those associations which are statistically significant. Further, in cohort study designs there is often the potential to look at numerous exposures and diseases, making data dredging exercises very possible.

Generally all the authors only considered one meta-analysis from each identified systematic review, often identified either on grounds of size or importance of the outcome. This is sensible practice, since multiple outcomes from the same review would generally be correlated since they would contain a proportion of the same primary studies, invalidating assessments since inferences regarding funnel asymmetry would not be independent of related meta-analyses. This does, however, potentially mask the problem of outcome reporting bias in meta-analysis (see Chapter 12) since outcomes which fewer studies report were less likely to be included in the assessments.

The assessments relating specifically to time-lag and language biases largely corroborate the results of the more general funnel asymmetry assessments. It should be realized that these forms of publication bias may contribute to funnel asymmetry and are assessed collectively in the more general assessments reported above (i.e., not just suppression of whole studies, or 'classic' publication bias, but all forms of publication bias lead to funnel asymmetry).

It is unclear whether the assessments currently carried out are broadly generalizable to other areas of scientific investigation, and thus it would be interesting to see similar assessments carried out in other areas of scientific inquiry. In the future, the Campbell Collaboration Database of Systematic Reviews (C2-SPECTR) may provide a framework for such an assessment of trials in the social sciences. In other areas of science, studies are so different in design from the ones used in the assessments reviewed here that levels of publication bias are bound to be different. As mentioned above, within the field of genetic epidemiology there is a growing concern that publication bias could be a huge problem due to analyses in which large numbers of hypothesis tests are considered (Colhoun *et al.*, 2003).

A lot of information has been gathered in the assessments summarized here, but what use are such assessments? I can think of three answers to this question:

1. They raise awareness of the problems of publication and related biases.
2. They identify particular published meta-analyses which need to be treated with caution due to excessive funnel asymmetry and reinforce the potential threat of publication bias when interpreting the results of any meta-analysis.
3. They raise awareness that an assessment of publication bias should be a required step in any meta-analysis.

There is evidence that publication bias is only considered in a fraction of meta-analyses carried out. For example, the 48 Cochrane systematic reviews, from which the 48 meta-analyses assessed by Sutton *et al.* (2000) were taken, were searched for any descriptions of the issue of publication bias or any steps taken to deal with it. Thirty (63 %) made no reference to publication bias; 5 examined a funnel plot and 3 used a regression test (Egger *et al.*, 1997a). Although many of these reviews were completed before the regression test was first described, and all had been completed before 'trim and fill' had been described in print, the numbers that made any assessment of publication bias were disappointing. Further, Professor Hannah Rothstein (personal communication) carried out a survey of all meta-analyses in *Journal of Applied Psychology* and *Personnel Psychology* (two leading journals in industrial organizational psychology) between 1985 and 2002. Fewer than 10 % considered publication bias at all (and of those that did few conducted any analyses to assess it). Most disappointingly, there was little evidence to suggest that there was any increase in the tendency to consider publication bias over time.

Further, under a Bayesian statistical framework, where prior information can be included in any quantitative analysis, there is a potential to actually utilize the information gained from the global assessments. For example, if one is assessing the likelihood of publication bias in a future meta-analysis, the data on rates of asymmetry from these assessments generally, or perhaps using more specific related meta-analysis, could be used to derive a prior distribution for bias parameters in any publication modelling, possibly using selection modelling as described in Chapter 9. For example, data on how asymmetric funnels are in meta-analyses of RCTs may contribute to the specification and estimation of relevant parameters of the weight function used in the selection model. Such an analysis would have similarities to analyses described previously where knowledge regarding levels of heterogeneity in previously carried out related meta-analyses have been used to form a prior distribution for heterogeneity in a further meta-analysis (Higgins and Whitehead, 1996).

ACKNOWLEDGEMENT

The author would like to thank Professors Hannah Rothstein and Doug Altman, whose comments improved this chapter considerably.

REFERENCES

Begg, C.B. and Mazumdar, M. (1994). Operating characteristics of a rank correlation test for publication bias. *Biometrics*, **50**, 1088–1099.

Cardiac Arrhythmia Suppression Trial Investigators (CAST). (1989). Preliminary report: Effect of encainide and flecainide on mortality in a randomised trial of arrhythmia suppression after myocardial infarction. *New England Journal of Medicine*, **321**, 406–412.

Cardiac Arrhythmia Suppression Trial II Investigators (CAST). (1992). Effect of the antiarrhythmic agent moricisine on survival after myocardial infarction. *New England Journal of Medicine*, **327**, 227–233.

Chalmers, I. (1990). Underreporting research is scientific misconduct. *Journal of the American Medical Association, 263*, 1405–1408.

Cochrane Database of Systematic Reviews (1998). [Computer Software]. Oxford: Cochrane.

Colhoun, H.M., McKeigue, P.M. and Davey Smith, G. (2003). Problems of reporting genetic associations with complex outcomes. *Lancet*, **361**, 865–872.

Copas, J.B. and Shi, J.Q. (2000). Reanalysis of epidemiological evidence on lung cancer and passive smoking. *British Medical Journal*, **320**, 417–418.

Cowley, A.J., Skene, A., Stainer, K. and Hampton, J.R. (1993). The effect of lorcainide on arrhythmias and survival in patients with acute myocardial-infarction – an example of publication bias. *International Journal of Cardiology*, **40**, 161–166.

DARE-Database of Abstracts of Reviews of Effects (n.d.). *Complied for the Centre for Reviews and Dissemination*. Retrieved from University of York website: http://agatha. york.ac.uk/welcome.htm.

Deeks, J.J. (2001). Systematic reviews of evaluations of diagnostic and screening tests. In M. Egger, G. Davey Smith and D.G. Altman (eds), *Systematic Reviews in Health Care: Meta-analysis in context*, 2nd edn (pp. 248–282). London: BMJ Publishing Group.

Deeks, J.J. and Altman, D.G. (2001). Effect measures for meta-analysis of trials with binary outcomes. In M. Egger, G. Davey Smith and D.G. Altman (eds), *Systematic Reviews in Health Care: Meta-analysis in Context*, 2nd edn (pp. 313–335). London: BMJ Publishing Group.

Egger, M. and Smith, G.D. (1995). Misleading meta-analysis. *British Medical Journal*, **310**, 752–754.

Egger, M., Smith, G.D., Schneider, M. and Minder, C. (1997a). Bias in meta-analysis detected by a simple, graphical test. *British Medical Journal*, **315**, 629–634.

Egger, M., Zellweger-Zahner, T., Schneider, M., Junker, C. Lengeler, C. and Antes, G. (1997b). Language bias in randomised controlled trials published in English and German. *Lancet*, **350**, 326–329.

Givens, G.H., Smith, D.D. and Tweedie, R.L. (1997). Publication bias in meta-analysis: A Bayesian data-augmentation approach to account for issues exemplified in the passive smoking debate (with discussion). *Statistical Science*, **12**, 221–250.

Gregoire, G., Derderian, F., Lelorier, J. and Le Lorier, J. (1995). Selecting the language of the publications included in a meta-analysis – is there a Tower-of-Babel bias? *Journal of Clinical Epidemiology*, **48**, 159–163.

Higgins, J.P.T. and Whitehead, A. (1996). Borrowing strength from external trials in a meta-analysis. *Statistics in Medicine*, **15**, 2733–2749.

ISIS-4 Collaborative Group (1995). ISIS-4: A randomised factorial trial assessing early oral captopril, oral mononitrate, and intravenous magnesium sulphate in 58050 patients with acute myocardial infarction. *Lancet*, **345**, 669–685.

Jennions, M.D. and Møller, A.P. (2002a). Publication bias in ecology and evolution: An empirical assessment using the 'trim and fill' method. *Biological Review*, **77**, 211–222.

Jennions M.D. and Møller, A.P. (2002b). Relationships fade with time: A meta-analysis of temporal trends in publication in ecology and evolution. *Proceedings of the Royal Society of London B*, **269**, 43–48.

Jüni, P., Holenstein, F., Sterne, J., Bartlett, C. and Egger, M. (2002). Direction and impact of language bias in meta-analyses of controlled trials: Empirical study. *International Journal of Epidemiology*, **31**, 115–123.

Little, J., Bradley, L.B.M.S., Clyne, M., Dorman, J., Ellsworth, D.L., Hanson, J., Khoury, M., Lau, J., O'Brien, T.R., Rothman, N., Stroup, D., Taioli, E., Thomas, D., Vainio, H., Wacholder, S. and Weinberg, C. (2002). Reporting, appraising, and integrating data on genotype prevalence and gene–disease associations. *American Journal of Epidemiology*, **156**, 300–310.

Macaskill, P., Walter, S.D. and Irwig, L. (2001). A comparison of methods to detect publication bias in meta-analysis. *Statistics in Medicine*, **20**, 641–654.

Moher, D., Pham, B., Klassen, T., Schulz, K., Berlin, J., Jadad, A. and Liberati, A. (2000). What contributions do languages other than English make on the results of meta-analyses? *Journal of Clinical Epidemiology*, **53**, 964–972.

Pham, B., Platt, R., McAuley, L., Klassen, T.P. and Moher, D. (2001). Is there a 'best' way to detect and minimize publication bias? An empirical evaluation. *Evaluation and the Health Professionals*, **24**, 109–125.

Song, F., Eastwood, A., Gilbody, S., Duley, L. and Sutton, A.J. (2000). Publication and other selection biases in systematic reviews. *Health Technology Assessment*, **4**, 1–115.

Song, F., Khan, K.S., Dinnes, J. and Sutton, A.J. (2002). Asymmetric funnel plots and publication bias in meta-analyses of diagnostic accuracy. *International Journal of Epidemiology*, **31**, 88–95.

Stern, J.M. and Simes, R.J. (1997). Publication bias: Evidence of delayed publication in a cohort study of clinical research projects. *British Medical Journal*, **315**, 640–645.

Sterne, J.A.C. and Egger, M. (2001). Funnel plots for detecting bias in meta-analysis: Guidelines on choice of axis. *Journal of Clinical Epidemiology*, **54**, 1046–1055.

Sterne, J.A.C., Gavaghan, D. and Egger, M. (2000). Publication and related bias in meta-analysis: Power of statistical tests and prevalence in the literature. *Journal of Clinical Epidemiology*, **53**, 1119–1129.

Sutton, A.J., Duval, S.J., Tweedie, R.L., Abrams, K.R. and Jones, D.R. (2000). Empirical assessment of effect of publication bias on meta-analyses. *British Medical Journal*, **320**, 1574–1577.

Sutton, A.J., Abrams, K.A. and Jones, D.R. (2002). Generalised synthesis of evidence and the threat of publication bias: The example of electronic fetal heart rate monitoring (EFM). *Journal of Clinical Epidemiology*, **55**, 1013–1024.

Tang, J.L. and Liu, J.L. (2000). Misleading funnel plot for detection of bias in meta-analysis. *Journal of Clinical Epidemiology*, **53**, 477–484.

Villar, J., Carroli, G. and Belizan, J.M. (1995). Predictive ability of meta-analyses of randomised controlled trials. *Lancet*, **345**, 772–776.

Yusuf, S., Koon, T. and Woods, K. (1993). Intravenous magnesium in acute myocardial infarction: An effective, safe, simple and inexpensive intervention. *Circulation*, **87**, 2043–2046.

CHAPTER 11

Software for Publication Bias

Michael Borenstein

Biostat, Inc., USA

KEY POINTS

- Various procedures for addressing publication bias are discussed elsewhere in this volume. The goal of this chapter is to show how these different procedures fit into an overall strategy for addressing bias, and to discuss computer programs that can be used to implement this strategy.
- To address publication bias the researcher should proceed through a logical sequence of analyses. First, forest plots and funnel plots provide a visual sense of the data. Then, rank correlation and regression procedures ask whether or not there is evidence of bias, while the failsafe N and its variants ask if the *entire* effect size may be attributed to bias. Finally, trim and fill and the cumulative forest plot offer a more nuanced perspective, and ask how the effect size would *shift* if the apparent bias were to be removed.
- The computer programs discussed are Comprehensive Meta Analysis, Stata, MetaWin, and RevMan. We show how the researcher would address publication bias using each of these programs in turn. The chapter closes with a subjective assessment of each program's strengths and weaknesses.
- The appendix includes a list of additional resources, including web links to data sets and to step-by-step instructions for running the publication bias procedures with each of the programs.

Publication Bias in Meta-Analysis – Prevention, Assessment and Adjustments Edited by H.R. Rothstein, A.J. Sutton and M. Borenstein © 2005 John Wiley & Sons, Ltd

INTRODUCTION

This chapter discusses computer programs for meta-analysis and publication bias. The author of this chapter is also one of the developers of Comprehensive Meta Analysis (CMA), one of the programs discussed below. While I have endeavored to make this chapter as accurate as possible, the chapter will reflect the same sensibilities that shaped the development of CMA, and as such is not entirely objective. Additional detail on this potential conflict of interest, and also of steps taken to ensure accuracy, are included at the end of the chapter.

Where other chapters in this volume discuss the rationale and interpretation of specific procedures for assessing publication bias, the goal of this chapter is to present an overall strategy for working with these different procedures, and then to discuss computer programs that can be used to implement this strategy. The overview that follows, and then the discussion of each program in turn, proceeds through a logical sequence of analyses: Forest plots and funnel plots provide a visual sense of the data. Rank correlation and regression procedures can test for the presence of bias. The failsafe *N* and its variants ask if we need to be concerned that the *entire* observed effect may be an artifact of bias. Finally, trim and fill and the cumulative forest plot offer a more nuanced perspective, and ask how the treatment effect (or other effect size) would *shift* if the apparent bias were to be removed.

Step-by-step instructions for each of the programs, and copies of the data sets, are available on the book's website, http://www.meta-analysis.com/publication-bias.

STATISTICAL PROCEDURES

Getting a sense of the data

Forest plots
A key element in any meta-analysis is the forest plot, which serves as the visual representation of the data. In this plot each study as well as the combined effect is depicted as a point estimate bounded by its confidence interval. The plot, as suggested by its appellation, allows the researcher to see both the forest and the trees. It shows if the overall effect is based on many studies or a few; on studies that are precise or imprecise; whether the effects for all studies tend to line up in a row, or whether they vary substantially from one study to the next. The plot puts a face on the statistics, helping to ensure that they will be interpreted properly, and highlighting anomalies such as outliers that require attention.

While the forest plot is more closely associated with the core meta-analysis than with publication bias, an examination of this plot is a logical first step in any analysis. Before turning to the funnel plot or statistical tests to look for bias, the researcher should study the forest plot to get a sense of the data.

Funnel plots
The funnel plot, in its traditional form, is a plot of a measure of study size on the vertical axis as a function of effect size on the horizontal axis (Chapter 5). Large studies appear toward the top of the graph, and tend to cluster near the mean

effect size. Smaller studies appear toward the bottom of the graph, and (since there is more sampling variation in effect size estimates in the smaller studies) will be dispersed across a range of values. This pattern tends to resemble a funnel, which is the basis for the plot's name (Light and Pillemer, 1984; Light *et al.*, 1994).

In the absence of publication bias, the studies will be distributed symmetrically about the combined effect size. By contrast, in the presence of bias, we would expect that the bottom of the plot would show a higher concentration of studies on one side of the mean than on the other. This would reflect the fact that smaller studies (which appear toward the bottom) are more likely to be published if they have larger than average effects, which makes them more likely to meet the criterion for statistical significance.

Sterne, Becker and Egger (Chapter 5) explain that the selection of an effect index (e.g., the odds ratio or the risk difference) will have an impact on the symmetry of the funnel plot. Therefore, the selection of an index should reflect the fundamental questions being asked in the analysis. They note also that the index used to represent study size will affect the way studies are dispersed on the plot, and that the proper index will facilitate the researcher's ability to detect bias. They recommend the use of the standard error (rather than its inverse) for this purpose. Finally, they suggest that it may be helpful to superimpose guidelines on the funnel plot to show the expected distribution of studies in the absence of bias. These guidelines can help to identify outliers and facilitate the process of detecting asymmetry (see also Egger *et al.*, 1997; Sterne and Egger, 2001).

Is there evidence of bias?

The funnel plot offers a visual sense of the relationship between effect size and precision, but the interpretation of the plot is largely subjective. Two tests are commonly used in an attempt to quantify the amount of bias captured by the funnel plot.

Begg and Mazumdar's rank correlation test

Begg and Mazumdar's rank correlation test reports the rank correlation (Kendall's tau) between the standardized effect size and the variances (or standard errors) of these effects. Tau would be interpreted much the same way as any correlation, with a value of zero indicating no relationship between effect size and precision, and deviations from zero indicating the presence of a relationship (Begg and Berlin, 1988; Begg and Mazumdar, 1994; Begg, 1994).

If asymmetry is caused by publication bias we would expect that high standard errors (small studies) would be associated with larger effect sizes. If larger effects are represented by low values (e.g., odds ratio for preventing lung cancer) tau would be positive, while if larger effects are represented by high values (e.g., risk ratio for passive smoking) tau would be negative. Since asymmetry could appear in the reverse direction however, the significance test is two-sided. (Kendall's tau should not be confused with the convention of using τ^2 to denote a variance component in random-effect meta-analysis models. There is no relationship between the two.)

Sterne and Egger (Chapter 6) caution against using the test unless the meta-analysis includes a range of study sizes, including at least one of 'medium' size. Otherwise, the result will be driven primarily by noise. They also note that the test has low power unless there is severe bias, and so a non-significant tau should not be taken as proof that bias is absent (see also Sterne *et al.*, 2000, 2001b, c).

Egger's regression

Egger's linear regression method (Egger *et al.*, 1997; Sterne *et al.*, 2001b, c), like the rank correlation test, is intended to quantify the bias captured by the funnel plot. This differs from Begg and Mazumdar's test in that Egger uses the actual values of the effect sizes and their precision, rather than ranks. In the Egger test, the standard normal deviate is regressed on precision, defined as the inverse of the standard error. The intercept in this regression corresponds to the slope in a weighted regression of the effect size on the standard error. As was true for the rank correlation test, the significance test should be two-tailed.

Sterne and Egger (Chapter 6) discuss this method in more detail. They report that power for this test is generally higher than power for the rank correlation method, but is still low unless there is severe bias or a substantial number of studies. As is true of all significance tests, the point estimate and confidence interval are more informative, and less likely to be misinterpreted, than the significance test. As was true for the rank correlation method, the Egger test should only be used if the analysis includes a range of study sizes and at least one of 'medium' size.

Can we be confident that the observed effect is not entirely an artifact of bias?

The failsafe *N*

Rosenthal suggested that we could compute the number of missing studies (with mean effect of zero) that would need to be added to the analysis before the combined effect would no longer be statistically significant. Rosenthal referred to this as a 'File drawer' analysis (this being the presumed location of the missing studies), and Harris Cooper (1979) suggested that the number of missing studies needed to nullify the effect should be called the 'failsafe *N*' (Rosenthal, 1979; Begg, 1994).

While Rosenthal's work was critical in focusing attention on publication bias, this approach is of limited utility for a number of reasons. First, it focuses on the question of statistical significance rather than clinical significance. That is, it asks how many hidden studies are required to make the effect not *statistically* significant, rather than how many hidden studies are required to reduce the effect to the point that it is not *clinically* important. Second, the formula assumes that the mean effect size in the hidden studies is nil, when in fact it could be negative (which would require fewer studies to nullify the effect) or positive. Third, the failsafe *N* is based on significance tests that combine *p*-values across studies, as was common at the time that Rosenthal suggested the method. Today, the common practice is to compute a *p*-value for the combined effect, and the failsafe algorithm would not work with this approach.

Therefore, as Becker (Chapter 7) suggests, the failsafe N might best be seen as a heuristic which can help to put the question of publication bias in perspective. If someone is concerned that the observed effect is entirely an artifact of publication bias, the statement that 'we would need 5000 studies with an effect size of zero to nullify the observed effect' may help to move the conversation along to more subtle and relevant issues about publication bias (see also Carson *et al.*, 1990).

The algorithm for the failsafe N involves computing a combined p-value for all studies in the analysis, and then determining how many additional studies with (an average) z of zero would be required to yield a non-significant p-value. The criterion p-value for non-significance could be either one-tailed or two-tailed. While this issue has not been addressed in discussions of this method, it seems that the use of a one- or two-tailed criterion for the failsafe N should match the criterion used in the meta-analysis.

Orwin's failsafe N

As noted, two problems with Rosenthal's method are that it focuses on statistical significance rather than clinical significance, and that it assumes that the mean effect size in the missing studies is nil.

Orwin (1983) proposed a variant on the Rosenthal formula which addresses both of these issues. First, Orwin's method allows the researcher to determine how many hidden studies would bring the overall effect to a specified level other than zero. The researcher could therefore select a value that would represent the smallest effect deemed to be clinically important, and ask how many missing studies it would take to bring the combined effect below *this* point. Second, it allows the researcher to specify the mean effect in the hidden studies as being some value other than nil. This would allow the researcher to model a series of other distributions for the missing studies (see Becker, Chapter 7, this volume; Begg, 1994).

What is the effect size after we adjust for bias?

Duval and Tweedie's trim and fill

As discussed above, the key idea behind the funnel plot is that in the absence of bias the plot would be symmetric about the summary effect. If there are more small studies on the right than on the left, our concern is that there may be studies missing from the left. The trim and fill procedure (Chapter 8) builds directly on this idea by imputing the missing studies, adding them to the analysis, and then recomputing the effect size.

Trim and fill uses an iterative procedure to remove the most extreme small studies from the positive side of the funnel plot, recomputing the effect size at each iteration, until the funnel plot is symmetric about the (new) effect size. While this 'trimming' yields the adjusted effect size, it also reduces the variance of the effects, yielding a too narrow confidence interval. Therefore, the algorithm then adds the original studies back into the analysis, and imputes a mirror image for each. This 'fill' has no impact on the point estimate but serves as a correction to the variance (Duval and Tweedie, 1998, 2000a, 2000b).

A major advantage of this approach is that it yields an effect size estimate that is adjusted for the funnel plot asymmetry. That is, rather than ask whether or not *any* bias exists, or whether the *entire* effect can be attributed to bias, it provides an estimate of the effect size after the bias has been taken into account. In many cases the adjusted effect will be essentially similar to the original effect. In other cases the magnitude of the effect size will shift but the core finding (e.g., that the treatment is, or is not, effective) will remain intact. In others, the core finding will be called into question. However, these are precisely the kinds of questions that we should be asking.

Another nice feature of this approach is that it lends itself to an intuitive visual display. The computer programs that incorporate trim and fill are able to create a funnel plot that includes both the observed studies and the imputed studies, so the researcher can see how the effect size shifts when the imputed studies are included. If this shift is trivial, then one can have more confidence that the reported effect is valid.

The problem with this approach is that it relies on a number of assumptions. The key assumption is that the observed asymmetry is due to publication bias rather than a 'small-study effect' (see below). If this assumption is incorrect, the idea of imputing the missing studies cannot be supported. Even if the assumption is correct, the remaining problem is that this procedure, like many of the others, assumes that publication bias in the real world follows a neat pattern, and uses this pattern to identify the number of missing studies. As Duval explains (Chapter 8), the algorithm seems to work well in many circumstances, but not all. Variants on the algorithm have been proposed, with one appearing to be more robust than the others (see Chapter 8).

Cumulative meta-analysis

A cumulative meta-analysis is a meta-analysis run with one study, then repeated with a second study added, then a third, and so on. Similarly, in a cumulative forest plot, the first row shows the effect based on one study, the second row shows the cumulative effect based on two studies, and so on. These procedures have traditionally been used to show shifts in the cumulative weight of the evidence over time. Typically, the studies were sorted chronologically, and the plot showed the effect size based on cumulative data though 1970, 1980, 1990, and so on (see, for example, Lau *et al.*, 1995).

The same mechanism, however, can be used to assess the potential impact of publication bias or of a small-study effect. For this purpose the studies would be sorted in the sequence of largest to smallest (or of most precise to least precise), and a cumulative meta-analysis performed with the addition of each study. If the point estimate has stabilized with the inclusion of the larger studies and does not shift with the addition of smaller studies, then there is no reason to assume that the inclusion of smaller studies had injected a bias (i.e., since it is the smaller studies in which study selection is likely to be greatest). On the other hand, if the point estimate does shift when the smaller studies are added, then there is at least a *prima facie* case for bias, and one would want to investigate the reason for the shift.

This approach also provides an estimate of the effect size based solely on the larger studies. And, even more so than trim and fill, this approach is entirely transparent: We compute the effect based on the larger studies and then determine if and how the effect shifts with the addition of the smaller studies (a clear distinction between larger and smaller studies will not usually exist, but is not needed).

An important caveat

Sterne and Egger (Chapter 6) discuss the fact that while these procedures may detect a relationship between sample size and effect size, they cannot assign a causal mechanism to it. That is, the effect size *may* be larger in small studies because we retrieved a biased sample of the smaller studies, but it is also possible that the effect size *really is* larger in smaller studies – perhaps because the smaller studies used different populations or different protocols than the larger ones. Sterne and Egger use the term 'small-study effect' to capture these and other potential confounds (see also Sterne *et al.* 2001b, c).

This caveat figures into the interpretation of all the procedures, but in different ways. Some procedures are based entirely on a model of *publication* bias and would have no meaning for a 'small-study effect'. These include the failsafe N and Orwin (which look for numbers of missing studies) and trim and fill (which imputes missing studies). Other procedures (funnel plot, rank correlation, regression) look for a relationship between sample size and effect size. These will 'work' for a small-study effect as well as for publication bias, but the caveat still applies. If the apparent bias is actually a small-study effect then the larger effect size in the smaller studies reflects legitimate heterogeneity in the effect sizes. Similarly, if the cumulative meta-analysis shows a shift in the effect sizes with the inclusion of smaller studies, this could reflect the fact that these studies used a different population or protocol than the larger ones. It may be possible to identify the nature of these populations or protocols by studying the moderator variables, and this would be a logical next step.

ILLUSTRATIVE EXAMPLE

This chapter uses data from the passive smoking meta-analysis that serves as one of the running examples throughout this book (Hackshaw *et al.*, 1997). The meta-analysis includes data from 37 studies that reported on the relationship between 'second-hand' (passive) smoking and lung cancer. The paper reported that exposure to 'second-hand' smoke increased the risk of lung cancer by about 20 %. There was concern that these 37 studies could have been a biased sample of all studies, since studies which had found a statistically significant relationship would have been more likely to be published than those which had found no (statistically significant) evidence of a relationship. The intent of the publication bias analyses is to assess the likely extent of the problem, and to determine what conclusions can be drawn despite the potential for bias.

In the original paper the data are presented in the form of a risk ratio and confidence interval for each study. Some of the programs will accept the data in this format, but others require that the data be used to compute a point estimate and standard error (or variance), which would then be entered into the program. As is standard practice, all programs run the analyses (both the core meta analyses and the publication bias analyses) using the log of the risk ratio. The analyses in the following examples use the inverse variance formula and a fixed-effect model, but can be extended to other computational options and to the random-effects model.

THE SOFTWARE

Four computer programs are discussed in this chapter. These programs – Comprehensive Meta Analysis, Stata, RevMan Analyses, and Metawin – were included because they are in widespread use for meta-analysis, and offer some of the best options for addressing publication bias. Other programs, including SPSS and SAS, are discussed briefly toward the end of the chapter. This chapter focuses on *what* the programs can do. Detailed instructions for running the analyses are available on the website (see the appendix to this chapter).

Comprehensive Meta Analysis (version 2.0)

Comprehensive Meta Analysis (CMA) is a stand-alone program for meta-analysis. CMA was developed in collaboration with people working in medicine, epidemiology, and the social sciences. This program is able to create a forest plot and a funnel plot, to compute the rank correlation test, Egger's test, the failsafe N and Orwin's variant, trim and fill, and to display a cumulative forest plot sorted by precision (Borenstein *et al.*, 2005).

Data entry and conversion
CMA will accept data in close to 100 formats including the number of events and sample size in each group, means and standard deviations, correlations, or point estimates and confidence intervals. The passive smoking data, provided as risk ratios and confidence intervals, can therefore be pasted directly into the program.

Forest plot
Figure 11.1 is a forest plot produced by CMA. The program will display the data using either the *log* relative risk or the relative risk (as in this example). The graphical part of the plot (the point estimate and confidence interval) is displayed toward the center of the screen, with points to the right of 1.0 indicating an increased risk for persons exposed to passive smoking. The overwhelming majority of studies show an increased risk, but the 95 % confidence intervals for most studies include the null value of 1.0, and therefore fail to meet the 0.05 criterion for statistical significance. The last row in the spreadsheet shows the summary data for the fixed-effect model. The risk ratio is 1.204 and the 95 % confidence interval is from 1.120

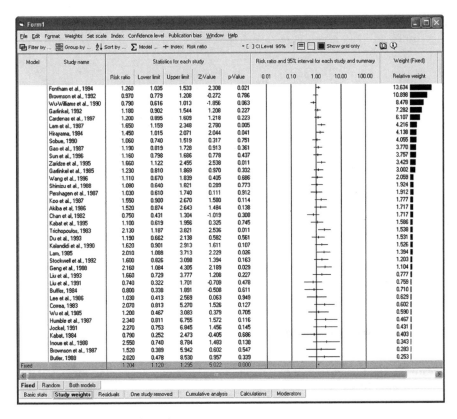

Figure 11.1 Forest plot (CMA).

to 1.295. Toward the left, the program shows the name and additional detail for each study. Toward the right, the program shows the relative weight assigned to each study.

In this plot, the studies have been listed from most precise to least precise, so that larger studies appear toward the top and smaller studies appear toward the bottom. This has no impact on the analysis, but allows us to get a sense of the relationship between sample size and effect size. Note that the point estimates become more widely dispersed but also shift toward the right as we move toward the bottom of the plot, where the smaller studies (note the wider confidence intervals) are located. This will be more evident in the funnel plot, which is designed to highlight this relationship.

Funnel plot

CMA will create a funnel plot of any effect size index on the X-axis by either the standard error (Figure 11.2) or precision (not shown) on the Y-axis. The program allows the user to include a vertical line at the summary effect, and guidelines for the 95 % confidence interval. In this example studies at the bottom are clustered

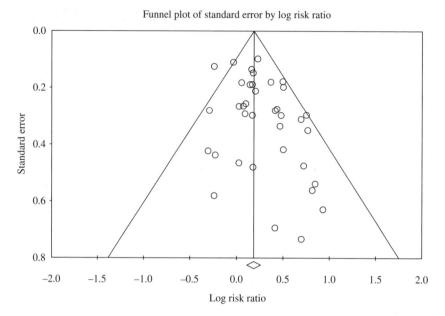

Figure 11.2   Funnel plot (CMA).

toward the right-hand side of the graph, this asymmetry suggesting the possibility of publication bias.

Statistical tests
Results for the various statistical tests are presented in tabular form by the program. For the rank correlation test, Kendall's tau is 0.144 with one-tailed $p = 0.102$. For Egger's test, the intercept (b) is 0.892, with a 95 % confidence interval from 0.127 to 1.657, and a one-tailed p-value of 0.024. The failsafe N is 398 using a one-tailed criterion (or 269 using a two-tailed criterion). Orwin's failsafe N is 103, assuming a mean risk ratio of 1.0 in the missing studies, with a 'trivial' effect defined as a risk ratio of 1.05.

The program creates a detailed text report summarizing all of these analyses. To save space, only the portion of this report which addresses trim and fill is shown in Figure 11.3.

Finally, the program redisplays the funnel plot, taking into account the trim and fill adjustment. Here, the observed studies are shown as open circles, and the observed point estimate in log units is shown as an open diamond at 0.185 (0.113, 0.258), corresponding to a risk ratio of 1.204 (1.120, 1.295). The seven imputed studies are shown as filled circles, and the imputed point estimate in log units is shown as a filled diamond at 0.156 (0.085, 0.227), corresponding to a risk ratio of 1.169 (1.089, 1.254). The 'adjusted' point estimate suggests a lower risk than the original analysis. Perhaps the key point, though, is that the adjusted estimate is fairly close to the original – in this context, a risk ratio of 1.17 has the same substantive implications as a risk ratio of 1.20.

Duval and Tweedie's Trim and Fill

If the meta analysis had captured all the relevant studies we would expect the funnel plot to be symmetric. If the funnel plot is actually asymmetric, with a relatively high number of small studies falling toward the right (representing a large treatment effect) and relatively few falling toward the left, we are concerned that these left-hand studies exist, but are missing from the analysis.

Duval and Tweedie developed a method that allows us to impute these studies. That is, we determine where the missing studies are likely to fall, add them to the analysis, and then recompute the combined effect. The method is known as 'Trim and Fill' as the method initially trims the asymmetric studies from the right side to identify the unbiased effect (in an iterative procedure), and then fills the plot by re-inserting the trimmed studies as well as their imputed counterparts.

In this case the method suggests that 7 studies are missing. Under the fixed effects model the point estimate and 95 % confidence interval for the combined studies is 1.204(1.120, 1.295). Using Trim and Fill the imputed point estimate is 1.169(1.089, 1.254).

Figure 11.3   Portion of text report created by CMA.

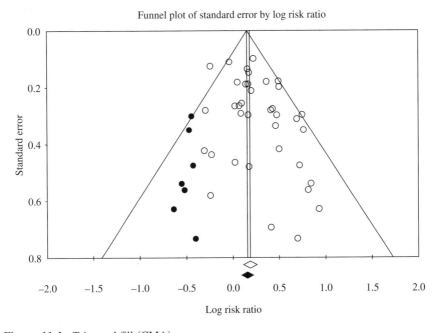

Figure 11.4   Trim and fill (CMA).

Cumulative forest plots

Figure 11.5 is a cumulative forest plot produced by CMA. Note the difference between the cumulative plot and the standard version shown earlier. Here, the

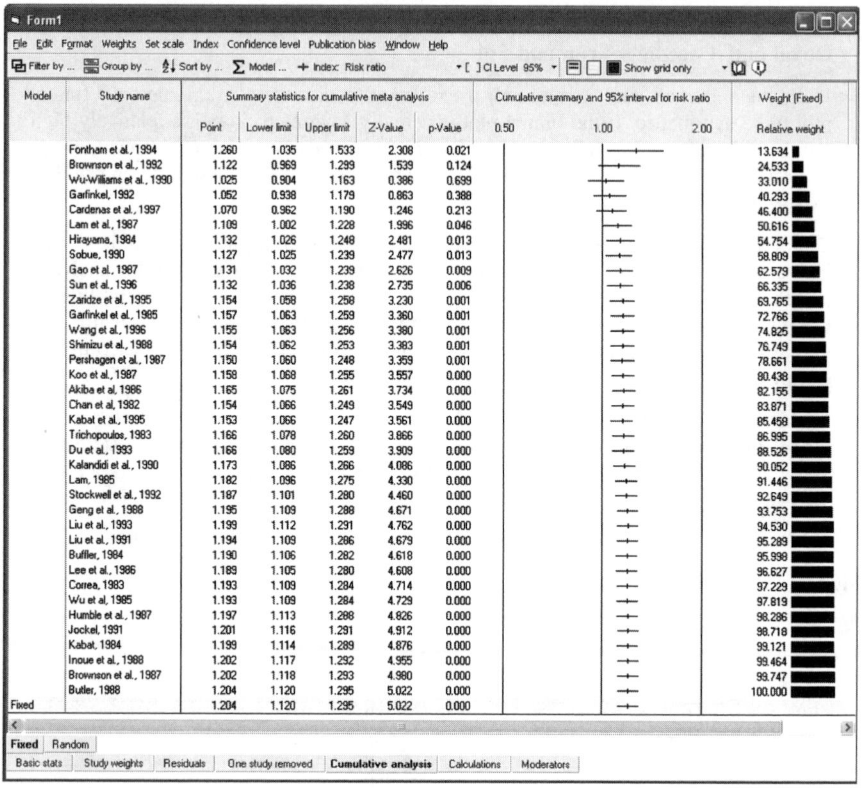

Model	Study name	Point	Lower limit	Upper limit	Z-Value	p-Value	Cumulative summary and 95% interval for risk ratio	Relative weight
	Fontham et al., 1994	1.260	1.035	1.533	2.308	0.021		13.634
	Brownson et al., 1992	1.122	0.969	1.299	1.539	0.124		24.533
	Wu-Williams et al., 1990	1.025	0.904	1.163	0.386	0.699		33.010
	Garfinkel, 1992	1.052	0.938	1.179	0.863	0.388		40.293
	Cardenas et al., 1997	1.070	0.962	1.190	1.246	0.213		46.400
	Lam et al., 1987	1.109	1.002	1.228	1.996	0.046		50.616
	Hirayama, 1984	1.132	1.026	1.248	2.481	0.013		54.754
	Sobue, 1990	1.127	1.025	1.239	2.477	0.013		58.809
	Gao et al., 1987	1.131	1.032	1.239	2.626	0.009		62.579
	Sun et al., 1996	1.132	1.036	1.238	2.735	0.006		66.335
	Zaridze et al., 1995	1.154	1.058	1.258	3.230	0.001		69.765
	Garfinkel et al., 1985	1.157	1.063	1.259	3.360	0.001		72.766
	Wang et al., 1996	1.155	1.063	1.256	3.380	0.001		74.825
	Shimizu et al., 1988	1.154	1.062	1.253	3.383	0.001		76.749
	Pershagen et al., 1987	1.150	1.060	1.248	3.359	0.001		78.661
	Koo et al., 1987	1.158	1.068	1.255	3.557	0.000		80.438
	Akiba et al., 1986	1.165	1.075	1.261	3.734	0.000		82.155
	Chan et al., 1982	1.154	1.066	1.249	3.549	0.000		83.871
	Kabat et al., 1995	1.153	1.066	1.247	3.561	0.000		85.458
	Trichopoulos, 1983	1.166	1.078	1.260	3.866	0.000		86.995
	Du et al., 1993	1.166	1.080	1.259	3.909	0.000		88.526
	Kalandidi et al., 1990	1.173	1.086	1.266	4.086	0.000		90.052
	Lam, 1985	1.182	1.096	1.275	4.330	0.000		91.446
	Stockwell et al., 1992	1.187	1.101	1.280	4.460	0.000		92.649
	Geng et al., 1988	1.195	1.109	1.288	4.671	0.000		93.753
	Liu et al., 1993	1.199	1.112	1.291	4.762	0.000		94.530
	Liu et al., 1991	1.194	1.109	1.286	4.679	0.000		95.289
	Buffler, 1984	1.190	1.106	1.282	4.618	0.000		95.998
	Lee et al., 1986	1.189	1.105	1.280	4.608	0.000		96.627
	Correa, 1983	1.193	1.109	1.284	4.714	0.000		97.229
	Wu et al., 1985	1.193	1.109	1.284	4.729	0.000		97.819
	Humble et al., 1987	1.197	1.113	1.288	4.826	0.000		98.286
	Jockel, 1991	1.201	1.116	1.291	4.912	0.000		98.718
	Kabat, 1984	1.199	1.114	1.289	4.876	0.000		99.121
	Inoue et al., 1988	1.202	1.117	1.292	4.955	0.000		99.464
	Brownson et al., 1987	1.202	1.118	1.293	4.980	0.000		99.747
	Butler, 1988	1.204	1.120	1.295	5.022	0.000		100.000
Fixed		1.204	1.120	1.295	5.022	0.000		

Fixed Random

Basic stats | Study weights | Residuals | One study removed | **Cumulative analysis** | Calculations | Moderators

Figure 11.5   Cumulative forest plot (CMA).

first row is a 'meta'-analysis based only on the Fontham *et al.* study. The second row is a meta-analysis based on two studies (Fontham *et al.* and Brownson *et al.*), and so on. The last study to be added is Butler, and so the point estimate and confidence interval shown on the line labeled 'Butler' are identical to that shown for the summary effect on the line labeled 'Fixed'.

The studies have been sorted from the most precise to the least precise (roughly corresponding to largest to smallest). With the 18 largest studies in the analysis (inclusive of Chan) the cumulative relative risk is 1.15. With the addition of another 19 (smaller) studies, the point estimate shifts to the right, and the relative risk is 1.20. As such, our estimate of the relative risk has increased. This could be due to publication bias or it could be due to small-study effects. Again, the key point is that even if we had limited the analysis to the larger studies, the relative risk would have been 1.15 (1.07, 1.25) and the clinical implications probably would have been the same.

Note also that the analysis that incorporates all 37 studies assigns nearly 84 % of its weight to the first 18 (see the bar graph in the right-hand column). In other words, if small studies are introducing a bias, we are protected to some extent by the fact that small studies are given less weight.

Stata (version 8.2)

Stata is a general purpose statistical package. While Stata itself does not include routines for meta-analysis, macros written by experts in meta-analysis are freely available (they can be located and downloaded by accessing the web from within the program) and provide a level of functionality comparable to the better stand-alone programs. With these macros in place Stata is able to create a forest plot and a funnel plot, to compute the rank correlation test, Egger's test, trim and fill, and to display a cumulative forest plot sorted by precision. Instructions for downloading, installing, and using the macros are on this book's website.

Data entry and conversion

Since the various macros were developed by different people, they tend to be somewhat idiosyncratic in the formats accepted for effect size data. Most will accept the number of events and sample size in each group, means and standard deviations, a point estimate and confidence interval, or a point estimate and standard error. The passive smoking data can therefore be pasted directly into the program.

Forest plot

Figure 11.6 is a forest plot produced by Stata. The user can elect to display the data using either the *log* relative risk or the relative risk (as in this example). Points to the right of 1.0 indicate an increased risk for persons exposed to passive smoking. The overwhelming majority of studies show an increased risk, but the 95 % confidence

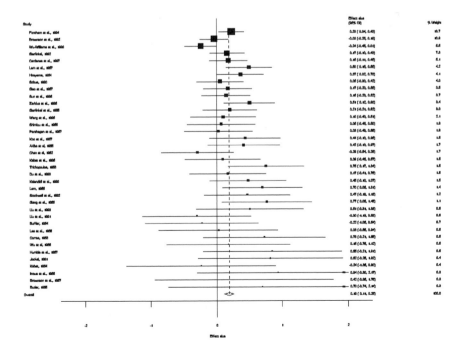

Figure 11.6 Forest plot (Stata).

intervals for most studies include the null value of 1.0, and therefore fail to meet the criterion for statistical significance. The last row in the spreadsheet shows the summary data for the fixed-effect model. The risk ratio is 1.204 and the 95% confidence interval is from 1.120 to 1.295.

In this plot, the studies have been listed from most precise to least precise, so that larger studies appear toward the top and smaller studies appear toward the bottom. This has no impact on the analysis, but allows us to get an initial sense of the relationship between sample size and effect size. Note that the point estimates become more widely dispersed but also shift toward the right as we move toward the bottom of the plot, where the smaller studies (note the wider confidence intervals) are located. This will be more evident in the funnel plot, which is designed to highlight this relationship.

Funnel plot

Stata will create a funnel plot of effect size on the *X*-axis by standard error on the *Y*-axis. In Figure 11.7, where larger studies are plotted at the top and smaller ones at the bottom, the latter cluster toward the right, suggesting the possibility of publication bias. There are several options within Stata for creating funnel plots, the most flexible being the metafunnel macro (Sterne and Harbord, 2004) which was used to create this figure as well as the figures in Chapter 5. This macro allows the user to customize the funnel plot by reversing the scale, and by including guidelines and annotations. It will also display studies from different subgroups in different colors/symbols, and will include the regression line corresponding to the Egger test.

For Kendall's test Stata reports a corrected *z*-value of 1.24 and *p*-value of 0.214. It reports Egger's test as 0.8922, standard error 0.3767, $t = 2.37$, $p = 0.024$, CI (0.127, 1.657).

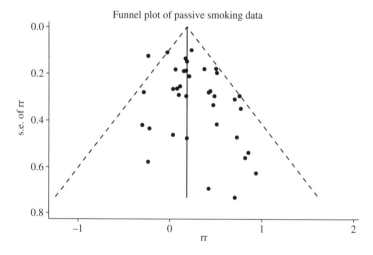

Figure 11.7   Funnel plot (Stata).

Duval and Tweedie's trim and fill

Stata runs the trim and fill algorithm and produces the results shown in Figure 11.8. In this example the trimming was based on the fixed-effect model. The original point estimate in log units had been 0.185, corresponding to a risk ratio of 1.20. The adjusted point estimate in log units is 0.156, corresponding to a risk ratio of 1.17. Perhaps the key point, though, is that the adjusted estimate is fairly close to the original – in this context, a risk ratio of 1.17 has the same substantive implications as a risk ratio of 1.20. The macro will allow the user to select any of the three trimming algorithms discussed in Duval and Tweedie's original paper (Duval and Tweedie, 2000a, 2000b) – linear, quadratic, and run. The results shown here are for the linear model, which is the default model in Stata and which Duval (personal communication) has reported as the most robust. The program also creates a funnel plot, with original points shown as circles and imputed points shown as circles encased in squares (see Figure 11.8).

Cumulative forest plot

Stata will produce a cumulative forest plot as shown in Figure 11.9. Note the difference between this cumulative forest plot and the standard version shown earlier. Here, the first row is a 'meta'-analysis based only on the Fontham *et al.* study. The second row is a meta-analysis based on two studies (Fontham *et al.* and Brownson *et al.*), and so on. The last study to be added is Butler, and so the point estimate and confidence interval shown on that line are identical to that shown for the summary effect.

The studies have been sorted from the most precise to the least precise (roughly corresponding to largest to smallest). With the 18 largest studies in the analysis (inclusive of Chan) the cumulative relative risk is 1.15. With the addition of another

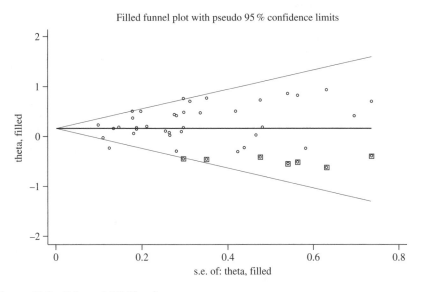

Figure 11.8 Trim and fill (Stata).

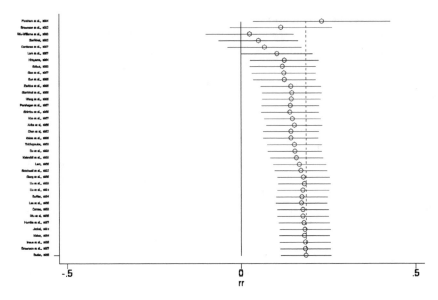

Figure 11.9 Cumulative forest plot (Stata).

19 (smaller) studies, the point estimate shifts to the right, and the relative risk is 1.20. As such, our estimate of the relative risk has increased. This could be due to publication bias or it could be due to small-study effects. Again, the key point is that even if we had limited the analysis to the larger studies, the relative risk would have been 1.15 (1.07, 1.25) and the clinical implications probably would have been the same.

This report used the procedures meta (Sharp, Sterne), metan (Bradburn, Deeks, Altman), metabias (Steichen, Egger, Sterne), metatrim (Steichen) metafunnel (Sterne and Harbord), metacum (Sterne), and meta-dialog (Steichen). Technical details for the Stata meta-analysis macros are given in Bradburn *et al.* (1999a, 1999b), Sharp and Sterne (1998a, 1998b, 1999), StataCorp. (2003), Steichen (1998, 2001a, 2001b, 2001c), Steichen *et al.* (1998, 1999), Sterne (1998), and Sterne and Harbord (2004). Perhaps the most comprehensive resource for using Stata in meta-analysis is Sterne *et al.* (2001a). The book that includes this chapter (Egger *et al.*, 2001) can be found at www.systematicreviews.com, and the chapter itself is available online as a PDF file at http://www.blackwellpublishing.com/medicine/bmj/systreviews/pdfs/chapter18.pdf.

Review Manager (version 4.2)

Review Manager (RevMan) is software provided by the Cochrane Collaboration for researchers whose reviews are intended for inclusion in the Cochrane Library. RevMan's statistical procedures are encapsulated in a module called RevMan Analyses (formerly MetaView) and invoked from within the larger program. RevMan will create a forest plot and a funnel plot but will not run any of the other procedures discussed in this chapter. Details for this program are provided in Clarke and Oxman, 2003.

Data entry and conversion

RevMan will accept data in two formats (the number of events and sample size in each group, means and standard deviations), as well as a generic format (point estimate and standard error). RevMan will not accept the passive smoking data in the format provided (risk ratios and confidence intervals), so the researcher would need to compute the log risk ratio and its standard error using another program such as Excel and then enter these data into RevMan.

Forest plot

Figure 11.10 is a forest plot produced by RevMan. The graphical part of the plot (the point estimate and confidence interval) is displayed toward the center of the screen, with points to the right of 1.0 indicating an increased risk for persons exposed to passive smoking. The overwhelming majority of studies show an increased risk, but the 95 % confidence intervals for most studies include the null value of 1.0,

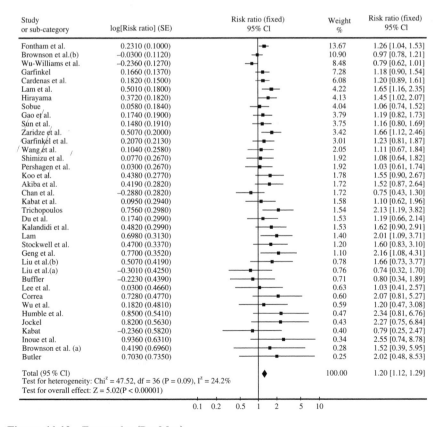

Study or sub-category	log[Risk ratio] (SE)	Risk ratio (fixed) 95% CI	Weight %	Risk ratio (fixed) 95% CI
Fontham et al.	0.2310 (0.1000)		13.67	1.26 [1.04, 1.53]
Brownson et al.(b)	−0.0300 (0.1120)		10.90	0.97 [0.78, 1.21]
Wu-Williams et al.	−0.2360 (0.1270)		8.48	0.79 [0.62, 1.01]
Garfinkel	0.1660 (0.1370)		7.28	1.18 [0.90, 1.54]
Cardenas et al.	0.1820 (0.1500)		6.08	1.20 [0.89, 1.61]
Lam et al.	0.5010 (0.1800)		4.22	1.65 [1.16, 2.35]
Hirayama	0.3720 (0.1820)		4.13	1.45 [1.02, 2.07]
Sobue	0.0580 (0.1840)		4.04	1.06 [0.74, 1.52]
Gao et al.	0.1740 (0.1900)		3.79	1.19 [0.82, 1.73]
Sun et al.	0.1480 (0.1910)		3.75	1.16 [0.80, 1.69]
Zaridze et al.	0.5070 (0.2000)		3.42	1.66 [1.12, 2.46]
Garfinkel et al.	0.2070 (0.2130)		3.01	1.23 [0.81, 1.87]
Wang et al.	0.1040 (0.2580)		2.05	1.11 [0.67, 1.84]
Shimizu et al.	0.0770 (0.2670)		1.92	1.08 [0.64, 1.82]
Pershagen et al.	0.0300 (0.2670)		1.92	1.03 [0.61, 1.74]
Koo et al.	0.4380 (0.2770)		1.78	1.55 [0.90, 2.67]
Akiba et al.	0.4190 (0.2820)		1.72	1.52 [0.87, 2.64]
Chan et al.	−0.2880 (0.2820)		1.72	0.75 [0.43, 1.30]
Kabat et al.	0.0950 (0.2940)		1.58	1.10 [0.62, 1.96]
Trichopoulos	0.7560 (0.2980)		1.54	2.13 [1.19, 3.82]
Du et al.	0.1740 (0.2990)		1.53	1.19 [0.66, 2.14]
Kalandidi et al.	0.4820 (0.2990)		1.53	1.62 [0.90, 2.91]
Lam	0.6980 (0.3130)		1.40	2.01 [1.09, 3.71]
Stockwell et al.	0.4700 (0.3370)		1.20	1.60 [0.83, 3.10]
Geng et al.	0.7700 (0.3520)		1.10	2.16 [1.08, 4.31]
Liu et al.(b)	0.5070 (0.4190)		0.78	1.66 [0.73, 3.77]
Liu et al.(a)	−0.3010 (0.4250)		0.76	0.74 [0.32, 1.70]
Buffler	−0.2230 (0.4390)		0.71	0.80 [0.34, 1.89]
Lee et al.	0.0300 (0.4660)		0.63	1.03 [0.41, 2.57]
Correa	0.7280 (0.4770)		0.60	2.07 [0.81, 5.27]
Wu et al.	0.1820 (0.4810)		0.59	1.20 [0.47, 3.08]
Humble et al.	0.8500 (0.5410)		0.47	2.34 [0.81, 6.76]
Jockel	0.8200 (0.5630)		0.43	2.27 [0.75, 6.84]
Kabat	−0.2360 (0.5820)		0.40	0.79 [0.25, 2.47]
Inoue et al.	0.9360 (0.6310)		0.34	2.55 [0.74, 8.78]
Brownson et al. (a)	0.4190 (0.6960)		0.28	1.52 [0.39, 5.95]
Butler	0.7030 (0.7350)		0.25	2.02 [0.48, 8.53]
Total (95 % CI)			100.00	1.20 [1.12, 1.29]

Test for heterogeneity: Chi² = 47.52, df = 36 (P = 0.09), I² = 24.2%
Test for overall effect: Z = 5.02(P < 0.00001)

0.1   0.2   0.5   1   2   5   10

Figure 11.10   Forest plot (RevMan).

and therefore fail to meet the criterion for statistical significance. The last row in the spreadsheet shows the summary data for the fixed-effect model. The risk ratio is 1.20 and the 95 % confidence interval is from 1.12 to 1.29. Toward the left, the program shows the name and log risk ratio (SE) for each study. Toward the right, the program shows the relative weight assigned to each study, the risk ratio and confidence interval.

In this plot, the studies have been listed from most precise to least precise, so that larger studies appear toward the top and smaller studies appear toward the bottom. This has no impact on the analysis, but allows us to get an initial sense of the relationship between sample size and effect size. Note that the point estimates become more widely dispersed but also shift toward the right as we move toward the bottom of the plot, where the smaller studies (note the wider confidence intervals) are located. This will be more evident in the funnel plot, which is designed to highlight this relationship.

Funnel plot
RevMan Analyses will display a funnel plot of effect size on the *x*-axis by standard error on the *y*-axis (Figure 11.11). It allows the user to include a vertical line at the summary effect, and guidelines for the 95 % confidence interval as shown. In RevMan these are based on the fixed-effect model. In this example studies toward the bottom of the plot tend to cluster toward the right, suggesting the possibility of publication bias.

Aside from the funnel plot, RevMan Analyses is not able to run any of the publication bias procedures nor to create a cumulative forest plot. The best option

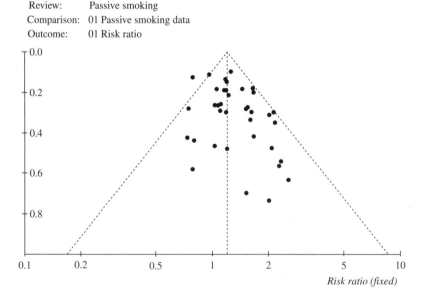

Figure 11.11 Funnel plot (RevMan).

for running additional analyses would therefore be to export the data to one of the other programs. RevMan Analyses does not support cut-and-paste of data using the Windows© clipboard, but does allow the user to export data to an ASCII (text) file. Step-by-step instructions are provided on the website.

MetaWin (version 2.0)

MetaWin is a stand alone program for meta-analysis. It was developed by people working in the field of ecology, but its features are general enough to have application in most other fields as well. MetaWin is able to create a forest plot and a funnel plot. It will compute the rank correlation test, the failsafe N and Orwin's variant, and display a cumulative forest plot sorted by precision (see Rosenberg *et al.*, 2000).

Data entry and conversion
MetaWin will accept data in three formats (the number of events and sample size in each group, means and standard deviations, correlations), as well as a generic format (point estimate and variance). MetaWin will not accept the passive smoking data in the format provided (risk ratios and confidence intervals), so the researcher would need to compute the log risk ratio and its variance using another program such as Excel and then cut and paste these data into MetaWin.

Forest plot
The forest plot (Figure 11.12) has been sorted according to each study's precision, with larger studies appearing toward the bottom and smaller studies appearing

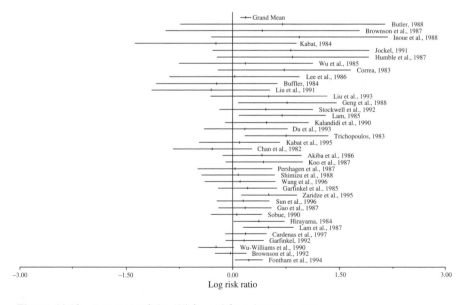

Figure 11.12   Forest plot (MetaWin), read from bottom to top.

toward the top (note that this is the reverse of the other programs). This has no impact on the analysis, but allows us to get an initial sense of the relationship between sample size and effect size. The summary effect of 0.186 (0.113, 0.259), corresponding to a risk ratio of 1.204 (1.120, 1.295) is shown at the top, and labeled 'Grand mean'. Note that the point estimates become more widely dispersed but also shift toward the right as we move toward the top of the plot, where the smaller studies (note the wider confidence intervals) are located. This will be more evident in the funnel plot, which is designed to highlight this relationship.

Funnel plot
In MetaWin the funnel plot (Figure 11.13) is shown with the effect size on the *Y*-axis and the variance (as in this example) or the sample size on the *X*-axis. In this example smaller studies (which appear toward the right) are clustered toward the top (representing larger effects). MetaWin does not offer the option of including a line to denote the summary effect, nor guidelines for the 95 % confidence interval. MetaWin will also let the user create a scatter plot of any two continuous variables, so it would be possible to plot the effect size against any other index of precision that the researcher had included in the data sheet.

Statistical tests
The failsafe *N* is given as 397.3. For computing Orwin's failsafe *N*, MetaWin assumes the mean value in the missing studies is zero but the user is allowed to define the value of a negligible effect. If this is defined as a log odds ratio of 0.05, Orwin's failsafe *N* is computed as 103, assuming a risk ratio of 1.0 in missing studies and a defining a trivial effect as a risk ratio of 1.05 (log risk ratio of 0.049).

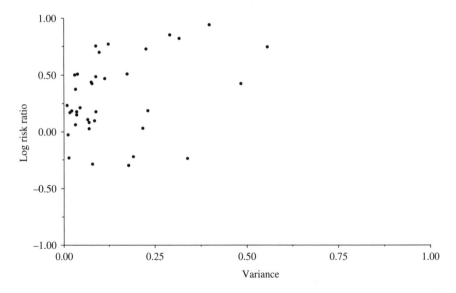

Figure 11.13 Funnel plot (MetaWin).

MetaWin reports Kendall's tau as 0.144, $z = 1.256$, p (one-tailed) $= 0.21$. MetaWin will not computer Egger's test of the intercept.

Cumulative forest plot

MetaWin will produce a cumulative forest plot as shown in Figure 11.14. While the program does include various options for sorting the data, these will not work with the cumulative forest plot. Therefore, the user must ensure that the studies have been entered into the data sheet in the sequence of most precise to least precise studies in order for this option to work.

As was true for the basic forest plot, MetaWin plots the studies from the bottom up. The meta-analysis for the first study (Fontham *et al.*) is omitted (since it would include only one study) and so the bottom row shows the analysis for the first two studies (Fontham *et al.* and Brownson *et al.*). The second row from the bottom is a meta-analysis based on three studies, and so on. The last study to be added is Butler, at the top.

The studies have been sorted from the most precise to the least precise (roughly corresponding to largest to smallest). With the 18 largest studies in the analysis (inclusive of Chan) the cumulative relative risk is 1.15 (the graph shows this in log units). With the addition of another 19 (smaller) studies, the point estimate shifts to the right, and the relative risk is 1.20. As such, our estimate of the relative risk has increased. This could be due to publication bias or it could be due to small-study effects. Perhaps the key point, though, is that even if we had limited the analysis to the larger studies, the relative risk would have been 1.15 (1.07, 1.25) and the clinical implications probably would have been the same.

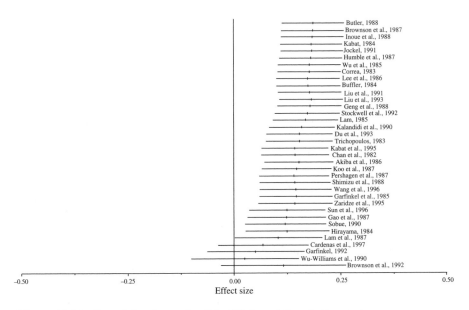

Figure 11.14 Cumulative forest plot (MetaWin), read from bottom to top.

Summary of the findings for the illustrative example

The various statistical procedures approach the problem of bias from a number of directions. One would not expect the results of the different procedures to 'match' each other since the procedures ask different questions. Rather, the goal should be to synthesize the different pieces of information provided by the various procedures.

Is there evidence of bias?

The funnel plot is noticeably asymmetric, with a majority of the smaller studies clustering to the right of the mean. This impression is confirmed by Egger's test which yields a statistically significant p-value. The rank correlation test did not yield a significant p-value, but this could be due to the low power of the test, and therefore does not contradict the Egger test. The body of evidence suggests that smaller studies reported a higher association between passive smoking and lung cancer than did the larger studies.

Is it possible that the observed relationship is entirely an artifact of bias?

The failsafe N is 398, suggesting that there would need to be nearly 400 studies with a mean risk ratio of 1.0 added to the analysis, before the cumulative effect would become statistically non-significant. Similarly, Orwin's failsafe N is 103, suggesting that there would need to be over 100 studies with a mean risk ratio of 1.0 added to the analysis before the cumulative effect would become trivial (defined as a risk ratio of 1.05). Given that the authors of the meta-analysis were able to identify only 37 studies that looked at the relationship of passive smoking and lung cancer, it is unlikely that nearly 400 studies (Rosenthal) or even 103 studies (Orwin) were missed. Therefore, while the actual risk of passive smoking may be smaller than reported, it is unlikely to be zero.

What is the risk ratio adjusted for bias?

We can attempt to classify the bias as falling into one of three categories: (a) the bias, if any, is probably trivial; (b) the bias is probably non-trivial, but we can have confidence in the substantive conclusions of the meta-analysis; or (c) the bias is potentially severe enough to call the substantive findings into question.

In this example the meta-analysis based on all studies showed that passive smoking was associated with a 20 % increase in risk of lung cancer. The cumulative meta-analysis showed that the increased risk in the larger studies was 16 %. Similarly, the trim and fill method suggested that if we trimmed the 'biased' studies, the increased risk would be imputed as 15 %. As such, this meta-analysis seems to fall squarely within category (b). There is evidence of bias, and the risk is probably somewhat lower than reported, but there is no reason to doubt the validity of the core finding, that passive smoking is associated with a clinically important increase in the risk of lung cancer.

WHERE DO WE GO FROM HERE?

As noted, the exaggerated effect in the small studies could be due to publication bias but could also be due to a small-study 'bias'. If the latter, this 'bias' reflects a source of heterogeneity among studies which, like any other heterogeneity, should be investigated.

COMPARING THE PROGRAMS

Table 11.1 shows the publication bias procedures incorporated in the four programs. As indicated in this table, three of the four programs will carry out most of the publication bias procedures. Aside from this, however, the programs have surprisingly little in common, regarding their approaches to data entry and to the core meta-analysis, and even the publication bias procedures are quite different. This section is meant to give the reader a better sense of the 'look and feel' of the four programs, and as such is bound to be subjective. Therefore, I want to remind the reader of the conflict of interest mentioned at the outset.

Comprehensive Meta Analysis (2.0) offers the advantage of being able to perform a wide range of tasks required for the analysis. It will accept data in virtually any format, automatically compute the effect size, and (where needed) convert data from different formats to a common effect size index. CMA allows the researcher to run the meta-analysis with one click, but allows the advanced user to select

Table 11.1   Publication bias procedures in meta-analysis programs.

	CMA (2.0)	Stata (8.2)	RevMan (4.2)	MetaWin (2.0)
Data entry formats				
Effect size and standard error	•	•	•	•
Events and sample size/non-events	•	•	•	•
Means and standard deviation	•	•	•	•
Point estimate and confidence interval	•	•		
Get a sense of the data				
Forest plot	•	•	•	•
Funnel plot	•	•	•	•
Is there evidence of bias?				
Begg and Mazumdar's rank correlation	•	•		•
Egger's regression	•	•		•
Can effect be entirely an artifact of bias?				
Failsafe N	•			•
Orwin's failsafe N	•			•
What is the adjusted effect size?				
Duval and Tweedie's trim and fill	•	•		
Cumulative forest plot	•	•		•
Miscellaneous				
Text report with detailed narrative	•			

from an array of computational options. The program features advanced options for graphics, including the ability to extensively customize the forest plot (including the use of proportionally sized symbols) and to export high-resolution plots to other programs such as Word and PowerPoint. The interface is similar to that of programs such as Excel™, so that the learning curve is minimal, and the functions are intuitive.

Stata (8.2) is a general purpose statistical program. To use Stata for meta-analysis the researcher would incorporate a series of macros (code segments) that have been written by experts in meta-analysis. These macros can be located and downloaded by accessing the web from within Stata. Stata differs from the other three programs in that it is command-driven (i.e., the user writes code, or uses a dialog box to create code), rather than being directly menu-driven, and therein lie both its advantages and disadvantages. For someone who is familiar with Stata, and who is relatively expert in meta-analysis, Stata offers the advantage that the user has complete control over every element of the analysis, from the computation of effect size to the precise weighting of each study. Code can be written from scratch, the provided macros can be viewed and modified, and all code can be saved and reused with other data sets. Someone who is not familiar with Stata, on the other hand, would need to learn the basic Stata language before proceeding to meta-analysis, and would encounter problems working with data formats that were not automatically handled by the Stata macros. One disadvantage of Stata, even for the advanced user, is that the forest plot may be difficult to read and relatively little customization is currently possible (see Figure 11.6, for example).

RevMan Analyses (4.2) is meant to function as part of a larger package whose goal is not only to run the meta-analysis but also to track other aspects of the systematic review and to ensure that the data can be uploaded to the Cochrane Library. The advantage of this program is that it is compatible with the Cochrane Library. The disadvantage is that data entry is not intuitive, and is limited to only a few formats (2×2 tables, means and standard deviations, and a generic point estimate and standard error). While the forest plot can be customized only slightly, the default format is well designed and incorporates all key data. Anyone working with the Cochrane Collaboration will need to enter their data into RevMan, so the question for these people is whether or not to use another program as well. If they want to produce higher quality graphics, or to run additional publication bias analyses, they should consider exporting the data to CMA or to Stata. RevMan does not allow copy and paste of data, but it does allow the user to export the entire data file in ASCII format, and this file can then be read by either CMA or Stata (the appendix to this chapter lists a website with instructions).

MetaWin 2.0 has taken the same general approach as CMA in that it offers a spreadsheet interface with menus and toolbars. However, MetaWin offers a much more limited version of this approach. The data input will work with only a few formats. The forest plot is relatively difficult to read and offers few options for customizing. A strength of this program is that it allows the user to work with the response ratio, an effect size index that is used in ecology and is not found in any of the other programs. The program also provides resampling methods (bootstrap and randomization) for Q-statistics and confidence intervals of cumulative effect sizes.

Computational accuracy

In choosing among programs, accuracy of results is not an issue. The programs all give the same results, both for the core meta-analyses and also for the publication bias analyses. The literature includes two versions of Egger's test, the original using a weighting scheme and the later version (at Egger's suggestion) omitting the weight. All programs that include this test use the newer version of the formula. Two of the programs reviewed here compute the failsafe N, and they both use the Stouffer method (Stouffer *et al.*, 1949), which is the one originally used by Rosenthal. Note, however, that the programs will yield the same results only if they are using the same options, and that different programs may use different computational options as the default. For example, the default in CMA is to use a two-tailed test for failsafe N, while the default in MetaWin is to use a one-tailed test. CMA offers a button to set all the defaults to match those used in RevMan. None of the programs implements the selection models discussed by Hedges and Vevea (Chapter 9).

Other programs

The discussion in this chapter does not include all programs used for meta-analysis. SPSS and SAS are general purpose statistical packages which have often been pressed into service for meta-analysis by researchers who write their own code or who incorporate the macros written by David Wilson (Lipsey and Wilson, 2001). Additional detail for these programs, including instructions for downloading the macros, are included on the book's website. These programs were not included in the discussion because there are no widely available macros to address publication bias (see also Wang and Bushman, 1999; Arthur *et al.*, 2001).

ACKNOWLEDGEMENTS

This chapter, this volume, and the computer program Comprehensive Meta Analysis, were funded by the following grants from the National Institute on Aging: Publication bias in meta-analysis (AG20052) and Combining data types in meta-analysis (AG021360).

The author would like to express his appreciation to the National Institute on Aging, and in particular to Dr. Sidney Stahl for making this endeavor possible.

The term 'Illustrative example' used in this chapter was coined by my mentor and friend, Jacob Cohen.

REFERENCES

Arthur, W., Bennett, W. and Huffcutt, A.J. (2001). *Conducting Meta-analysis Using SAS*. Mahwah, NJ: Lawrence Erlbaum Associates.

Begg, C.B. (1994). Publication bias. In H.M. Cooper and L.V. Hedges (eds), *The Handbook of Research Synthesis*. New York: Russell Sage Foundation.

Begg, C.B. and Berlin, J.A. (1988). Publication bias: A problem in interpreting medical data. *Journal of the Royal Statistical Society, Series A*, **151**, 419–463.

Begg, C.B. and Mazumdar, M. (1994). Operating characteristics of a rank correlation test for publication bias. *Biometrics*, **50**, 1088–1101.

Borenstein, M., Hedges, L., Higgins, J. and Rothstein, H. (2005) *Comprehensive Meta Analysis, Version 2*. Englewood, NJ: Biostat.

Bradburn, M.J., Deeks, J.J. and Altman, D.G. (1999a). Metan – an alternative meta-analysis command. *Stata Technical Bulletin Reprints*, **8**, 86–100.

Bradbury, M.J., Deeks, J.J. and Altman, D.G. (1999b). Correction to funnel plot. *Stata Technical Bulletin Reprints*, **8**, 100.

Cooper, H.M. (1979). Statistically combining independent studies: A meta-analysis of sex differences in conformity research. *Journal of Personality and Social Psychology*, **37**, 131–146.

Clarke, M. and Oxman, A.D. (eds) (2003). Cochrane Reviewers' Handbook 4.2.0 [updated March 2003]. In *The Cochrane Library, Issue 2, 2003*. Oxford: Update Software. Updated quarterly.

Carson, K.P., Schriesheim, C.A. and Kinicki, A.J. (1990). The usefulness of the 'fail-safe' statistic in meta-analysis. *Educational and Psychological Measurement*, **50**, 233–243.

Duval, S.J. and Tweedie, R.L. (1998). Practical estimates of the effect of publication bias in meta-analysis. *Australasian Epidemiologist*, **5**, 14–17.

Duval, S. and Tweedie, R. (2000a) A nonparametric 'trim and fill' method of accounting for publication bias in meta-analysis. *Journal of the American Statistical Association*, **95**, 89–99.

Duval, S.J. and Tweedie, R.L. (2000b) Trim and fill: A simple funnel-plot-based method of testing and adjusting for publication bias in meta-analysis. *Biometrics*, **56**, 455–463.

Egger, M., Davey Smith, G., Schneider, M. and Minder, C. (1997). Bias in meta-analysis detected by a simple, graphical test. *British Medical Journal*, **315**, 629–634.

Egger, M., Davey Smith, G., and Altman, D. (eds) (2001). *Systematic Reviews in Health Care: Meta-analysis in Context*. London: BMJ.

Hackshaw, A.K., Law, M.R. and Wald, N.J. (1997). The accumulated evidence on lung cancer and environmental tobacco smoke. *British Medical Journal*, **315**, 980–988.

Lau, J., Schmid, C.H. and Chalmers, T.C. (1995). Cumulative meta-analysis of clinical trials: builds evidence for exemplary medical care. *Journal of Clinical Epidemiology*, **48**, 45–57.

Light, R. & Pillemer, D. (1984). *Summing Up: The Science of Reviewing Research*. Cambridge, MA: Harvard University Press, Cambridge.

Light, R.J., Singer, J.D. and Willett, J.B. (1994). The visual presentation and interpretation of meta-analyses. In H.M. Cooper and L.V. Hedges (eds), *The Handbook of Research Synthesis*. New York: Russell Sage Foundation

Lipsey, M.W. and Wilson, D.B. (2001) *Practical Meta-analysis*. Thousand Oaks, CA: Sage.

Orwin, R.G. (1983). A fail-safe *N* for effect size in meta-analysis. *Journal of Educational Statistics*, **8**, 157–159.

RevMan Analyses (2002) [Computer program]. Version 1.0 for Windows. In Review Manager (RevMan) 4.2. Oxford, England. The Cochrane Collaboration.

Rosenberg, M.S., Adams, D.C. and Gurevitch, J. (2000) *MetaWin: Statistical Software for Meta-Analysis Version 2.0*. Sunderland, MA: Sinauer Associates, Inc.

Rosenthal, R. (1979). The 'file-drawer problem' and tolerance for null results. *Psychological Bulletin*, **86**, 638–641.

Sharp, S.J. and Sterne, J. (1998a). Meta-analysis. *Stata Technical Bulletin Reprints*, **7**, 100–106.

Sharp, S.J. and Sterne, J. (1998b). New syntax and output for the meta-analysis command. *Stata Technical Bulletin Reprints*, **7**, 106–108.

Sharp, S.J. and Sterne, J. (1999). Corrections to the meta-analysis command. *Stata Technical Bulletin Reprints*, **8**, 84.

StataCorp. (2003). Stata Statistical Software: Release 8.0. College Station, TX: StataCorp LP.

Steichen, T.J. (1998). Tests for publication bias in meta-analysis. *Stata Technical Bulletin Reprints*, **7**, 125–133.

Steichen, T.J. (2001a). Update of tests for publication bias in meta-analysis. *Stata Technical Bulletin Reprints*, **10**, 70.

Steichen, T.J. (2001b). Tests for publication bias in meta-analysis: erratum. *Stata Technical Bulletin Reprints*, **10**, 71.

Steichen, T.J. (2001c). Update to metabias to work under version 7. *Stata Technical Bulletin Reprints*, **10**, 71–72.

Steichen, T.J., Egger, M. and Sterne, J. (1998) Modification of the metabias program. *Stata Technical Bulletin*, **STB-44**(sbe19.1), 3–4.

Steichen, T.J., Egger, M. and Sterne, J. (1999). Tests for publication bias in meta-analysis. *Stata Technical Bulletin Reprints*, **8**, 84–85.

Sterne, J. (1998). Cumulative meta-analysis. *Stata Technical Bulletin Reprints*, **7**, 143–147.

Sterne, J. A. & Egger, M. (2001). Funnel plots for detecting bias in meta-analysis: guidelines on choice of axis, *Journal of Clinical Epidemiology*, **54**, 1046–1055.

Sterne, J.A. and Harbord, R.M. (2004) Funnel plots in meta analysis. *Stata Journal*, **4**, 127–141.

Sterne, J.A.C., Gavaghan, D. and Egger, M. (2000) Publication and related bias in meta-analysis: Power of statistical tests and prevalence in the literature. *Journal of Clinical Epidemiology*, **53**, 1119–1129.

Sterne, J. A., Bradburn, M. J., & Egger, M. (2001a). Meta-Analysis in Stata. In M. Egger, G. Davey Smith and D. G. Altman (eds.), *Systematic Reviews in Health Care: Meta-analysis in Context* (pp. 347–369). London: BMJ Books.

Sterne, J. A. C., Egger, M. and Davey Smith, G. (2001b). Investigating and dealing with publication and other biases. In M. Egger, G. Davey Smith and D. G. Altman (eds), *Systematic Reviews in Health Care: Meta-analysis in Context.*, 2nd edn. London: BMJ Books

Sterne, J. A., Egger, M., and Davey Smith, G. (2001c). Systematic reviews in health care: Investigating and dealing with publication and other biases in meta-analysis, *British Medical Journal*, **323**, 101–105.

Stouffer, S.A., Suchman, E.A., DeVinney, L.C., Star, S.A. and Williams, R.M., Jr. (1949). *The American Soldier: Adjustment during Army Life* (Vol. 1). Princeton, NJ: Princeton University Press.

Wang, M.C. and Bushman, B.J. (1999). *Integrating Results through Meta-analytic Review Using SAS Software*. Cary, NC: SAS Institute Inc.

APPENDIX

Updates and data sets

Updates to this chapter and also the data sets used as examples in this volume, will be posted on the book's website, http://www.meta-analysis.com/publication-bias.

Conflicts of interest

The development of Comprehensive Meta Analysis was funded by the National Institutes of Health in the United States. The editors of this book, as well as many of the chapter authors, have been involved in the development of that program, and some have a financial interest in its success. The author of this chapter is the lead developer on the program and has a financial interest in its success.

To help ensure the accuracy of the information in this chapter, developers and users of the various programs were asked to review a draft of this chapter. Jonathan Sterne, who collaborated on the development of the Stata macros and has written a chapter on using Stata for meta-analysis, provided feedback on Stata. Julian Higgins, a member of the RevMan Advisory Group, provided comments on RevMan and on Stata. Dean Adams, one of the developers of MetaWin, provided comments on that program. Alex Sutton, Hannah Rothstein, and Steven Tarlow provided extensive and very helpful comments on earlier drafts of this chapter.

Trademarks

Comprehensive Meta Analysis is a trademark of Biostat, Inc. MetaWin is a trademark of that program's developers. RevMan is a trademark of the Cochrane Collaboration. Stata is a trademark of StataCORP.

Contacts

Comprehensive Meta Analysis
Biostat, Inc., 14 North Dean Street, Englewood, NJ 07631, USA
Phone + 1 201 541 5688
Fax + 1 201 541 5526
MBorenstein@Meta-Analysis.com
www.Meta-Analysis.com

MetaWin

Sinauer Associates, Inc., 23 Plumtree Road, P.O. Box 407, Sunderland, MA 01375-0407, USA
Phone + 1 413 549 4300
Fax + 1 413 549 1118
www.metawinsoft.com

Stata

4905 Lakeway Drive, College Station, TX 77845, USA
Phone + 1 979 696 4600 (USA 800 782 8272)
Fax + 1 979 696 4601
www.Stata.com

RevMan

http://www.cochrane.org/software/download.htm

Advanced and Emerging Approaches

CHAPTER 12

Bias in Meta-Analysis Induced by Incompletely Reported Studies

Alexander J. Sutton
Department of Health Sciences, University of Leicester, UK

Therese D. Pigott
School of Education, Loyola University Chicago, USA

KEY ISSUES

- Incomplete data reporting may exist at various levels other than the suppression of whole studies. For example, specific outcomes, subgroups or covariates may not be available for meta-analysis.
- In some instances, it may be possible to use standard missing-data methods to address missing-data problems in meta-analysis. However, this will not always be possible because data will often not be missing at random – a necessary assumption of such methods.
- Specific sensitivity-type methods are starting to be developed to deal with such incomplete data reporting in meta-analysis, although these will often need adapting before they can be implemented in specific situations.
- There is a need for further development of specific methods to deal with incomplete study reporting in meta-analysis generally, and specifically a framework to deal simultaneously with data missing on different levels.

Publication Bias in Meta-Analysis – Prevention, Assessment and Adjustments   Edited by H.R. Rothstein, A.J. Sutton
and M. Borenstein   © 2005 John Wiley & Sons, Ltd

INTRODUCTION

The success of a meta-analysis depends on the quality and completeness of the primary studies the review intends to synthesize. In some applications of meta-analysis, the primary studies are essentially replications: similar methods, treatments and measures are collected and reported, facilitating the review. In other areas, a more heterogeneous range of methods, treatments and measures are employed. Complicating the review further, the studies may differ in how completely and thoroughly they report data.

In a sense, this entire book deals with one major source of missing data in a research synthesis, that of missing whole studies. This chapter will examine issues related to missing data within studies, specifically those related to incomplete and biased reporting in primary research. The chapter begins with an overview of possible reasons why data may be missing. For example, social scientists often use different methods or measures in a study; not all of the primary authors report all the details that a reviewer finds important. In medicine, though the primary authors may use similar methods, they may fail to report information about treatment failures (selective reporting of individuals), about subgroup analyses, or about particular outcomes measured in the study. Empirical evidence is beginning to emerge on the last of these issues, and this is reviewed. We then provide a brief review of current methods for missing data generally, and discuss their utility in handling typical missing-data issues in meta-analysis. Then the chapter considers novel methods to deal with biased or incomplete reporting in primary studies. A discussion, including thoughts on needed future developments, concludes this chapter.

LEVELS OF MISSING STUDY DATA

Before we consider the reasons and mechanisms why data from studies may be missing, and potential approaches to dealing with the missingness, a categorization of the different levels of potentially missing data from a study is presented.

- *Missing whole studies.* Absolutely no results from a study exist, and hence, often the study is not even known by the systematic reviewer/meta-analyst.
- *Missing outcomes.* Studies often collect and analyse data on more than one outcome measure. However, not all of these outcomes may be reported.
- *Missing subgroup-specific analyses.* Investigators may do secondary analyses on the data they collect. For example, in a clinical trial, there may be interest in whether the intervention under trial works equally well for all subjects, or whether there are subgroups of subjects for whom differential effects can be established. Obviously if many such analyses are conducted there may be a temptation to only report the 'interesting' subgroups.
- *Missing study data required for the meta-analysis.* Data may not be provided in study reports required for the meta-analysis. Examples of such data include missing standard errors of the effect size outcome measure, and missing study or subject covariates.

- *Missing data on individual study subjects.* In some medical meta-analyses, the reviewer analyses the individual-level subject data from the primary study. Data may be missing or excluded from analyses on these individual study units. The missing data may include outcome data, or other data such as subject characteristics. Missing data on individual subjects are a less common occurrence in the social sciences than in medicine since individual-level data are rarely used in a meta-analysis in the social science context. If it were possible to obtain the raw data from the individual studies – for example, the scores on reading assessments for each child in a given study – then the reviewer could use the raw data to obtain effect size estimates that would then be used in a meta-analysis. This type of meta-analysis is much more common in medical applications, and is called individual patient data meta-analysis (see Chapter 14).

Illustration of levels of missing data in a meta-analysis

The different levels of missing data possible in a meta-analysis can be illustrated in a review of phonics programmes by Ehri *et al.* (2001). Since the reviewers focused only on studies published in peer-reviewed journals, we could conceptualize the review as missing whole studies in the form of unpublished reports, doctoral dissertations, or published studies in non-peer-reviewed outlets. The review examined various measures of reading outcomes presented in the studies. While most studies presented findings from word identification tasks (reading words out of context), only a subset of studies report results on decoding tasks or spelling assessments. It is plausible that primary study authors did collect results on decoding and spelling tasks but chose not to report them, resulting in missing outcomes in the analysis. Missing subgroup analyses may also apply to this review. For example, only half of the studies in the review report on outcomes analysed by the socioeconomic status of the child. A related issue is missing study data required for the analysis. Ehri *et al.* indicate that they planned on examining how characteristics of the phonics programme related to effect size, but that insufficient data were given in most reports to sustain this analysis.

Empirical evidence of outcome-reporting bias in medical randomized controlled trials

Although little attention has been given to the problem of incomplete reporting of studies relative to that given to selective publication of whole studies, extensive empirical investigations into the problem of incomplete and biased outcome reporting in medical randomized controlled trials (RCTs) have been published (Chan *et al.*, 2004a, 2004b). Both these studies used the same approach by identifying RCTs through Danish ethical committees (Chan *et al.*, 2004a) and those approved for funding by the Canadian Institutes of Health Research (Chan *et al.*, 2004b) and comparing the protocols of these trials with all available published reports of the trials. Fifty per cent of efficacy and 65 % of harm outcomes were not reported completely enough in the Danish trials to enable inclusion in a meta-analysis. The corresponding figures in the Canadian trials were 31 % and 59 %. Importantly, there was also evidence of biased as statistically significant outcomes had higher odds of being fully reported compared to non-significant ones (efficacy Danish trials odds ratio $= 2.4$, 95% CI (1.4, 4.0); harm Danish trials odds ratio $= 4.7$, 95% CI (1.8, 12.0); efficacy Canadian trials odds ratio $= 2.7$, 95% CI (1.5, 5.0)). Such serious findings should bring more attention to the potential biases induced by incomplete outcome reporting in the future.

REASONS FOR MISSING DATA IN A META-ANALYSIS

Missing data occur in a meta-analysis when a primary author fails to report information that a reviewer wishes to use. Missing data could be at one or more of the levels specified above. As in any data analysis, we can imagine a variety of mechanisms that could cause missing data. It is helpful to consider these mechanisms as they will help guide any strategy for dealing with the missingness. Three mechanisms that could cause missing data are discussed below under established missing-data terminology (Rubin, 1976).

Missing completely at random

We might assume that the missing information in a study is missing completely at random (MCAR), so that the data available at each level are actually a random subsample of all the data that exists.

What types of missing meta-analytic data are more likely to be MCAR? In many research reviews, primary studies are not strict replications of one another. We might assume that decisions made by primary authors about what to include in a study report approximate a random sampling mechanism. However, this mechanism may not be easy to imagine. Much early meta-analysis research demonstrated a bias in the published literature towards statistically significant results (much of this literature is reviewed in Chapter 2). It is not difficult to foresee that much incomplete reporting may be less innocent and follow a similar pattern.

Missing at random

A second classification of missing data is described as missing at random (MAR). Conceptually, MAR data occur when the reasons for the missing data depend only on completely observed data in a sample rather than on the value of the missing observations themselves. For example, a review in the social sciences often includes primary studies whose authors have differing disciplinary affiliations. Sociologists may differ systematically from psychologists in how they design, measure and report a study. Academic discipline may be related to missing information on subjects' academic ability, for example, if psychologists tend to report ability information whereas sociologists tend to report academic achievement. If this relationship were true, and academic discipline were known, it would be reasonable to assume academic ability and academic achievement could be assumed MAR.

While theoretically we can devise examples of MAR data, we can never gather empirical evidence that our missing data in a meta-analysis are MAR. There is some theoretical research on possible response mechanisms operating in meta-analysis. For example, Orwin and Cordray (1985) first discussed a number of reporting practices that may operate in the social sciences, an example of how meta-analytic data could be MAR. More recently, Lipsey and Wilson (2001) describe many intervention studies as having 'personalities' or clusters of interrelated substantive and methodological characteristics that may be confounded with study outcomes. If reporting practices hold true in certain disciplines, or if clusters of studies with similar 'personalities' exist, then missing data in these clusters or groups of studies could be MAR.

MAR data may not seem at first to hold much promise for missing data in a meta-analysis. However, when missing data are MAR, a reviewer could use maximum likelihood and multiple imputation techniques as alternatives to analysing only the complete cases. Assuming MAR data allows a reviewer to examine the sensitivity of analyses to differing hypotheses about the reasons for missing data.

Not missing at random

The more likely reasons for missing data result from a primary author's biased reporting of results. For example, primary authors may fail to report information on treatment failures, making the outcome data for the treatment successes more positive than they should be. In this situation, the missing data are not missing at random (NMAR); the missing data are missing because the individual did not progress through the treatment, and would be likely to score low on measures of treatment effectiveness.

Methods for dealing with data missing by these three mechanisms are considered below. It is crucial to know that it is impossible to test the assumptions made by the three different missing-data mechanisms for a particular data set. Due to this difficulty, it may often be sensible to err on the side of caution and assume data are NMAR, a topic that will be taken up later in the chapter. In Box 12.1 the likely reasons why data are missing for each level of missing data and the implications for such missingness, as outlined above, are considered.

Box 12.1   Likely reasons why data are missing for each level of missing data

Missing whole studies
If whole studies are missing, much evidence reviewed elsewhere in this book would suggest that this is likely to occur because study results were not statistically significant or were uninteresting. Since usually no details about these studies are available, NMAR is the only reasonable assumption that can be made. However, to complicate matters further, since we often do not know how many studies are missing, the extent of this type of missing data is not known. The implications of missing whole studies are considered extensively throughout this book, and empirical evidence concerning the consequences is reviewed in Chapter 10.

Box 12.1   (Continued)

Missing outcomes

Outcomes may be missing because only those which are interesting or statistically significant are selected for publication. Alternatively, secondary outcomes may be omitted simply due to lack of journal space or reported in a format not compatible with the other studies included in the meta-analysis. In a clinical setting, adverse event outcomes are particularly prone to selective reporting. Hence, situations potentially exist where outcomes are MCAR, MAR (since other covariates and outcomes of the study will be known and could be predicted on the basis of the set of fully observed variables) or NMAR. Hence, unless MCAR, missing outcomes may potentially lead to biased pooled effect size estimates in a very similar way to missing whole studies, and are thus a serious form of missing data.

Missing subgroup-specific analyses

Similar to missing outcomes, we often can assume that subgroup analyses may not be reported due to space considerations; or more likely, that they are not reported because their results were not interesting. Again MCAR, MAR or NMAR are all potentially possible. If subgroups are explored in a meta-analysis, and data are not MCAR, then pooled effect sizes are likely to be biased and thus this is potentially a very serious problem.

Missing study data required for the meta-analysis

It is easy to foresee that data such as covariates and standard errors may not be published due to poor reporting practice, or lack of journal space. Hence MCAR and MAR are possible and arguably more realistic than for the other data types discussed above. That is not to say that some selection process that cannot be modelled by available subject covariates is acting and the mechanism is, in fact, NMAR. If the data that are missing are required for meta-analysis, for example standard errors of effect sizes, and these are NMAR, then this will have serious implications for a meta-analysis because pooling only the studies where data are available will potentially produce biased results. If the data missing are less essential, for example study-level moderator variables, then this will limit the analyses that can be performed, such as subgroup analyses or meta-regression.

Missing data on individual study subjects

Data may be missing on study subjects for a number of reasons. It could be due to seemingly random processes such as accidental loss of data. However, there may exist more systematic reasons why data are missing. One important example exists in the clinical trial setting where subjects are randomized into two different treatments. If a subject stops taking their treatment during the trial, perhaps counterintuitively, these persons should still be included in the statistical analysis. This is called an intention-to-treat analysis and gives a better estimate of an intervention in routine practice since treatment failures are an aspect of 'real life'. However, sometimes such patients are excluded from the analysis, which will often have the effect of making the treatment with subject attrition seem more beneficial on average than is actually the case as all those who stopped the treatment have not been accounted for. Hence MCAR, MAR or NMAR are potentially possible when data are missing on individual study subjects.

STANDARD MISSING-DATA METHODS AND WHEN THEY CAN AND CANNOT BE APPLIED TO META-ANALYSIS

Currently used methods for missing data can be divided into two sets: (1) *ad hoc*, simple-to-implement methods; and (2) model-based methods often using maximum likelihood techniques. In general, the more easily implemented methods do not apply to all forms of missing data, and require stronger assumptions than model-based methods. *Ad hoc* methods that could be applied to missing data in meta-analysis include complete-case analysis, available-case analysis and single-value imputation. Model-based methods employing maximum likelihood include the use of the EM algorithm and multiple imputation.

Simple-to-implement methods

Complete-case analysis is probably the most commonly used and most easily implemented method. All commercially available statistical analysis packages utilize complete-case analysis as the default procedure; if any cases are missing even a single variable, that case is deleted from the analysis. Complete-case analysis provides unbiased estimates only when the data are missing completely at random. With MCAR data, the complete cases are, in fact, a random sample of the originally identified data.

Another commonly employed method involves replacing a missing value with a 'likely' value such as the complete-case mean for some variable. Many statistical packages such as SPSS (1999) provide mean substitution as a strategy for handling missing data. Little (1992) points out, however, that mean imputation leads to underestimating the variance in a particular variable. Filling in all the missing values with exactly the same value (i.e., the mean) does not replicate the variance that would be present if these missing values were observed; it is unlikely that each missing value would take on exactly the same value if they were observed. This underestimation of the variance is a problem because it leads to an underestimation of standard errors, and inferences based on them will be biased. In addition, filling in the missing values allows the use of all of the data, a larger sample than the complete cases. Larger samples have smaller variances, all things being equal.

Many other strategies have been employed to address the problem of replacing a single value for missing observations on a variable. Buck (1960) suggested a regression technique, using predicted values for the missing values given the other variables observed in a particular case. Hot-deck techniques are another example; 'similar' cases are used to impute the missing value in a random way. As Sinhary *et al.* (2001) point out, all single-imputation methods fail to account for the uncertainty in the missing data and will underestimate the true variability in the data set.

A third method for treating missing data is available-case analysis, or pairwise deletion. Many statistical packages provide this method as an option. In available-case analysis, all relevant data are used, so that means of variables are computed on all cases that observe that variable, and not just on those cases with complete data on all variables. While available-case analysis does maximize the amount of data used in the analysis, different subsets of the data provide estimates of the parameters of interest. For example, two different subsets of studies may provide cases for estimating the means of two variables, while a third subset provides cases for the covariance of these two variables. Available-case analysis does produce unbiased estimates if each subset of data is MCAR. However, Fahrbach (2001) studied the performance of available-case methods for missing data in meta-analysis through simulations and found the procedure much less efficient than maximum likelihood methods discussed below.

When data are MAR, Little and Rubin (1987) indicate that Buck's (1960) method provides consistent estimates of the means particularly when the data are normally distributed. Buck's method fills in the missing values using regression estimates; the missing values are estimated from the predicted values for a regression of the missing variable on the other variables in the data. The sample covariance matrix from the filled-in data using Buck's method underestimates the variances and covariances. In general, the more easily implemented methods will not produce unbiased estimates when the data are MAR.

Model based methods

Model-based methods such as maximum likelihood and multiple imputation are alternative methods for missing data. Both of these methods provide valid and unbiased estimates when the data are MAR, a less restrictive assumption than MCAR. Both methods assume that the data are multivariate normal. Space considerations do not permit a full treatment of these methods. Little and Rubin (1987) and Schafer (1997) discuss maximum likelihood and multiple imputation methods for missing data. The application of these methods for missing data on covariates in meta-analysis is given in detail elsewhere (Pigott, 2001).

When standard methods can and cannot be used

Complete-case analysis for missing data may be adequate if data are MCAR. When data are MAR, maximum likelihood and multiple imputation are the preferred methods. Thus, missing observations at any of the levels described in Box 12.1 could employ complete-case analysis when data are MCAR, and maximum likelihood or multiple imputation when data are MAR. These analyses procedures also apply when outcomes or subgroup analyses are missing. With missing effect sizes, often the standard errors of the effect sizes will need to be assumed MAR, and, for example, imputed from the data.

Although data missing on individual subjects may be MCAR or MAR, application of standard methods may be limited if summary statistics based on individual subjects are being considered. Meta-analyses of individual subject data are not considered at length in this chapter, since their benefits with respect to publication biases are discussed in Chapter 14. It is worth noting here that standard missing-data techniques could be applied to each data set (in a standardized way) in a meta-analysis of individual subject-level data before combining results. Additionally, individual subject data deliberately missing could also be reinstated allowing intention-to-treat analyses to be performed. See Lambert *et al.* (1997) for an application of a Bayesian hierarchical missing-data model to an individual subject-level meta-analysis.

Before moving on to methods developed specifically to deal with missing data due to incomplete reporting, it is worth spending a few moments considering exactly why we may wish to employ methods to deal with missing data. There are two quite distinct reasons that should be made explicit. Firstly, if data is MCAR then, as explained above, the missingness will not introduce bias into the analysis if only complete-case analysis is carried out. In such instances, the only benefits of employing missing-data methods are related to improved efficiency in the estimation of the quantities of interest. Since this chapter is on bias associated with missing data, such issues are technically beyond its scope but included for completeness. Secondly, and usually more importantly, if the missing data are not MCAR, complete-case analysis will often be biased and missing-data methods should be employed to try and correct for the selection procedure.

As discussed earlier, missing data in meta-analysis often result from incomplete or biased reporting, a mechanism that is NMAR. In such instances the standard missing-data methods described above do not fully address the mechanisms by which data are missing, and hence the application of such methods cannot account for the associated bias. The remainder of this chapter considers methods that have been developed to address data missing through biased reporting specifically in a meta-analysis context. Each level of missing data (as outlined above) is considered in turn below.

DEALING WITH SELECTIVELY MISSING WHOLE STUDIES

In essence, the whole of Part B of this book considers the statistical techniques that have been developed to deal with missing whole studies and thus they will not be considered here. This is the area in which most research has been carried out to date. Note that it may be possible to apply the methods developed for missing whole studies to other levels of missing data; for example the use of trim and fill (see Chapter 8) is considered for assessing selectively missing study outcomes below. Selection modelling (see Chapter 9) may also be a strategy for dealing with missing data at other levels of a meta-analysis.

DEALING WITH SELECTIVELY MISSING STUDY OUTCOMES

Hutton and Williamson (2000) develop a sensitivity analysis method which considers patterns of missing data across multiple outcomes to assess the impact of potential multiple testing of outcomes in a study followed by selective reporting based on size of effect and statistical significance. For each potentially reported outcome, different adjustments are made for studies in which: the outcome of interest is not reported; and the outcome of interest is reported but other outcomes are not reported, suggesting selectivity may have taken place.

The adjustments made are described in detail elsewhere (Hutton & Williamson, 2000) but are based on the premise that the most extreme, with respect to statistical significance, of a fixed number of identically distributed independent outcomes are reported. Under these (probably unrealistically strict) assumptions, inferences can be estimated regarding the degree of overestimation in the suspected selectively reported outcomes and the expected magnitude in the unreported outcomes conditional on those reported.

Although the assumptions made in this sensitivity analysis are restrictive and rather extreme, they do allow the assessment of whether conclusions are likely to be robust to the missing data. Hence, if conclusions are threatened by such an analysis, then this adds strength to the time-consuming notion of trying to acquire missing outcome data directly (in summary or in the form of individual subject data (Chapter 14)) from the original investigators.

A sensitivity analysis by the approach of Hutton and Williamson (2000) considers patterns of missing data across multiple outcomes. Another way to address the problem is to consider each outcome separately, and use methods for addressing publication bias in whole studies. An illustrative example of this is given below.

Example of aortic rupture

The data set of interest comes from a systematic review of success rates of a surgical procedure to repair a ruptured abdominal aortic aneurysm over a 50-year period (Bown *et al.*, 2002). The literature base was large, with 171 study reports included in the meta-analysis. One of the aims of this analysis was to assess if survival had improved over time since the operation was introduced in the 1950s. Several outcomes were of interest, including overall mortality (up to 30 days after surgery) and mortality in surgery. Mortality during surgery consists of a subset of the total mortality events. Both outcomes were of interest, since dissecting location of death would help attribute any improvements in mortality over time to improved procedure in the operating theatre or to postoperative intensive care. It was possible to extract an overall mortality rate in all 171 studies, but mortality during surgery was only reported in 77 of these 171 reports. There is a concern that likelihood of reporting mortality during surgery may be related to outcome, and hence an outcome-reporting bias exists.

Funnel plots (Chapter 5) for both outcomes plotted on the log odds scale (defined as $\log_e$(number of deaths/(total number of patients − number of deaths)) are shown in Figure 12.1(a)–(b). Interpretation of these funnels would appear to be relatively straightforward (in part due to the large number of studies they include). The plot for overall mortality would appear to be symmetric, while that for mortality in surgery would appear highly asymmetric with a large 'gap' in the bottom right-hand corner relating to small and moderately sized studies in which mortality was high. Hence, there is a suspicion that surgeons with high mortality rates in surgery

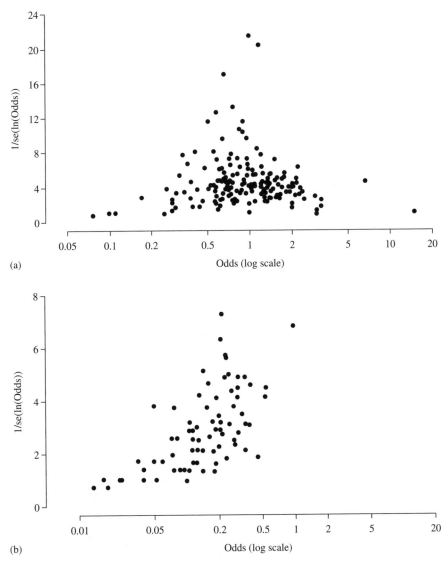

(a)

(b)

Figure 12.1 (a) Funnel plot of total mortality for all 171 studies included in analysis; (b) mortality during surgery for 77 studies which reported it; (c) intra-operative mortality showing those studies in the analysis (black filled circles, $n = 77$) and, in addition, those predicted to be missing by the trim and fill method (open triangles, $n = 29$).

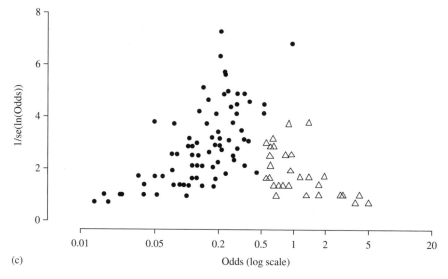

(c)

Figure 12.1 (Continued).

selectively chose not to disclose these figures and hence a meta-analysis of the published literature may underestimate true levels of mortality during surgery.

A trim and fill analysis (see Chapter 8) is carried out on the surgery mortality funnel, although any of the methods described in Part B could be applied. Applying the L_0 estimator to this data set using a random effect model, we estimate that 29 studies are missing (see Figure 12.1(c)) which changes the surgery mortality rate from 15 % (95 % CI (13 %, 17 %) to 20 % (18 %, 23 %). Hence it is concluded that mortality is likely to be underestimated by around 5 % due to outcome reporting bias.

We note that although this analysis is informative, perhaps it is using the information available suboptimally. Using 'standard' methods for publication bias means that information on the other outcomes is being ignored (other than to add weight to the conclusion that the funnel asymmetry observed is indeed due to publication bias rather than other influences). In this example, we know there are (at least) 171 studies, and due to the nature of the outcome, mortality during surgery exists for all of them. Additionally, we know the total number of patients in each study, and it would appear reasonable that some relationship between total (or post-surgery) mortality and mortality in surgery may exist which could be used to help predict missing outcomes. At the time of writing both custom imputation and selection models (Chapter 9) were being explored to better utilize the data (Jackson *et al.*, 2005).

DEALING WITH SELECTIVELY MISSING SUBGROUPS OF INDIVIDUALS

Hahn *et al.* (2000) develop an approach for dealing with selectively reported subgroup analyses similar in spirit to that of Hutton and Williamson (2000) for selective outcome reporting. Since potentially many subgroups could be compared there is the possibility that only those which are most 'interesting' or statistically significant are reported. The method developed is intended to be used as a form of sensitivity analysis and involves data imputation for missing subgroup data under the assumption that the results which are not published tend to be non-significant. Details of the method are reported elsewhere for studies with a binary endpoint (Hahn *et al.*, 2000), but the general approach can be summarized as follows. For each study in which subgroup analyses have not been reported, limits on the range of possible values they could take can be derived and analyses imputing all possible values for each study in the meta-analysis can be conducted. Following this, those data sets which give $p > 0.05$ for the subgroups of interest are selected (i.e., it is assumed they would be published if $p < 0.05$). One can obtain an estimate of the pooled effect size by reanalysing the meta-analysis using each of these selected imputed data sets in turn, and averaging the results over all data sets as if all studies had reported the subgroups of interest.

As for outcomes, potential bias in reported subgroups could also be explored using 'standard' funnel plot methods, but, for reasons discussed above, the utilization of data would be suboptimal and methods would lack power unless many studies were included in the meta-analysis.

DEALING WITH MISSING STUDY DATA REQUIRED FOR THE META-ANALYSIS

When certain pieces of data are missing, such as effect size standard errors or study-level covariates, it may be necessary to use *ad hoc* methods as a sensitivity analysis to assess whether the omission of such data is likely to threaten the conclusions of the meta-analysis, although as discussed previously the assumption that such data are MCAR or MAR may be reasonable and standard missing-data methods may suffice in some situations.

Historically, when effect sizes are missing, the idea of imputing the value of zero for each one has been advocated (Whitley *et al.*, 1986) as a sensitivity analysis method to check the robustness of the results to missing data (assumptions about the standard error may also have to be made). This procedure has similarities with the idea behind the 'file drawer' method (Chapter 7). When information such as the direction of effect and p-value are given, a researcher may instead calculate the range in which the effect must fall and carry out sensitivity analyses imputing extreme values.

Bushman and Wang (1995, 1996) describe more sophisticated methods to combine studies where data to calculate an effect size and its standard deviation are not available from all studies. These procedures combine information about study size, and the direction of the observed effect from the studies where no effect size is given, with effect sizes from studies from which they are known. Hence, the methods combine sample effect sizes and vote counts (Sutton *et al.*, 2000) to estimate the population mean effect. Such methods are efficient as they allow studies to be included in the meta-analysis even if data usually required are not available. It should be stressed that such methods assume missing data are MCAR and hence no bias mechanism is influencing which data are missing.

Abrams *et al.* (2000) consider a situation where a continuous outcome is being considered but in some studies responses are reported separately in terms of outcome measure and standard deviation for baseline and follow-up. The desired outcome is the difference between baseline and follow-up, but its standard deviation cannot be calculated since the correlation between individuals' responses is required but not reported. A sensitivity analysis approach is presented which imputes a range of values for the unknown correlations, and a Bayesian solution is presented which incorporates the uncertainty in the unknown correlation into the analysis.

Example when variances are not known

A good example of the use of *ad hoc* methods to assess the impact of missing data is by Song *et al.* (1993) who meta-analysed 53 RCTs which had measured depression change following treatment using a standard scale. Only 20 of these trials reported a standard deviation for the change outcome, meaning weights for a meta-analysis could not be calculated directly for 33 RCTs. Values were imputed for the missing standard deviations based on average values in those which did report it allowing all 53 studies to be meta-analysed. Since results for a meta-analysis of the 20 studies in which the standard deviations were known, and a meta-analysis of all 53 studies imputing standard deviations where necessary, produced similar results, this strengthened the conclusions the review reported.

DEALING WITH SELECTIVELY MISSING SUBJECTS

Little has been written on selectively missing subjects within individual studies. Clearly an individual subject data meta-analysis would provide most scope to assess this issue (see Chapter 14).

On the specific issue of wanting an intention-to-treat analysis, where only treatment compliers are reported, it may be possible to carry out an *ad hoc* sensitivity analysis. One would first impute extreme results for the treatment non-compliers in order to establish a range of possible values for the treatment effect in each study, and hence by meta-analysing the extremes, would obtain a range of values for the overall pooled estimate.

If one is working with binary outcomes in a clinical trials setting, sensitivity analysis methods have recently been proposed to assess the potential impact of missing-outcome data (Preston and Hollis, 2002; Higgins *et al.*, 2003).

DISCUSSION

This chapter has provided an overview of the problems of incomplete reporting of primary investigations for meta-analysis and potential methods to address such problems. Methods for missing data generally have a reasonably long history, but are of limited use in a meta-analysis context since assumptions about the missingness of the data required by such methods will often not be satisfied. Methods specific to meta-analysis do exist, but they are currently underdeveloped. It is hoped that as methods for dealing with missing whole studies become established, problems concerning missing data within studies will receive more attention. Indeed, the first empirical evidence of the problem of within-study selective reporting within clinical research is now available which should help to publicize the problem and need for solutions (Hahn *et al.*, 2002; Chan *et al.*, 2004a, b). However, from the work carried out so far, it would seem that there are issues with generalization across applications and often it will be necessary to take a somewhat *ad hoc* approach, with existing methods needing custom tailoring to specific applications.

In some applications it will be possible to ascertain that data are definitely missing, and hence it will be obvious that steps should be taken to assess this likely impact (although the best method of doing this may be far from obvious). In other instances, data may be suspected missing but this is not known for certain. For example, patterns of outcome or subgroup reporting may look suspicious, but it cannot be proven that selectivity has gone on (note this was not an issue in the aortic rupture example because a figure for mortality in theatre must exist – even if it is zero). It may be that certain outcomes were truly not measured in some studies. This raises the question whether it is desirable to treat all incomplete data as missing if there is uncertainty that they were ever measured/recorded. This question requires further consideration.

One area which has received minimal attention is the need for a unified approach to dealing with data missing at the different levels. For example, it can be envisaged that in a particular application area there may be missing whole studies, missing outcomes in some of the known studies, and individual patients may have been incorrectly excluded from the endpoints of a proportion of the known studies. A sensitivity analysis framework addressing the potential impact of all three types of missing data simultaneously would be valuable. Bayesian Markov chain Monte Carlo methods (Gilks *et al.*, 1996) may provide the flexibility required for the development of such a framework

ACKNOWLEDGEMENT

The authors would like to thank Dr David B. Wilson whose comments improved this chapter considerably.

REFERENCES

Abrams, K.R., Lambert, P.C., Sanso, B., Shaw, C. and Marteau, T.M. (2000). Meta-analysis of heterogeneously reported study results – a Bayesian approach. In D.K. Stangl and D.A. Berry (eds), *Meta-analysis in Medicine and Health Policy*. New York: Marcel Dekker.

Bown, M.J., Sutton, A.J., Nicholson, M.L. and Bell, R.F. (2002). A meta-analysis of 50 years of ruptured abdominal aortic aneurysm repair. *British Journal of Surgery*, **89**, 1–18.

Buck, S.F. (1960). A method of estimation of missing values in multivariate data suitable for use with an electronic computer. *Journal of the Royal Statistical Society, Series B*, **22**, 302–306.

Bushman, B.J. and Wang, M.C. (1995). A procedure for combining sample correlation coefficients and vote counts to obtain an estimate and a confidence interval for the population correlation coefficient. *Psychological Bulletin*, **117**, 530–546.

Bushman, B.J. and Wang, M.C. (1996). A procedure for combining sample standardized mean differences and vote counts to estimate the population standardized mean difference in fixed effect models. *Psychological Methods, 1*, 66–80.

Chan, A.-W., Hróbjartsson, A., Haahr, M.T., Gøtzsche, P.C. and Altman, D.G. (2004a). Empirical evidence for selective reporting of outcomes in randomized trials. Comparison of protocols to published articles. *Journal of the American Medical Association,* **291**, 2457–2465.

Chan, A.-W., Krleza-Jeric, K., Schmid, I. and Altman, D.G. (2004b) Outcome reporting bias in randomized trials funded by the Canadian Institutes of Health Research. *Canadian Medical Association Journal,* **17**, 735–740.

Ehri, L.C., Nunes, S., Stahl, S. and Willows, D. (2001). Systematic phonics instruction helps students learn to read: Evidence from the National Reading Panel's meta-analysis. *Review of Educational Research,* **71**, 393–448.

Fahrbach, K.R. (2001). An investigation of methods for mixed-model meta-anlaysis in the presence of missing data. Unpublished doctoral dissertation, Michigan State University.

Gilks, W.R., Richardson, S. and Spiegelhalter, D.J. (1996). *Markov Chain Monte Carlo in Practice.* London: Chapman & Hall.

Hahn, S., Williamson, P.R., Hutton, J.L., Garner, P. and Flynn, E.V. (2000). Assessing the potential for bias in meta-analysis due to selective reporting of subgroup analyses within studies. *Statistics in Medicine,* **19**, 3325–3336.

Hahn, S., Williamson, P.R. and Hutton, J.L. (2002). Investigation of within-study selective reporting in clinical research: Follow-up of applications submitted to a local research ethics committee. *Journal of Evaluation in Clinical Practice,* **8**, 353–359.

Higgins, J.P.T., White, I.R. and Wood, A.M. (2003) Dealing with missing outcome data in meta-analysis of clinical trials [abstract]. *Controlled Clinical Trials,* **24**, 88S.

Hutton, J.L. and Williamson, P.R. (2000). Bias in meta-analysis due to outcome variable selection within studies. *Applied Statistics,* **49**, 359–370.

Jackson, D., Copas, J. and Sutton, A. (2005). Modelling reporting bias: The operative reporting rate for ruptured abdominal aortic aneurysm repair. *Journal of the Royal Statistical Society, Series A* (in press)

Lambert, P.C., Abrams, K.R., Sanso, B. and Jones, D.R. (1997). Synthesis of incomplete data using Bayesian hierarchical models: An illustration based on data describing survival from neuroblastoma. Technical Report 97-03, Department of Epidemiology and Public Health, University of Leicester, UK.

Lipsey, M.W. and Wilson, D.B. (2001). The way in which intervention studies have 'personality' and why it is important to meta-analysis. *Evaluation and the Health Professions,* **24**, 236–254.

Little, R.J.A. (1992). Regression with missing X's: A review. *Journal of the American Statistical Association,* **87**, 1227–1237.

Little, R.J.A. and Rubin, D.B. (1987). *Statistical Analysis with Missing Data.* New York: John Wiley & Sons, Inc.

Orwin, R.G. and Cordray, D.S. (1985). Effects of deficient reporting on meta-analysis: A conceptual framework and reanalysis. *Psychological Bulletin,* **97**, 134–147.

Pigott, T.D. (2001). Missing predictors in models of effect size. *Evaluation and the Health Professions,* **24**, 277–307.

Preston, C. and Hollis, S. (2002) Method of investigating the effect of missing data on meta-analysis. Paper presented at the 4th Symposium on Systematic Reviews: Pushing the boundaries, Oxford, UK, July. Abstract available at http://www.ihs.ox.ac.uk/csm/pushingtheboundaries/poster/20_pres.html accessed May 2005.

Rubin, D.B. (1976). Inference and missing data. *Biometrika,* **63**, 581–592.

Rubin, D.B. (1987). Multiple imputation for nonresponse in surveys. New York: John Wiley.

Schafer, J.L. (1997). *Analysis of Incomplete Multivariate Data.* New York: Chapman & Hall.

Sinhary, S., Stern, H.S. and Russell, D. (2001). The use of multiple imputation for the analysis of missing data. *Psychological Methods,* **6**, 317–329.

Song, F., Freemantle, N., Sheldon, T.A., House, A., Watson, P. and Long, A. (1993) Selective serotonin reuptake inhibitors: Meta-analysis of efficacy and acceptability. *British Medical Journal*, **306**, 683–687.

SPSS (1999). *SPSS for Windows (Version 9.0)*. Chicago.

Sutton, A.J., Abrams, K.R., Jones, D.R., Sheldon, T.A. and Song, F. (2000) *Methods for Meta-analysis in Medical Research*. Chichester: John Wiley & Sons, Ltd.

Whitley Jr., B.E., McHugh, M.C. and Frieze, I.H. (1986). Assessing the theoretical models for sex differences in causal attributions of success and failure. In J.S. Hyde and M.C. Linn (eds), *The Psychology of Gender*. Baltimore, MD: Johns Hopkins University Press.

CHAPTER 13

Assessing the Evolution of Effect Sizes over Time

Thomas A. Trikalinos and John P.A. Ioannidis

Department of Hygiene and Epidemiology, University of Ioannina School of Medicine, Greece, and Tufts-New England Medical Center, USA

KEY POINTS

- Evidence is never static. Current knowledge may change in the light of newer evidence.
- When *time-lag bias* operates, the first published studies show systematically greater effect sizes compared to subsequently published investigations. Their summary effect size would then diminish over time, as data accumulate.
- However, such a waning temporal pattern in the summary effect size may be due to chance or genuine heterogeneity.
- One may empirically assess the anticipated uncertainty in the future evolution of the summary effect size when additional relevant studies emerge. The future uncertainty diminishes drastically as more studies accumulate.

INTRODUCTION

Meta-analyses of randomized controlled trials or epidemiological studies have recently emerged as a key source of evidence across medical disciplines (Chalmers and Lau, 1996; Lau *et al.*, 1995; Mosteller and Colditz, 1996). However, evidence is

---

Publication Bias in Meta-Analysis – Prevention, Assessment and Adjustments   Edited by H.R. Rothstein, A.J. Sutton and M. Borenstein   © 2005 John Wiley & Sons, Ltd

never static. New studies may be published and their results may agree or disagree with those of previously published research. Occasionally, later published studies fail to replicate early estimates in terms of the magnitude of effect sizes, formal statistical significance or even identification of the favoured arm (Hirschhorn *et al.*, 2002; Ioannidis *et al.*, 2001; Lohmueller *et al.*, 2003). In several cases the estimated effect sizes change considerably as more evidence accumulates (Ioannidis *et al.*, 1999).

We shall briefly discuss the factors that influence the evolution of effect sizes over time, using examples from randomized research and epidemiological studies, in particular genetic epidemiology. We shall also refer to methodologies that assess the differences between early and later published studies: evaluations of discrepancies between initial and subsequent evidence, cumulative meta-analysis, recursive cumulative meta-analysis and domain-wide graphs. For descriptive purposes, we shall analyse the data set on passive smoking and lung cancer in women and the teacher expectancy data set, along with other examples to illustrate the described methodologies.

FACTORS UNDERLYING THE EVOLUTION OF EFFECT SIZES OVER TIME

Sometimes initial studies show an impressive estimate of effect size that diminishes or vanishes in subsequent research. This waning pattern could be explained by operating biases (systematic errors that pose a threat to the validity of a meta-analysis); true heterogeneity in the effect size across various population subgroups; spurious findings due to chance; and, hopefully not often, flawed research. We shall briefly comment on these factors in the following paragraphs.

Time-lag bias

When time-lag bias operates, the time to completion and publication of a study are dependent on its results. Time to publication may be longer for statistically non-significant ('negative') studies, compared to formally significant ('positive') ones (Stern and Simes, 1997; Ioannidis, 1998). At least in randomized clinical trials, empirical evidence suggests that this could be attributed to investigator delay, rather than editorial preference (Dickersin *et al.*, 2002; Olson *et al.*, 2002). In molecular medicine, impressively strong genetic associations tend to be published earlier (Goring *et al.*, 2001; Ioannidis *et al.*, 2001). Theoretically, the net effect is the attenuation of the optimistically biased initial estimates towards more modest effect sizes.

Proteus phenomenon

In scientific fields with rapid data turnaround, the results of the first study may influence the publication order of subsequent studies. Highly contradictory findings are most tantalizing and attractive to investigators and editors, and may have an advantage for rapid publication. Thus, one often encounters in the early literature a succession of extreme opposite results. Studies with intermediate and potentially less spectacular findings may be published at a slower pace, filling the gap between the early extremes. This rapid, early succession of extreme findings has been called the *Proteus phenomenon* after the mythological god who rapidly metamorphosed himself into very different figures (Ioannidis and Trikalinos, 2005). As stated, the Proteus phenomenon is more likely to occur when data production is rapid and copious. This may be more typical of retrospective study designs where data can be produced rapidly. Conversely, this rapid succession of extremes is less likely in prospective research where studies take considerable time to perform and their conduct is spread over a longer time span. This bias highlights the importance of performing further validation research and of systematically appraising the evolving accumulated evidence on each research question.

Publication bias

Publication bias may be conceived as an extreme form of time-lag bias. 'Negative' studies that are never published may never have an opportunity to impact on the evolving summary effect size. Thus under conditions that favor the existence of publication bias, a relatively constant effect size over time does not guarantee that this effect is true. It may well be that a filter against the publication of 'negative' studies is continuously operating in the field. Sometimes a comprehensive meta-analysis manages to unearth the available unpublished information. The incorporation of these data in the calculations as the latest available evidence – along with other unpublished recent results – would diminish the previously postulated effect size.

True heterogeneity

Heterogeneity is common both in randomized (Lau *et al.*, 1998) and epidemiologic research (Ioannidis *et al.*, 2001, 2003; Lohmueller *et al.*, 2003), and meta-analysis can accommodate it (Lau *et al.*, 1998). Subpopulations (defined by 'race', gender, age or other parameters) that are at higher risk levels may obtain more benefit (or experience more harm) from an intervention than low-risk subjects (Glasziou and Irwig, 1995). When efficacy is time-limited, the treatment effect would be smaller when assessed at longer follow-up periods, in later published studies (Ioannidis *et al.*, 1997). Thus the different effect sizes encountered over time may be ascribed to true heterogeneity rather than biases. Meta-analysis of individual patient (or participant) data is a useful tool in modeling true heterogeneity, as risk is estimated at the individual level (Stewart and Clarke, 1995; Stewart and Parmar, 1993).

True heterogeneity could justify the lack of replication of initial estimates in meta-analyses, provided that in the initially examined subpopulation(s) the association is extreme. Usually, however, there is no clearly discernible reason for heterogeneity alone to determine the differential order of study publication and we would therefore expect no distinctive pattern in the evolution of the effect size in the majority of meta-analyses. Chapter 15 discusses in more detail the differentiation of biases from genuine heterogeneity.

Other biases

Other potential biases also pose a threat to meta-analyses: language bias (not identified reports in various non-English languages) (Jüni *et al.*, 2002) and database bias (reports not indexed in the searched databases) (Hopewell *et al.*, 2002) among others. Their potential impact on the evolution of effect sizes has not been studied. In addition, methodological flaws may yield peculiar results. Including studies of poor quality in a meta-analysis is an example (Moher *et al.*, 1998). Effect sizes may be inflated in investigations of low quality, and sometimes earlier studies may be qualitatively inferior to more recent ones. Moreover, completely flawed studies may still get published, eluding peer review and editorial scrutiny (Hey and Chalmers, 2000). Finally, one can never ignore the potential action of financial interests that may wish to prolong the apparent maximal efficacy for a treatment in the published literature, or even in meta-analyses thereof, by withholding the publication of important 'negative' data in company archives for as long as possible.

Chance

When it comes to epidemiologic associations, researchers may screen hundreds and thousands of putative risk factors as potential markers of a disease. In the sea of tested possible associations, we expect 5 % of them to be 'positive' by chance, at the 5 % significance level. Moreover, one could speculate that meta-analyses are more likely to be conducted on topics where large, 'positive' effect sizes have been found by the first studies, as such topics may seem more attractive research areas. Hence, first studies that are not replicated by subsequent research could represent incidental spurious findings due to type I error in a setting of multiple comparisons (Ioannidis, 2003). For example, empirical data suggest that only approximately 30 % of the proposed genetic associations are replicated by subsequent research (Ioannidis *et al.*, 2001, 2003; Lohmueller *et al.*, 2003). This may reflect a high infiltration of the literature with chance findings.

There are no empirical data regarding the frequency of chance findings in randomized research. Some highly surprising results, especially in small studies that show big survival differences that are not eventually replicated, may be due to chance and/or manipulation of endpoints (Ioannidis and Lau, 1997).

INITIAL VERSUS SUBSEQUENT EVIDENCE

Methodology

Theoretically, in the presence of time-lag bias, the initial estimates of treatment effects would be optimistically inflated. Hence, the tendency for diminishing effect sizes over time may easily be detected by comparing early studies with subsequent evidence. In the typical scenario, the study (or studies) published during the earliest calendar year is considered 'first'; the rest comprise 'subsequent research'. Other definitions of 'first' and 'subsequent research' are also feasible. Sometimes it is possible to separate out which is the first study among several published in the same calendar year, but for practical purposes this is typically not easy. We should caution that the order may change, if we also consider meeting presentations, and that it is very difficult to ensure the comprehensive scrutiny of all meeting presentations. Therefore the calendar year of publication is usually the best available proxy for determining study order. The first studies may be thought of as 'hypothesis-generating' ones, while subsequent ones play the role of validation experiments.

Consider a meta-analysis with n_1 studies published during the first calendar year and n_2 subsequent studies; let T_1, v_1 and T_2, v_2, respectively, be the effect sizes and their variances (T_i, v_i pertain to the effect size, and its variance, of either a single study, or the proper synthesis of several studies). T_1 and T_2 stand for generic effect size estimates that follow an approximate normal distribution; if, for example, the odds ratio (OR) is the metric of choice, we would employ the transformation $OR_i \rightarrow \ln(OR_i) = T_i$. Considering T_1 and T_2 to be independent, the extent of their discrepancy is assessed by the standardized z score,

$$z = \frac{T_1 - T_2}{\sqrt{v_1 + v_2}}, \tag{13.1}$$

that follows the typical normal distribution. Values of $|z| > 1.96 \leftrightarrow p < 0.05$ reflect statistically significant differences at the 5 % significance level.

Most often T_2 represents the synthesis of many studies and may be derived by fixed or random effects. To be consistent, we have to employ the same synthetic method for the early results as well, in case there is more than one first study. When there is significant between-study heterogeneity, random-effects syntheses should be preferred. In full random-effects approaches (Moses *et al.*, 2002), the variance v_1 of a large first study is inflated by τ_2^2, the DerSimonian and Laird estimate of between-study variance for the subsequent studies (DerSimonian and Laird, 1986),

$$z = \frac{T_1 - T_2}{\sqrt{(v_1 + \tau_2^2) + v_2}}, \tag{13.2}$$

leading to even smaller estimates of the prevalence of statistically significant first versus subsequent discrepancies. Note that τ_2^2 has already been used in the calculation of v_2.

Empirical overview

We have previously examined the discrepancies between first and subsequent studies for 36 meta-analyses of associations in genetic epidemiology (Ioannidis *et al.*, 2001), using the definition of the z score described by (13.1). The odds of finding a statistically significant discrepancy between first and subsequent research increased with the number of studies available, when the sample size of the first study (or studies) was small, and when there was only a single first publication that had highlighted a clearly defined genetic contrast. In 21 out of the 36 studied

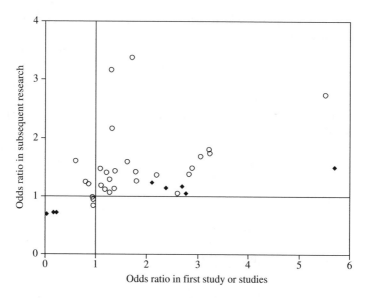

Figure 13.1   Initial versus subsequent evidence. Correlation between the odds ratio in the first study or studies and in subsequent research for 36 meta-analyses of genetic association studies. $OR > 1$ suggests predisposition towards disease, whereas $OR < 1$ the opposite. Diamonds denote meta-analyses with statistically significant discrepancies beyond chance between initial and subsequent studies. Triangles stand for non-significant differences. All calculations were performed with random-effects modeling. By fixed-effects modeling formal significant discrepancies would appear in two additional instances. (Reproduced with permission (Ioannidis *et al.*, 2001).)

associations (with random-effects syntheses), the first study (or studies) tended to give more impressive results. This was true for all eight cases where first and subsequent evidence differed beyond chance (Figure 13.1). In all but one of these eight instances, statistically significant between-study heterogeneity was present. Results based on fixed-effects modeling were similar.

Worked examples: passive smoking and teacher expectancy data sets

For the passive smoking data set, the first study (Garfinkel, 1981) showed $OR_1 = 1.18(95\% \text{ CI } 0.90-1.54) \rightarrow \ln(OR_1) = 0.166$ and $v_1 = 0.0188$. The random-effects synthesis of the subsequent studies yielded $OR_2 = 1.25(95\% \text{ CI } 1.13-1.37) \rightarrow \ln(OR_2) = 0.219$ and $v_2 = 0.0025$. Using (13.1), the z score $= -0.37 \rightarrow p = 0.71$. Similarly, in the teacher expectancy data set, the first study was that by Flowers, showing standardized difference and corresponding variance $d_1 = 0.180$ and $v_1 = 0.0498$, respectively, and the random-effects synthesis of the subsequent studies yielded $d_2 = 0.081$, $v_2 = 0.0032$. Using (13.1) we get a z-score of $0.43 \rightarrow p = 0.67$. Hence, there was no statistically significant difference between the first and the subsequently published studies in these examples.

Interpretation

In scientific fields where the production of data is rapid and copious, such as genetic epidemiology, the first published reports are very often optimistically biased (Ioannidis *et al.*, 2001, 2003; Lohmueller *et al.*, 2003), and their results may not be replicated by subsequent research. The phenomenon has been described as the 'winner's curse', and it should be taken into account to avoid overinterpreting early results. However, the extrapolation of such findings to other scientific areas is at least precarious. Genetic epidemiology is a special case mainly because of the large amount of data that are produced with relative ease (Ioannidis, 2003). There are no empirical data on how often first and subsequent evidence disagree in different settings: in the rest of epidemiology, in randomized research, or across different scientific areas (i.e., health care and social sciences).

CUMULATIVE META-ANALYSIS

Description and methodology

In cumulative meta-analysis one re-evaluates the summary estimates for the measured outcome every time a new study is added in a series of studies, so that updated summary effect sizes are recorded. The semantic differential from a simple meta-analysis update is that cumulative meta-analysis highlights the history of the successive summary estimates. It is straightforward to inspect for trends in the evolution of the summary effect size and to assess the impact of a specific study on the overall conclusions.

For the evaluation of the temporal evolution of the measured outcomes, the (ascending) year of publication defines the study order (Lau *et al.*, 1995). Because cumulative meta-analysis revises information in the light of new information, it is naturally amenable to a Bayesian analysis. The knowledge before the first study (or studies) represents the prior probability distribution of the effect size, and could be represented by a subjective non-informative distribution. The knowledge available after the consideration of each set of studies represents the posterior probability distribution, which will in turn become the prior probability distribution when the next set of studies is considered (Lau *et al.*, 1995). However, from a classical (frequentist) perspective, cumulative meta-analysis suffers from the problem of repeated testing and an inflated overall type I error. Some techniques have been proposed regarding the correction for serially re-estimated *p*-values in cumulative meta-analysis of binary outcomes (Pogue and Yusuf, 1997), but they have inherent problems (Egger *et al.*, 1998). A technique for adjustments for multiple views in cumulative meta-analysis of continuous outcomes has also been described (Lan *et al.*, 2003). Despite all these, typically, no *p*-value or confidence interval adjustments for serial considerations have been applied to cumulative meta-analyses (Lau *et al.*, 1995).

Every re-evaluation of the summary estimates in cumulative meta-analysis may be considered as an 'information step'. Since it may be unrealistic to update meta-analyses more often than once a year, all studies published in the same year may be thought to belong by convention to the same information step (Ioannidis and Lau, 2001). Figure 13.2 shows cumulative meta-analyses for the passive smoking and

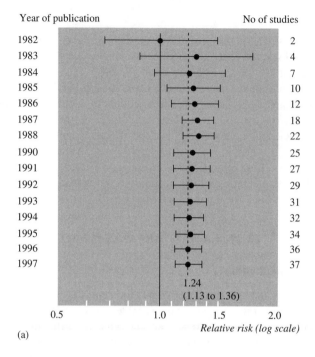

(a)

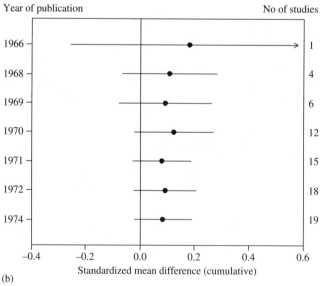

(b)

Figure 13.2 (a) Cumulative meta-analysis of the pooled odds ratio estimate for lung cancer in non-smokers and exposure to environmental tobacco smoke. The addition of new studies does not materially change the estimate, so that the final 1997 odds ratio estimate of 1.24 seems robust (Hackshaw *et al.*, 1997). (Reprinted from *British Medical Journal* (1997) Vol. 315, 980–988, with permission from the BMJ Publishing Group.) (b) Cumulative meta-analysis of the standardized mean difference for the teacher expectancy data set.

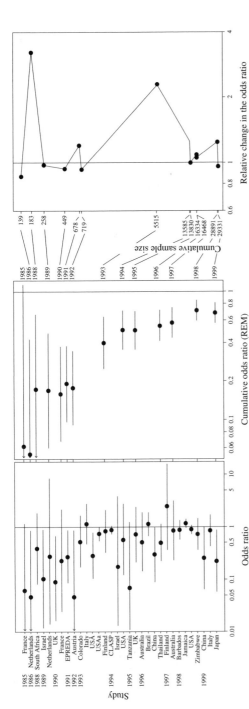

Figure 13.3 Outline of cumulative and recursive cumulative meta-analysis. This meta-analysis from Knight *et al.* (2000) addresses the application of antiplatelet agents for prevention of pre-eclampsia. The outcome is proteinuric pre-eclampsia. There are a total of 32 studies ordered chronologically. The odds ratio was chosen as a metric of the effect size. The left-hand part of this rugplot is a classic meta-analysis 'forest graph', without the final synthesis. The second part is a cumulative meta-analysis diagram, where the pooled odds ratio (by random-effects modeling) is re-estimated at the end of each calendar year, that is, after each information step. The little dashes connecting the two parts correspond to each information step. The third part is a recursive cumulative meta-analysis plot, where the relative changes in the odds ratio are plotted against the cumulative sample size. The first information step is used in the calculation of the first relative change. Relative changes away from 1 imply great fluctuations in the pooled odds ratio estimate. In most cases, when the cumulative sample size increases, the relative changes become smaller. This meta-analysis is an exception, as a large fluctuation in the summary effect size was observed at a relatively large cumulative sample size (see text). (Adopted and updated from Ioannidis and Lau (2001). Reprinted from PNAS, Vol. 98, Evolution of treatment effects over time: Empirical insight from recursive cumulative meta-analyses, 831–836 (© 2001), with permission from National Academy of Sciences, USA.)

teacher expectancy data sets. The first two panels of Figure 13.3 outline the meta-analysis and cumulative meta-analysis of the administration of antiplatelet agents for the prevention and treatment of pre-eclampsia (Knight *et al.*, 2000). This has been a controversial topic where small early studies suggested very large protective effects that were not validated in subsequent larger trials.

Empirical overview

In the absence of biases and significant heterogeneity, the cumulative meta-analysis plot should exhibit a fairly constant estimate of treatment effects over time with some fluctuation due to chance. However, every possible temporal pattern may be encountered as a result of operating biases or heterogeneity (Figures 13.4 and 13.5). Meta-analyses may transiently reach or lose formal significance, or may even completely reverse the initial association. Examples are given in Figure 13.5. The time of reaching statistical significance in a meta-analysis and the uncertainty thereof may be appropriately modeled (Berkey *et al.*, 1996).

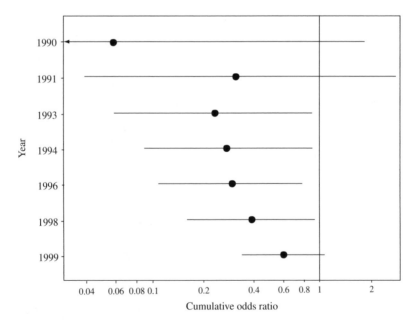

Figure 13.4 Example of meta-analysis of controlled trials that transiently reached formal significance. This is a cumulative meta-analysis graph constructed with data from a Cochrane review on vitamin E versus placebo for neuroleptic induced tardive dyskinesia (Soares and McGrath, 2001). For a time-span of 6 years the application of vitamin E seemed to be considered a statistically significant matter. However, all seven studies are of small size (range 10–158) and there is room for major fluctuations in the overall effect size, should a new study be published.

Interpretation

The conduct of a meta-analysis resembles a snapshot of the available evidence at a more or less arbitrary time. When does a team decide to perform a meta-analysis? The timing is probably dependent on the emergence of a meaningful number of relevant studies, but it often depends also on how imperative the addressed matters are (Lau *et al.*, 1995). One suspects that meta-analyses are often performed, because the issues at hand are 'hot'. Nevertheless, as shown in the trajectories in Figures 13.4 and 13.5, various temporal patterns may be encountered and the complete picture may evolve over time. Hence, especially for meta-analyses on

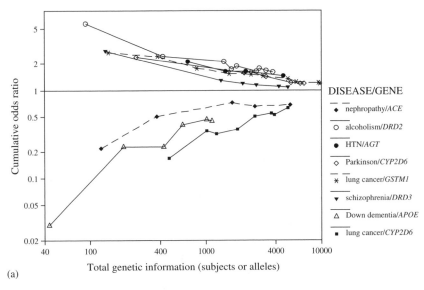

(a)

Figure 13.5 Cumulative meta-analysis trajectories for 16 genetic associations. The strength of the association is shown as an estimate of the odds ratio without confidence intervals. (a) Eight topics, for which the results of the first study (or studies) differed beyond chance ($p < 0.05$) when compared to the results of the subsequent research. (b) For eight genetic associations that finally reached formal significance, despite the fact that the first study (or studies) estimate was not formally significant. The horizontal axis corresponds to the cumulative amount of information gathered (genotyped people or alleles) (Ioannidis *et al.*, 2001). Note the different types of the trajectories of the addressed topics, including increasing, constant and decreasing odds ratios, and even a case where formal significance is gained, yet for the opposite arm than originally thought. *ACE*: gene encoding angiotensin converting enzyme; *AGT*: gene encoding angiotensinogen; *APOE*: gene encoding apolipoprotein E; *CYP*: gene(s) encoding cytochrome P450; *DRD2/DRD3*: genes encoding dopamine receptor D2/D3; *GSTM1*: gene encoding glutathione-S-transferase M1; HTN: hypertension; ICVD: ischemic cerebrovascular disease; IHD: ischemic heart disease; *KIR6.2-BIR*: K$^+$ inwardly rectifying channel; MI: myocardial infarction; MTHFR: gene encoding methylenetetrahydrofolate reductase; *NAT2*: gene encoding N-acetyltransferase 2; NIDDM: non-insulin dependent diabetes mellitus; NTD: neural tube defects. (Reproduced with permission (Ioannidis *et al.*, 2001).).

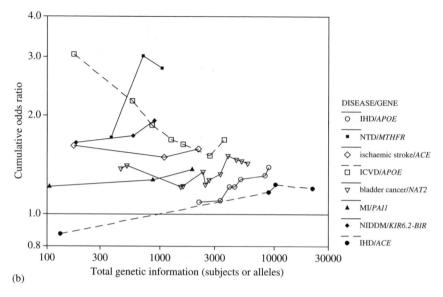

(b)

Figure 13.5   (Continued).

controversial matters (Figure 13.3) or meta-analyses with a small total number of subjects, any conclusions should be drawn with caution. An effort should be made to try to keep updating meta-analyses on a regular basis, instead of considering meta-analyses as fixed, unaltered publications. The regular updating of meta-analyses is consistent with the philosophy of the Cochrane Collaboration (Bero and Rennie, 1995; Chalmers and Haynes, 1994; Clarke and Langhorne, 2001). Moreover, we need to gather more evidence on how frequently formal statistical significance is lost during the update of a meta-analysis and what the optimal timing interval should be for updates. The counterargument is that spending time and effort to keep updating highly conclusive meta-analyses may be a poor investment of resources.

RECURSIVE CUMULATIVE META-ANALYSIS

p-Values and uncertainty in the future evolution of the effect size

By amassing large quantities of information, meta-analysis sometimes has the power to reveal even negligible effect sizes and raise them into statistical significance. Yet, formal statistical significance does not lend, for example, clinical meaning to the intervention or association examined. In clinical medicine, a very small effect size would not actually make a difference in everyday practice, no matter how small its *p*-value were. Moreover, formal statistical significance, especially when marginally reached, does not ensure that current knowledge will persist in the light of new evidence.

In the presence of time-lag bias, early conclusions are uncertain because they are derived from biased selection of data; this is not directly reflected by any *p*-value.

Thus, it is useful to get a glimpse of a distinct quantity, the extent of the anticipated uncertainty in the future evolution of the effect size, if additional studies are to appear in subsequent years.

Increments and decrements in the summary effect size

We expect that, in the absence of bias, the number of information steps where the summary effect size increases should be more or less equal to the number of information steps where the summary effect size decreases. Any differences between the number of information steps where the summary effect size increases or decreases can be statistically assessed with a non-parametric test, such as the sign test. For example, for the passive smoking data set, the summary odds ratio increased in 7 and decreased in 8 information steps, giving an exact $p = 1.0$. Similarly, for the teacher expectancy dataset, the standardized mean difference, d, increased in 4 information steps and decreased in 2, giving an exact $p = 0.69$. If there were a statistically significant difference, one of the factors described in the section on 'Factors underlying the evolution of effect sizes over time' might be responsible.

Methodology outline for recursive cumulative meta-analysis

In recursive cumulative meta-analysis (Ioannidis *et al.*, 1999; Ioannidis and Lau, 2001) one estimates the relative changes in the odds ratio, RC, for every new information step. Let OR_t denote the estimate of the summary odds ratio at the tth information step. Then the RC at the $(t + 1)$th information step would be

$$RC_t = \frac{OR_{t+1}}{OR_t}. \tag{13.3}$$

The advantage of using the odds ratio in recursive cumulative meta-analysis lies in the fact that it is symmetric, while other multiplicative effect sizes, such as the relative risk (RR), are not. Thus, the overall assessment would yield similar inferences if OR_t/OR_{t+1} were used instead of OR_{t+1}/OR_t. This would not be true for RR_{t+1}/RR_t vs. RR_t/RR_{t+1} or other, non-symmetric effect sizes.

Apart from the above, it might be informative for someone to calculate the relative changes for any metric, making modifications according to his/her needs: for example, one could use transformed correlation coefficients (e.g. the Fisher transformation), or even change equation (13.3) to capture differences instead of quotients per information step. However, no such examples exist in the literature yet.

Interpretation

Relative changes reflect the observed uncertainty of the treatment effects. It is informative to plot the RC against the cumulative sample size N of the meta-analysis. One expects that great fluctuations in the estimate of treatment effects should be observed early in a meta-analysis when few data have accumulated, while relative changes should be smaller when excessive data have been amassed. The typical behavior under conditions where there is no bias resembles the oscillations of a pendulum with decreasing oscillation amplitude time due to dampening. In the case of antiplatelet agents for pre-eclampsia we encounter an exception: as evidence accumulates the deviations of relative changes from 1 are not always minor (Figure 13.3, last panel). Over time there is a gradual decrease of the treatment effect. Note that the relative changes do not inform on the formal statistical significance of an intervention or an association, or even on the favored arm. Rather, they stress the observed fluctuations of the summary odds ratios.

ZOOMING OUT: DOMAIN-WIDE RECURSIVE CUMULATIVE META-ANALYSIS

Domain-wide plots and the uncertainty of the effect sizes

By plotting the relative changes in the odds ratio against the cumulative sample size of several meta-analyses that cover a wider medical domain, we can generate a domain-wide scatter plot. This assesses visually the evolution of the uncertainty of the summary effect size in the specific domain (Ioannidis and Lau, 2001). What we are trying to do is to obtain an empirical answer to the question: given a certain amount of already available evidence (number of randomized subjects), how big a change in the summary odds ratio should one anticipate in the future, if more studies become available?

Intuitive approach: graphical overview

Figure 13.6 is an example of a domain-wide plot for 45 meta-analyses of interventions in pregnancy and perinatal medicine (Ioannidis and Lau, 2001). The relative changes for all 45 meta-analyses are plotted in the same plot as a function of the previously accumulated sample size. Evidently, the relative changes' dispersion decreases impressively as more evidence accumulates. There is room for large fluctuations in the effect size when only a few people have been randomized. Conversely, a greater accumulation of evidence yields more robust results that are less prone to change.

Mathematical approach

The expected values of the relative changes for any given cumulative sample size N, along with their 95 % confidence boundaries, can be calculated based on a linear regression of the form

$$\log_{10}(RC) = \beta_1 \log_{10}(N) + \beta_0, \qquad \text{weighted by } w(N) = N^g,$$

as the dispersion of the RC-values (and consequently their logarithms) is dependent on N. The extent of this dependency is accounted for by the weight function $w(N)$, where g is the power that optimizes the log-likelihood function in this

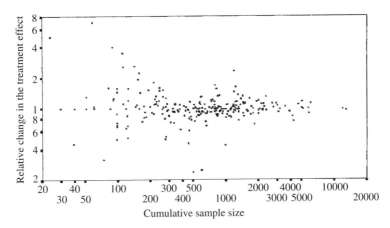

Figure 13.6   Scatter plot of the relative change in the pooled odds ratio as a function of the cumulative number of patients in previously published randomized trials in perinatal medicine and pregnancy (random-effects modeling) (Ioannidis & Lau. 2001). Note the drastically decreased uncertainty in the odds ratio (as conveyed by the *RC* dispersion) with accumulating patients. (Reprinted from PNAS, Vol. 98, Evolution of treatment effects over time: Empirical insight from recursive cumulative meta-analyses, 831–836 (© 2001), with permission from National Academy of Sciences, USA.)

regression and is derived by an iterative algorithm that is available in most statistical computer packages (Ioannidis and Lau, 2001). The expected values of the *RC*s for a given *N*, along with their 95 % confidence boundaries, can be derived from this regression. Variants of this approach or proper bootstrapping could also be used for the same purpose. Although the points in the domain plot are not independent, this methodology provides an approximation for the boundaries of the scatter of the relative changes.

Empirical overview

Empirical evaluations with meta-analyses in the domains of myocardial infarction and pregnancy/perinatal medicine have shown that threefold relative changes in the odds ratio, in either direction, are not surprising when only 100 patients have been randomized (Ioannidis and Lau, 2001). This uncertainty is drastically curtailed when more patients accumulate. With 2000 patients, we would expect with 95 % certainty a less than 20–25 % change in the odds ratio, should an additional study be published (Table 13.1). The predicted uncertainties, for given cumulative sample sizes, were fairly similar in the two medical domains examined (Ioannidis and Lau, 2001).

An empiric scrutiny of a larger set of meta-analyses focused on mental health-related interventions. Mental health is a domain dominated by small trials. The estimated uncertainties for different levels of amassed information were similar to those described above. The observations were consistent for all examined efficacy

Table 13.1   Ninety-five per cent prediction intervals for the relative change (RC) in the odds ratio for different numbers of accumulated patients (cumulative sample size, N) (Ioannidis and Lau 2001).[a]

Patients, N	Pregnancy/perinatal		Myocardial infarction	
	Fixed effects	Random effects	Fixed effects	Random effects
100	0.37–2.78	0.32–3.13	0.18–5.51	0.23–4.43
500	0.59–1.71	0.56–1.71	0.60–1.67	0.63–1.58
1 000	0.67–1.49	0.65–1.53	0.74–1.35	0.76–1.32
2 000	0.74–1.35	0.73–1.37	0.83–1.21	0.84–1.20
15 000	0.85–1.14	0.86–1.15	0.96–1.05	0.96–1.05

Reprinted from PNAS, Vol. 98, Evolution of treatment effects over time: Empirical insight from recursive cumulative meta-analyses, 831–836 (© 2001), with permission from National Academy of Sciences, USA.
[a] Predicted values were derived from domain-wide regressions (for fixed and random effects models) of the form

$$\log_{10}(RC) = \beta_1 \log_{10}(N) + \beta_0, \qquad \text{weighted by } w(N) = N^g.$$

For example, if the treatment effect of a given drug in the treatment of myocardial infarction is provided by a fixed-effects odds ratio of 0.60 based on $N = 2000$ randomized to date in published trials until now, then we can be 95 % certain that the odds ratio, should new trials appear, would be between $0.60 \times 0.83 = 0.50$ and $0.60 \times 1.21 = 0.73$.

outcomes (treatment failures, relapses, and deaths) as well as for dropouts, an outcome that combines toxicity and efficacy components (Trikalinos *et al.*, 2004).

Early extreme fluctuations in the treatment effects

The studentized residuals of the RCs – derived from the domain-wide regressions – reflect how far from its predicted value an observed RC is. The greater the absolute value of the residual, the more extreme the observed fluctuation in the odds ratio. When the studentized residual is very large, one may claim that the observed fluctuation is too large compared to what one would expect, given the accumulated sample size.

A challenging question is whether large fluctuations in the evolution of evidence early on may be signals for additional subsequent fluctuations in the future. When the summary odds ratio for an evolving meta-analysis shows a large change, is this a signal that it may change again disproportionately in the future? If this is true, one would have to be extra cautious with meta-analyses that show disproportionately large fluctuations. The situation resembles the attempt to predict a major earthquake based on the appearance of earlier earthquakes.

Some empirical evidence exists that such a prediction of future 'earthquakes' in the evolution of the summary effect size may be possible. In the medical domains of myocardial infarction and pregnancy/perinatal medicine that we discussed in the previous examples, extreme fluctuations in the effect size were identified for seven medical interventions. Interestingly, in three of these (namely magnesium in myocardial infarction, and calcium and antiplatelet agents for prevention of pre-eclampsia), more recent mega-trials eventually completely contradicted the prior respective meta-analyses. Several years before the publication of these mega-trials, extreme fluctuations in the effect size had been observed as evidence accumulated from early small trials. Moreover, in a fourth case (nitrates for myocardial infarction) where a similarly refuting mega-trial (1995) had been performed, the a priori specified criteria for extreme fluctuations were marginally missed among the early published trials. In another four topics with extreme fluctuations, no mega-trials had ever been performed. In all of these four cases, early large effect sizes were dissipated when more data appeared, even without mega-trials (Ioannidis & Lau, 2001).

Interpretation

In the presence of limited cumulative evidence, treatment effects must be examined with extra caution. The obtained empirical confidence intervals for the anticipated future fluctuations in the effect size pertain to the whole medical domain rather than a specific meta-analysis. An intriguing possibility is to predict future extreme fluctuations in the treatment effects by early ones, and more empirical evidence is needed to address the generalizability of this finding. The domain-wide plots have their limitations: it is not always easy to define what constitutes a domain and it is not always straightforward why a certain number of meta-analyses should dictate how the next one should behave. These analyses require a large amount of data from many meta-analyses. Such data are nevertheless likely to become increasingly available in the future. Finally, the above considerations may be particularly applicable to randomized research, and more empirical evidence is needed to understand the uncertainty in the evolution of effect sizes for non-randomized research. The Proteus phenomenon (as discussed earlier in this chapter) may be more prominent in non-randomized research.

SAFETY OUTCOMES

There are limited empirical data on the evolution of the effect sizes for safety outcomes (Trikalinos *et al.*, 2004). Sometimes safety outcomes may carry a significant efficacy component. For example, withdrawals may be considered as a safety measure, as patients may withdraw from a study due to toxicity. However, withdrawals may also be due to lack of efficacy: patients are less likely to remain on an impotent intervention, or may refuse treatment if they feel very well. The net effect may be difficult to predict. Regardless, it would be most useful if cumulative meta-analysis and recursive cumulative meta-analysis could be preformed side-by-side for both efficacy and safety outcomes. This would allow evaluation of the evolving risk–benefit ratios, and such information would be very helpful for the rational updating of evidence-based guidelines and recommendations.

REFERENCES

Berkey, C.S., Mosteller, F., Lau, J. and Antman, E.M. (1996). Uncertainty of the time of first significance in random effects cumulative meta-analysis. *Controlled Clinical Trials*, **17**, 357–371.

Bero, L. and Rennie, D. (1995). The Cochrane Collaboration. Preparing, maintaining, and disseminating systematic reviews of the effects of health care. *Journal of the American Medial Association*, **274**, 1935–1938.

Chalmers, I. and Haynes, B. (1994). Reporting, updating, and correcting systematic reviews of the effects of health care. *British Medical Journal*, **309**, 862–865.

Chalmers, T.C. and Lau, J. (1996). Changes in clinical trials mandated by the advent of meta-analysis (with discussion). *Statistics in Medicine*, **15**, 1263–1272.

Clarke, M. and Langhorne, P. (2001). Revisiting the Cochrane Collaboration. Meeting the challenge of Archie Cochrane – and facing up to some new ones. *British Medical Journal*, **323**, 821.

DerSimonian, R. and Laird, N. (1986). Meta-analysis in clinical trials. *Controlled Clinical Trials*, **7**, 177–188.

Dickersin, K., Olson, C.M., Rennie, D., Cook, D., Flanagin, A., Zhu, Q. *et al.* (2002). Association between time interval to publication and statistical significance. *Journal of the American Medical Association*, **287**, 2829–2831.

Egger, M., Smith, G.D. and Sterne, J.A. (1998). Meta-analysis: Is moving the goal post the answer? *Lancet*, **351**, 1517.

Garfinkel, L. (1981). Time trends in lung cancer mortality among nonsmokers and a note on passive smoking. *Journal of the National Cancer Institute*, **66**, 1061–1066.

Glasziou, P.P. and Irwig, L.M. (1995). An evidence based approach to individualizing treatment. *British Medical Journal*, **311**, 1356–1359.

Goring, H.H., Terwilliger, J.D. and Blangero, J. (2001). Large upward bias in estimation of locus-specific effects from genomewide scans. *American Journal of Human Genetics*, **69**, 1357–1369.

Hackshaw, A.K., Law, M.R. and Wald, N.J. (1997). The accumulated evidence on lung cancer and environmental tobacco smoke. *British Medical Journal*, **315**, 980–988.

Hey, E. and Chalmers, I. (2000). Investigating allegations of research misconduct: The vital need for due process. *British Medical Journal*, **321**, 752–755.

Hirschhorn, J.N., Lohmueller, K., Byrne, E. & Hirschhorn, K. (2002). A comprehensive review of genetic association studies. *Genetics in Medicine*, **4**, 45–61.

Hopewell, S., Clarke, M., Lusher, A., Lefebvre, C. and Westby, M. (2002). A comparison of handsearching versus MEDLINE searching to identify reports of randomized controlled trials. *Statistics in Medicine*, **21**, 1625–1634.

Ioannidis, J.P. (1998). Effect of the statistical significance of results on the time to completion and publication of randomized efficacy trials. *Journal of the American Medical Association*, **279**, 281–286.

Ioannidis, J.P. (2003). Genetic associations: False or true? *Trends in Molecular Medicine*, **9**, 135–138.

Ioannidis, J.P. and Lau, J. (1997). The impact of high-risk patients on the results of clinical trials. *Journal of Clinical Epidemiology*, **50**, 1089–1098.

Ioannidis, J. and Lau, J. (2001). Evolution of treatment effects over time: Empirical insight from recursive cumulative meta-analyses. *Proceedings of the National Academy of Sciences of the United States*, **98**, 831–836.

Ioannidis, J.P.A. and Trikalinos, T.A (2005). Early extreme contradictory estimates may appear in published research: The Proteus phenomenon in molecular genetics research and randomized trials. *Journal of Clinical Epidemiology*, **58**, 543–549.

Ioannidis, J.P., Cappelleri, J.C., Sacks, H.S. and Lau, J. (1997). The relationship between study design, results, and reporting of randomized clinical trials of HIV infection. *Controlled Clinical Trials*, **18**, 431–444.

Ioannidis, J.P., Contopoulos-Ioannidis, D.G. and Lau, J. (1999). Recursive cumulative meta-analysis: A diagnostic for the evolution of total randomized evidence from group and individual patient data. *Journal of Clinical Epidemiology*, **52**, 281–291.

Ioannidis, J.P., Ntzani, E.E., Trikalinos, T.A. and Contopoulos-Ioannidis, D.G. (2001). Replication validity of genetic association studies. *Nature Genetics*, **29**, 306–309.

Ioannidis, J.P., Trikalinos, T.A., Ntzani, E.E. and Contopoulos-Ioannidis, D.G. (2003). Genetic associations in large versus small studies: An empirical assessment. *Lancet*, **361**, 567–571.

Jüni, P., Holenstein, F., Sterne, J. Bartlett, C. and Egger, M. (2002). Direction and impact of language bias in meta-analyses of controlled trials: Empirical study. *International Journal of Epidemiology*, **31**, 115–123.

Knight, M., Duley, L., Henderson-Smart, D.J. and King, J. F. (2000). Antiplatelet agents for preventing and treating pre-eclampsia (Cochrane Review). In *The Cochrane Library*, Issue 4, 2002. Oxford: Update Software.

Lan, K.K.G., Hu, M. and Cappelleri, J.C. (2003). Applying the law of iterated logarithm to cumulative meta-analysis of a continuous end-point. *Statistica Sinica*, **13**, 1135–1145.

Lau, J., Schmid, C.H. and Chalmers, T.C. (1995). Cumulative meta-analysis of clinical trials builds evidence for exemplary medical care. *Journal of Clinical Epidemiology*, **48**, 45–57; discussion 59–60.

Lau, J., Ioannidis, J.P. and Schmid, C.H. (1998). Summing up evidence: One answer is not always enough. *Lancet*, **351**, 123–127.

Lohmueller, K.E., Pearce, C.L., Pike, M., Lander, E.S. and Hirschhorn, J.N. (2003). Meta-analysis of genetic association studies supports a contribution of common variants to susceptibility to common disease. *Nature Genetics*, **33**, 177–182.

Moher, D., Pham, B., Jones, A., Cook, D.J., Jadad, A.R., Moher, M. *et al.* (1998). Does quality of reports of randomised trials affect estimates of intervention efficacy reported in meta-analyses? *Lancet*, **352**, 609–613.

Moses, L. E., Mosteller, F. & Buehler, J. H. (2002). Comparing results of large clinical trials to those of meta-analyses. *Statistics in Medicine 21*, 793–800.

Mosteller, F. and Colditz, G.A. (1996). Understanding research synthesis (meta-analysis). *Annual Review of Publiclic Health*, **17**, 1–23.

Olson, C.M., Rennie, D., Cook, D., Dickersin, K., Flanagin, A., Hogan, J.W. *et al.* (2002). Publication bias in editorial decision making. *Journal of the American Medical Association*, **287**, 2825–2828.

Pogue, J.M. and Yusuf, S. (1997). Cumulating evidence from randomized trials: Utilizing sequential monitoring boundaries for cumulative meta-analysis. *Controlled Clinical Trials*, **18**, 580–593; discussion 661–666.

Soares, K.V. and McGrath, J.J. (2001). Vitamin E for neuroleptic-induced tardive dyskinesia (Cochrane Review). In *The Cochrane Library*, Issue 4, 2002. Oxford: Update Software.

Stern, J.M. and Simes, R.J. (1997). Publication bias: Evidence of delayed publication in a cohort study of clinical research projects. *British Medical Journal*, **315**, 640–645.

Stewart, L.A. and Clarke, M.J. (1995). Practical methodology of meta-analyses (overviews) using updated individual patient data. Cochrane Working Group. *Statistics in Medicine*, **14**, 2057–2079.

Stewart, L.A. and Parmar, M.K. (1993). Meta-analysis of the literature or of individual patient data: Is there a difference? *Lancet*, **341**, 418–422.

Trikalinos, T.A., Churchill, R., Ferri, M., Leucht, S., Tuunainen, A., Wahlbeck, K., Ioannidis, J.P.A., for the EU-PSI Project (2004). Effect sizes in cumulative meta-analyses of mental health randomized trials evolved over time. *Journal of Clinical Epidemiology*, **57**, 1124–1130.

CHAPTER 14

Do Systematic Reviews Based on Individual Patient Data Offer a Means of Circumventing Biases Associated with Trial Publications?

Lesley Stewart, Jayne Tierney and Sarah Burdett
MRC Clinical Trials Unit, London, UK

KEY POINTS

- Individual patient data (IPD) meta-analyses centrally collect, verify and reanalyse the original 'raw' data directly from researchers.
- In doing so, they avoid dependence on study publications and provide a way of circumventing a number of reporting biases including publication bias, duplicate reporting bias, outcome-reporting bias, patient-exclusion bias and follow-up bias. Conversely, the approach could introduce availability bias if data were available only from particular types of trial.
- In all systematic reviews an active decision should be taken about what approach is most appropriate, giving particular consideration to the likely possible sources of bias.
- The fact that a systematic review uses IPD does not necessarily imply that it is of high quality. When done well, IPD meta-analyses of randomized controlled trials are likely to offer a 'gold standard' of research synthesis.

Publication Bias in Meta-Analysis – Prevention, Assessment and Adjustments Edited by H.R. Rothstein, A.J. Sutton and M. Borenstein © 2005 John Wiley & Sons, Ltd

INTRODUCTION

There are, as discussed throughout this book, many potential biases associated with whether trials are published, where they are published, how often they are published and in the way that data are presented in publications. It is clear that the publication process can distort the underlying findings of research. Overt or subconscious pressures such as the wish to bolster research ratings, the need to sell journals or simply the desire to deliver good news can lead to results being presented in an over-favourable light, or to publishing only those papers that will deliver a message of progress and improvement. This is of course potentially very damaging, and in the context of systematic review it is important that we do all we can to minimize sources of potential bias, including those associated with publication and reporting. Methods have been developed to assess the likelihood of publication bias, to predict its effect and correspondingly adjust outcome measures within individual systematic reviews and meta-analyses. An alternative approach is to bypass dependence on published reports and obtain data directly from researchers, as is done in individual patient data (IPD) meta-analyses.

Individual patient data meta-analysis of randomized controlled trials

Often described as a 'gold standard', the IPD approach to systematic review involves obtaining the original 'raw' research data from those responsible for trials addressing a common research question. The approach can be used both prospectively as described in Chapter 3 by Berlin and Ghersi on prospective meta-analyses, or more commonly, in the context of a prospectively defined, but nonetheless retrospective systematic review. It is the latter that we consider specifically in this chapter, although many of the issues that we discuss are applicable also to prospective meta-analyses. Typically data are sought from all trials irrespective of publication status and are then centrally collated, checked and reanalysed. Usually, a great deal of effort is spent in identifying relevant trials and in ensuring that the review is as complete as possible. The processes involved are the same as for any other well-conducted systematic review: a protocol is written, systematic searches for trials are done, and trials are critically appraised and assessed for eligibility most commonly the individual trials are reanalysed and the *results* of these trials are then combined in stratified analyses to provide overall estimates of effect. The IPD approach differs only in terms of data collection, checking and analysis. IPD meta-analyses are usually carried out as collaborative projects whereby all trialists' contributing information from their trials, together with those managing the project, become part of an active collaboration. The projects are usually managed by a small local organizing group or secretariat, which may be aided in important and strategic decision-making by a larger advisory group. Results are usually published in the name of the collaborative group and the secretariat may also organize a meeting of all collaborators, to bring individuals together to discuss the preliminary results. Potential advantages of the IPD approach stem from improved data quality, from the ability to perform more detailed and sensitive analyses, as well as from reviewing evidence as an international multidisciplinary team (Stewart and Tierney,

2002). The main disadvantages are that they are more time-consuming and costly than other types of systematic review.

There are, of course, differences in the ways that IPD meta-analyses are conducted. The methods and approaches used differ according to the subject area and outcomes explored, as well as by the techniques favoured by the groups responsible for coordinating them. The fact that a systematic review is done using IPD does not necessarily imply it is of high quality. Potentially, some IPD projects may not exploit the full advantages offered by the approach and, like all research projects, there may be cases where they have not been done most appropriately (Simmonds *et al.*, 2005). However, when done well, we believe that IPD meta-analyses of randomized controlled trials (RCTs) do offer a 'gold standard' in research synthesis.

History of IPD meta-analyses in health care

In the health-care field, IPD meta-analyses have an established history in cardiovascular disease and cancer, where the methodology has been developing steadily since the late 1980s. In cancer, for example, there are now more than 50 IPD meta-analyses of screening and treatment across a wide range of solid tumour sites and haematological malignancies (Clarke *et al.*, 1998). More recently, IPD has also been used in systematic reviews in many other fields, including HIV infection (HIV Trialists Collaborative Group, 1999), Alzheimer's disease (Qizilbash *et al.*, 1998), epilepsy (Marson *et al.*, 2002), depression (Storosum *et al.*, 2001), malaria (Artemether-Quinine Meta-analysis Study Group, 2001), hernia (EU Hernia Trialists Collaboration, 2000a, 2000b) addiction (Barnett *et al.*, 2001) and asthma (Holt *et al.*, 2001).

IPD meta-analyses in the social sciences and education

As far as is known, IPD methodology has not yet been adopted in the social sciences or education fields. However, there is no reason why it should not be. Indeed, the approach may offer a valuable tool in assessing complex interventions and allow in-depth exploration of individual characteristics. Although the methodology has developed around RCTs, which are less prevalent, and in some circumstances less relevant, in social sciences and education than in health care, many of the issues that we discuss are relevant also to non-randomized evidence. For example, the ability to harness the collaborative approach and the direct involvement of those responsible for the primary studies to help identify studies, clarify ambiguities of study design and weed out duplicate publications. It allows checking and verification of data and more detailed and powerful analyses, including those that deal with heterogeneity and confounding factors. As in most meta-analyses, direct comparisons are only ever made between individual studies and data pooling is stratified by study. If individual studies have used different tools or scoring systems to measure the same or similar outcomes it may not be possible to combine aggregate data. However, access to data at the individual participant level may enable rescoring according to a common system and thereby make it possible to pool data across studies that could not otherwise be combined. It also facilitates inclusion of data that may have been excluded from the primary analyses and the conduct of common standard analyses across studies. Although the examples that follow are drawn from meta-analyses of RCTs in the health-care setting, in particular from the cancer field, the underlying principles are likely to be equally applicable to research synthesis in other fields.

Using IPD to explore bias

The collection of updated IPD from all randomized trials (published and unpublished) and all randomized patients included in those trials provides a rare chance to circumvent publication and associated biases. Because the IPD approach to systematic review does not rely on data presented in the literature or on results of pre-existing analyses, it also provides an opportunity to further explore the underlying data sets obtained to try to tease out some general issues relating to the publication of clinical trials. With IPD, we are able to hold certain factors constant and explore the impact of other factors. A great deal of effort is spent on trying to identify all the relevant trials. Where such projects have collected a very large proportion of the research evidence, the data set can be thought of as the population of trial data from which we can draw various samples in order to explore the effect of different sampling frames, for example published trials only compared to all trials. Of course, defining the population of trials may be easier in some circumstances than others.

This chapter differs from others in that our examples are not based on the same standard data sets used throughout this book, for the obvious reason that the standard data sets are based on study level data. Rather, we draw on the results of empirical research that has used the data collected for IPD meta-analyses to explore methodological issues. This relates particularly to the impact of potential publication biases and confounding influences in the meta-analysis setting. We begin by giving examples of overall comparisons of the results of the IPD approach to systematic review with the results of those that extract data from published reports. We then discuss the ways in which the IPD approach may help avoid or reduce the possible impact of various types of reporting bias, including publication bias, duplicate-publication bias, trial-selection bias, outcome-reporting bias, patient-exclusion bias and follow-up bias. We also present the results of a project exploring whether the reporting of IPD meta-analyses might themselves be subject to publication bias and offer some guidance on deciding whether obtaining IPD might be valuable in a given review.

Comparing results of the IPD versus other approaches to meta-analysis

Given the extra resources and commitment required to undertake an IPD-based review, it is reasonable to question the need for the IPD approach and whether the potential advantages offered actually materially alter the results, conclusions and interpretation of the resulting review. There is evidence that meta-analyses based only on data extracted from published papers may give different estimates of treatment effects and of their significance than IPD meta-analyses (Jeng *et al.*, 1995; Pignon and Arriagada, 1993; Stewart and Parmar, 1993).

Examples

A comparison of an IPD meta-analysis in advanced ovarian cancer – comparing treatment by a more aggressive combination of drugs that included cisplatin versus treatment with a less-aggressive single drug (not cisplatin) (Advanced Ovarian Cancer Trialists Group, 1991) – with a similar analysis using only data that could be extracted from the published papers (Table 14.1), found that the IPD analysis gave less encouraging results (Stewart and Parmar, 1993).

In this study the aim was to compare the overall differences in outcome between the two approaches; comparing the policy of extracting data from published reports with the policy of collecting IPD, doing the best that was practically possible with each approach and reflecting what was/would have been done in practice. The analysis of summary data extracted from publications, which included 8 trials and 788 patients, favoured the combination of drugs with an estimated increase in survival at 30 months of 7.5 %. This was marginally significant at conventional levels ($p = 0.027$). The IPD meta-analysis, which was based on 11 trials and 1329 patients, suggested only a 2.5 % improvement and this did not reach conventional levels of significance ($p = 0.30$). Further investigation revealed that the differences between the two approaches were attributable to a number of factors. These included: trials that could not be included either because they were unpublished or because they did not publish the required information; patients that had been excluded from the published analyses; the point in time at which analyses using data from publications were based; the method of analysis that was possible with each approach; and the additional long-term follow-up which was available with the IPD. None of these factors was very much more influential than the others. Rather, they each contributed cumulatively to the difference. The analysis based on the summary data available from published reports not only provided a conventionally

Table 14.1 Comparison of meta-analysis approaches using an example in advanced ovarian cancer (Stewart and Parmar, 1993; Stewart, 2003).

	Individual patient data from trialists	Data extracted from publications
Trials	11	8
Patients	1329	788
Odds ratio	–	0.71
Hazard ratio	0.93	–
95 % confidence interval	0.83–1.05	0.52–0.96
p-value	0.30	0.027
Absolute benefit at 30 months	2.5 %	7.5 %
Comments	Median follow-up 6.5 years	Point estimate at 30 months

statistically significant result, but also gave an estimated benefit three times larger than that suggested by the IPD. Given the poor prognosis for women with advanced ovarian cancer and the limited treatment options available at the time, when balanced against other factors such as toxicity, ease of use and financial cost, there were likely to have been different clinical interpretations of the results from the two approaches.

A similar investigation, comparing the results of an IPD review with those obtained using data from published papers of trials comparing ovarian ablation (removal of the ovaries or other suppression of ovarian function) with control in the treatment of early breast cancer, also found differences in the results obtained by the two approaches (Clarke and Godwin, 1998). However, in this case, it was the IPD that gave the strongest evidence in favour of the intervention. The meta-analysis of data extracted from the publications available in 1990 was inconclusive about the effect of ovarian ablation. However, the meta-analysis that was carried out in 1990 using IPD showed a significant reduction in the annual odds of death with an absolute improvement in overall survival of 10.2 % at 15 years for women randomized to ovarian ablation. This was found to be mostly attributable to the increased follow-up resulting from the trialists bringing their data up to date for submission to the collaborative meta-analysis.

In contrast, an investigation of meta-analyses of trials exploring selective decontamination of the digestive tract (a form of antibiotic prophylaxis for patients in intensive care units that aimed to reduce infections) found that there was little difference in outcome between three types of approach (Table 14.2; D'Amico *et al.*, 1998). This comparison used data from 17 trials for which three methods of systematic review (data extracted from publications, tabular summary data provided by trialists, or IPD provided by trialists) could be used, out of a total of 25 trials (4310 patients) identified as relevant to the therapeutic question.

In this case the authors concluded that, despite the advantages of IPD, the approach is costly and that analysis of summary data *provided by study investigators* is a valid alternative to IPD, when categorical data are used and censoring is not relevant. Here the authors aimed to compare the differences attributable only to the data type and used exactly the same trials in each analysis. It is not clear how the most comprehensive and inclusive analysis including the additional eight trials (where data were not available in all three formats) would have affected the outcome and estimates achieved with each of the three approaches.

Table 14.2 Comparison of meta-analysis approaches using an example in selective decontamination of the digestive tract (D'Amico *et al.*, 1998; Stewart, 2003).

Type of meta-analysis	Deaths/Patients	Odds ratio	95 % CI
Data extracted from publications	762/3142	0.87	0.74–1.03
Summary data obtained from trialists	975/3564	0.92	0.80–1.08
IPD obtained from trialists	829/3357	0.89	0.76–1.05

Reproduced from *Clinical Trials in Cancer: Principles and Practice*, David Girling *et al.* (eds) (2003). Reprinted by permission of Oxford University Press.

REPORTING BIAS

Can the IPD approach help avoid reporting biases in systematic review?

There are unquestionably limits to relying only on data presented in published reports, and systematic reviews based only on data and information presented in published reports risk incorporating and potentially compounding any biases associated with individual publications. Unpublished trials are not included, data may be inconsistent or incompatible across trials, papers frequently present inadequate information and can be guilty of selective reporting of patients or outcomes. Consequently, systematic reviews based only on the information that can be extracted from published reports are often able to include only a proportion of the published trials and a lower still proportion of the totality of evidence. Collecting IPD also enables analyses to be done centrally and methods to be applied consistently across trials. If there have been problems with the published analyses, particularly those that may lead to biased assessment, these can often be corrected. For example, intention-to-treat analyses can be carried out even if this was not done originally or reported in the trial publication.

Publication bias: bias from unpublished trials and those published in the grey literature

There is good evidence that trials with more promising results in favour of a new or test intervention are more likely to be published than those with negative or inconclusive results (Dickersin, 1990; Dickersin *et al.*, 1992; Easterbrook *et al.*, 1991). Systematic reviews based on trial publications, by definition, do not include unpublished trials. Also trials published in the grey literature, particularly those presented in abstract form, often do not present sufficient information to be included in a systematic review. Consequently, any quantitative analyses are likely to be based on only a subset of the relevant evidence, and that evidence is likely to be biased towards the positive. IPD meta-analyses usually seek data from all trials irrespective of whether or where they are published, although data may not always be obtained for all trials. The collaboration and direct contact with the specific research community that is integral to the IPD approach is extremely valuable in identifying unpublished material. When inviting participation in an IPD project, trialists are usually asked to supplement a provisional list of trials with any other trials that they are aware of, including those that are unpublished or are reported in obscure places. These inquiries supplement other mechanisms of locating relevant trials, including bibliographic searches, consulting trial registers, where available (see Chapter 3 on prospective meta-analyses), hand-searching, contacting manufacturers and experts in the field (where appropriate) and often 'advertising' the IPD meta-analysis through a short journal communication that includes a request for information on any relevant trials that have been missed. Furthermore, because the original trial data are sought, there is essentially no difference between trials that are unpublished, published in the grey literature or fully published – the data obtained are in the same format, and dealing with unpublished material requires no special provision in terms of data handling.

Concerns have been raised about the quality of data from unpublished sources as they have not been subject to peer review. However, obtaining the IPD and study protocol enables much more rigorous checking, review and appraisal of the study design and the actual data set by the project secretariat and advisory group than is generally possible in peer review for a journal submission. Conversely, publication of a well-written paper in a respected journal with a good peer-review mechanism does not necessarily guarantee the quality of the underlying data.

Example

We explored the impact of data from unpublished sources and from the grey literature using the information collected for 13 completed IPD meta-analyses in the treatment of cancer, coordinated by our group (Burdett *et al.*, 2003). Our aim was to compare the results of a comprehensive systematic review that included all trials irrespective of source with the more common type of systematic review that is restricted to those trials fully published in English-language journals only.

In each of the original IPD reviews, we attempted to locate all relevant RCTs irrespective of whether they were published or not. This was done by conducting electronic bibliographic searches, hand-searching relevant review articles and conference proceedings and the bibliographies of trial reports, searching trial registers, and consulting widely with trialists and other relevant experts. Two of our meta-analyses included no unpublished or grey literature trials, therefore we investigated 11 IPD reviews (Advanced Ovarian Cancer Trialists Group, 1991, 1998; Advanced Bladder Cancer Overview Collaboration, 1995; Non-small Cell Lung Cancer Collaborative Group, 1995; PORT Meta-analysis Trialists Group, 1998; Sarcoma Meta-analysis Collaboration, 1997). As IPD were used in all analyses, any differences between the trial data sets attributable to factors such as differential follow-up, data maturity, patient exclusion and inappropriate analyses were minimized. The data checking procedures that were done as part of the original IPD projects also ensured that all trial data could be considered to be of high quality.

Survival results for (a) RCTs fully published in English-language journals and (b) RCTs fully published in English-language journals plus the grey literature plus unpublished trials, henceforth termed 'grey+' data, were compared. For each meta-analysis, time-to-event analyses were carried out. Pooled hazard ratios were calculated in a stratified analysis using the trial log rank observed-minus-expected number of events $(O - E)$ and variances calculated from individual patients' survival times within each trial. The hazard ratio of hazard ratios (HRHR) was calculated by pooling summated $O - E$ and variances from each individual meta-analysis.

The 11 IPD reviews included a total of 120 RCTs and 18 377 patients. Forty-five RCTs, including 6221 patients (34 % of all patients) were from unpublished or grey sources. These 'grey+' data were made up of unpublished data (53 %), abstracts (38 %), book chapters (7 %) and RCTs published in non-English-language journals (2 %). Results for all meta-analyses are shown in Figure 14.1.

In general, analyses based only on RCTs that were fully published in English-language journals tended to give more favourable results than those that also included 'grey+' data. Most observed differences were relatively modest. The HRHR for the results of the fully published data is 0.93 (95 % CI 0.90–0.97),

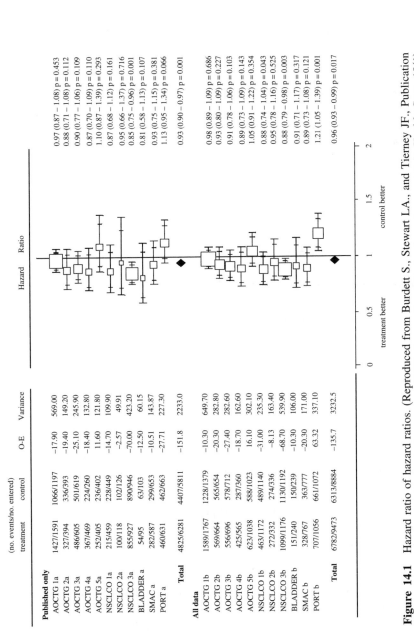

	(no. events/no. entered) treatment	control	O-E	Variance	Hazard Ratio
Published only					
AOCTG 1a	1427/1591	1066/1197	−17.90	569.00	0.97 (0.87 − 1.08) p = 0.453
AOCTG 2a	327/394	336/393	−19.40	149.20	0.88 (0.71 − 1.08) p = 0.112
AOCTG 3a	486/605	501/619	−25.10	245.90	0.90 (0.77 − 1.06) p = 0.109
AOCTG 4a	367/469	224/260	−18.40	132.80	0.87 (0.70 − 1.09) p = 0.110
AOCTG 5a	252/405	236/402	11.60	121.80	1.10 (0.87 − 1.39) p = 0.293
NSCLCO 1a	215/459	228/449	−14.70	109.90	0.87 (0.68 − 1.12) p = 0.161
NSCLCO 2a	100/118	102/126	−2.57	49.91	0.95 (0.66 − 1.37) p = 0.716
NSCLCO 3a	855/927	890/946	−70.00	423.20	0.85 (0.75 − 0.96) p = 0.001
BLADDER a	54/95	63/103	−12.50	60.15	0.81 (0.58 − 1.13) p = 0.107
SMAC a	282/587	299/653	−10.51	143.87	0.93 (0.75 − 1.15) p = 0.381
PORT a	460/631	462/663	−27.71	227.30	1.13 (0.95 − 1.34) p = 0.066
Total	4825/6281	4407/5811	−151.8	2233.0	0.93 (0.90 − 0.97) p = 0.001
All data					
AOCTG 1b	1589/1767	1228/1379	−10.30	649.70	0.98 (0.89 − 1.09) p = 0.686
AOCTG 2b	569/664	565/654	−20.30	282.80	0.93 (0.80 − 1.09) p = 0.227
AOCTG 3b	556/696	578/712	−27.40	282.60	0.91 (0.78 − 1.06) p = 0.103
AOCTG 4b	425/565	287/360	−18.70	162.60	0.89 (0.73 − 1.09) p = 0.143
AOCTG 5b	623/1038	588/1023	16.10	302.10	1.05 (0.91 − 1.22) p = 0.354
NSCLCO 1b	463/1172	489/1140	−31.00	235.30	0.88 (0.74 − 1.04) p = 0.043
NSCLCO 2b	272/332	274/336	−8.13	163.40	0.95 (0.78 − 1.16) p = 0.525
NSCLCO 3b	1099/1176	1130/1192	−68.70	539.90	0.88 (0.79 − 0.98) p = 0.003
BLADDER b	151/240	150/239	−10.30	106.00	0.91 (0.71 − 1.17) p = 0.317
SMAC b	328/767	363/777	−20.30	171.00	0.89 (0.73 − 1.08) p = 0.121
PORT b	707/1056	661/1072	63.32	337.10	1.21 (1.05 − 1.39) p = 0.001
Total	6782/9473	6313/8884	−135.7	3232.5	0.96 (0.93 − 0.99) p = 0.017

treatment better   control better

Figure 14.1 Hazard ratio of hazard ratios. (Reproduced from Burdett S., Stewart LA., and Tierney JF., Publication bias and meta-analyses – A practical example (2003). *International Technology Assessment in Health Care*, 19(1) 129–34.)

$p = 0.001$, compared with the HRHR for all data of 0.96 (95 % CI 0.93–0.99), $p = 0.017$ (Figure 14.1).

Including the 'grey+' data added more patients (a third more over all the IPD reviews), increased statistical power and led to improved confidence in the results. We found that in nine individual reviews the estimated treatment effect moved from a positive result towards a null result (e.g. AOCTG 5). The most extreme example of this was noted in a review that included only a small number of RCTs (Bladder), perhaps reflecting the instability of results based on small numbers of trials and individuals. In one review (PORT) adding the 'grey+' data led to a more extreme estimate of harm from the experimental treatment, making the result even more negative. This is interesting, as the results from unpublished RCTs were statistically significant, perhaps reflecting a subconscious desire to publish the encouraging results rather than discouraging ones. In one case adding 'grey+' data gave more encouraging results (SMAC).

The overall trend is similar to that found by other studies (McAuley *et al.*, 2000; Sterne *et al.*, 2000). The observed effect is less pronounced than reported by McAuley *et al.* (2000), possibly because a number of potential biases and differences between RCTs, such as differential follow-up and patient exclusion, have been minimized by the IPD approach.

Another barrier to including all trials, and so a potential source of bias when relying on published data, is that trials do not necessarily report sufficient information with which to calculate the necessary statistics to estimate the effect of interventions. If positive trials report their findings in more detail than those with null results, which may simply state that there was no significant difference between interventions, then this could introduce bias to a meta-analysis based on published data. About 85 % of the 138 trials included in our 14 completed and published meta-analysis were published in some form. Only about 60 % of these published trials (about half of all trials) provided sufficient data on the primary endpoint of survival to enable a meta-analysis to be carried out. For secondary outcomes of interest the inability to include eligible trials is likely to be more pronounced.

Duplicate publication bias: bias from including trials more than once in consequence of multiple publications

Although there may be good reasons for publishing a trial more than once – for example, reporting the results of long-term outcomes or presenting results in a particular format that is more accessible to a certain audience – there should be clear cross-referencing between such reports. It is important to actively ensure that each eligible trial is included only once in a systematic review. However, in practice it can sometimes be difficult to distinguish multiple publications of the same trial because authors change, different numbers of patients are reported or the trial design is described differently. This is particularly true of abstracts because they may report interim or immature data and the author may be the individual attending a particular conference rather than the principal investigator or trial group. If covert duplication leads to trials being included more than once in a systematic review it may compound publication bias. It is likely to be trials with impressive results that will be the focus of more than one publication and revisited over time, leading to overrepresentation and overestimation of the treatment effect (Tramèr *et al.*, 1997). Once again direct communication with trialists involved in an IPD meta-analysis can resolve these issues, ensuring that only the appropriate trials are included in the systematic review – trialists are not likely to provide the same data set more than once and it would be easy to tell from the original data if the same trial was supplied twice. For reviews based on data presented in publications there is a dilemma in using data presented in meeting abstracts: on the one hand, to avoid potential publication bias associated with excluding the grey literature we would wish to include such data; but on the other hand, to avoid duplicate reporting bias we might wish to be cautious about including it.

Example

Prior to an IPD meta-analysis in soft tissue sarcoma – a type of tumour that can affect the connective tissues in almost any area of the body, but most commonly occurs in the torso and limbs (Sarcoma Meta-analysis Collaboration, 1997) – we carried out a preliminary systematic review and meta-analysis of data from trial publications (Tierney *et al.*, 1995). We identified publications that appeared to be pertaining to 15 trials. It was only when we sought IPD and entered into discussion with the individuals and groups responsible for the trials that we were able to ascertain that a publication that seemed to report the results of single trial (Chang *et al.*, 1988), actually described the combined results of two separate trials. In addition, the results of another trial were published separately for two different subsets of patients as though they were two distinct RCTs (Glenn *et al.*, 1985a, 1985b) and in one subset there was a borderline significant result. This means that the results of the meta-analysis of data extracted from publications were potentially biased and that the statistical methods had not been applied appropriately.

Selection bias: bias from inclusion of inappropriate studies

Systematic reviews should specify a series of criteria that prospectively define the types of trials to be included. This then sets the scope of the review, and objectively applying these criteria can provide a safeguard against selection bias that could be associated with *ad hoc* inclusion/exclusion of certain studies (see Chapter 3 on prospective meta-analyses). To ensure the validity of the review, it is important that only eligible trials are included.

For example, the majority of systematic reviews of health-care interventions are limited to RCTs because there is evidence that results of randomized and non-randomized studies differ (Chalmers *et al.*, 1997; Sacks *et al.*, 1982; Diehl and Perry, 1986). However, not all trials published as such are in fact randomized. On further inquiry, it can transpire that allocation has been done by quasi-random methods such as date of birth, date of clinic visit or alternate allocation. In each of these the treatment is known before the decision to enter the trial and allocation is made. They are therefore potentially open to conscious or subconscious subversion whereby particular types of individual are selected for particular treatments. There is evidence that trials allocating interventions by such methods are likely to have more promising results than those that allocate treatments using appropriate methods of concealment (Schulz *et al.*, 1995). Consequently, inclusion of such trials in a systematic review and meta-analysis could potentially bias the results. The direct communication with trialists involved in an IPD meta-analysis allows us to establish the method of allocation and concealment and restrict the review to those trials that were 'properly randomized'. This dialogue, which takes place early in the project, also means that any other issues relating to uncertain eligibility can be clarified and only those trials that fully meet the inclusion criteria are included.

Implementation bias: bias from trial conduct

The data checking procedures that are integral to many IPD meta-analyses can sometimes reveal aspects of trial implementation or conduct which could potentially introduce bias to the analyses and which in some cases can be corrected.

Example

In an IPD meta-analysis in multiple myeloma (a cancer of plasma cells), as part of the normal data checking procedures, plotting the randomization sequence of the IPD supplied from one unpublished trial from India, comparing radiotherapy with chemotherapy, revealed the pattern shown in Figure 14.2. We see that from the start of the trial until early 1985 there are approximately equal numbers of patients randomized in each arm and the curves cross frequently. However, in the middle of 1985 the curves diverge – the cumulative accrual to the chemotherapy arm continues to rise, whereas the radiotherapy arm remains flat for a period, and the curves remain divergent for the rest of the accrual period. On further inquiry, it

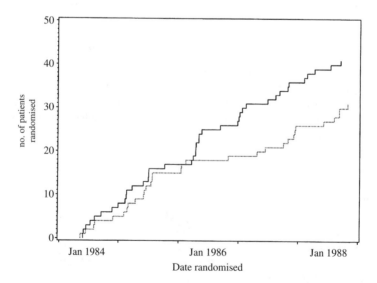

Figure 14.2   Cumulative randomization plot for a trial with a non-random phase during recruitment. (Reproduced with permission from Stewart and Clarke, 1995.)

transpired that in this trial the radiotherapy equipment was unavailable for 6 months, but during this time patients continued to enter the chemotherapy arm. In effect this was a trial in three sections, with the middle section being non-randomized. As they had the IPD, the meta-analysis secretariat, with agreement from the trialists, were able to remove those patients who had entered the trial during the middle non-randomized phase and analyse only those that had been randomized.

Patient-exclusion bias: bias from differential exclusion of patients

The use of random allocation in a trial should ensure the unbiased assignment and comparison of therapeutic groups. However, this can only be guaranteed if *all* randomized patients are analysed according to the treatments initially assigned by an intention-to-treat approach (Altman, 1991; Lachin, 2000; Schulz and Grimes, 2002). Yet trial investigators frequently exclude patients from trial analyses for reasons of ineligibility, protocol violations, pre-treatment outcomes, or losses to follow-up. If patients are excluded from analyses for reasons that are related to treatment and outcome, this may bias estimates of the effect of treatment in unpredictable ways. For example, if patients died from acute adverse effects associated with a new treatment and were subsequently excluded as 'early deaths' then the estimate of effect would be biased in favour of the new treatment. Moreover, combining these estimates in a meta-analysis could aggregate and compound any such biases (Lewis and Machin, 1993). Exclusion of participants does not necessarily introduce bias, but without information on such participants we are unable to assess their influence. Meta-analyses of IPD provide the opportunity to gather information on all randomized patients, even those excluded from the investigators' own analyses. This then allows an analysis of all randomized patients, which is the most robust and least biased approach. No value judgments are made as to whether individual patient exclusions are appropriate. It also allows us to assess the impact of patient exclusions.

Example

We have examined the effect of excluding patients after randomization on the results of our group's completed and published IPD meta-analyses in cancer (Tierney and Stewart, 2005). Because we collect all trials (published and unpublished), update follow-up and collect the same outcomes for each trial, we were able to do this in a context where other potential sources of bias such as publication bias, follow-up bias and outcome-reporting bias are limited. The 14 systematic reviews and meta-analyses of IPD addressing therapeutic questions used (Advanced Ovarian Cancer Trialists Group, 1991, 1998; Advanced Bladder Cancer Overview Collaboration, 1995; Arnott *et al.*, 1998; Glioma Meta-analysis Trialists Group, 2002; Non-small Cell Lung Cancer Collaborative Group, 1995; PORT Meta-analysis Trialists Group, 1998; Sarcoma Meta-analysis Collaboration, 1999) varied by cancer site, type and timing of the research treatment, number of trials, number of patients and the direction of overall treatment effect for survival, the primary endpoint. In all, 133 RCTs and 21 905 patients were included (Tierney and Stewart, 2005).

For each RCT and meta-analysis, we compared an intention-to-treat analysis of survival for all randomized patients, with an analysis based on just those patients included in the investigators' analysis. We looked at both hazard ratio estimates of effect and the associated heterogeneity statistics. Ninety-two trials (69 %) excluded between 0.3 % and 38 % of patients randomized. These exclusions varied substantially by meta-analysis, but with a tendency for more patients to be excluded from the treatment arm. Comparing the trial analyses based on 'included' patients with analyses of all randomized patients, there was no clear indication that the exclusion of patients altered the results more in one direction than another ($t = 1.537$, $p = 0.13$). In contrast, comparing the 14 pooled meta-analysis results, there was a tendency for the hazard ratios for 'included' patients to be more in favour of the research treatment than the HRs based on all patients ($t = 2.401$, $p = 0.03$; Figure 14.3).

The differences tended to be small and generally would not have altered the interpretation of the results. However, in the soft tissue sarcoma meta-analysis (Sarcoma Meta-analysis Collaboration, 1997), the HR changed from 0.90 ($p = 0.157$) to 0.85 ($p = 0.056$) when patients were excluded, and this approached conventional levels of significance (Table 14.3). Such a change in the size of the effect could have had an impact on the perception of the results of this particular meta-analysis. Also, the level of inconsistency was modestly increased as a result of the investigators' exclusions in 8 of the 14 meta-analyses. In fact there were some instances where an overall estimate of treatment effect in a meta-analysis of 'included' patients might not have been considered justified because of the level of inconsistency among trials, whilst a meta-analysis of all patients probably would have been.

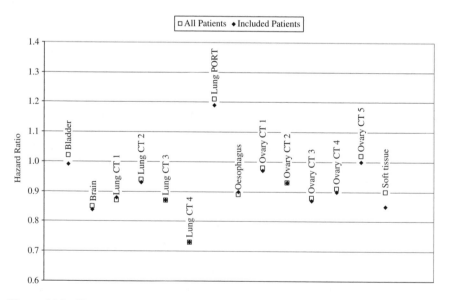

Figure 14.3 Hazard ratios of 'included' patients and of all patients for individual IPD meta-analyses. (Reproduced from 'Investigating patient exclusion bias in meta-analysis', Jayne F. Tierney and Lesley A. Stewart, in *International Journal of Epidemiology*. Reprinted by permission of Oxford University Press.)

There may be good clinical reasons for excluding certain types of patients from a trial or meta-analysis and comparing this secondary analysis to the primary intention-to-treat analysis. If there is agreement between the two, this could strengthen the conclusions drawn. However, to limit bias in the meta-analysis setting, any deliberate exclusions should be prespecified in the meta-analysis protocol and applied objectively and consistently across all trials. Ideally, their impact should then be assessed by sensitivity analyses (including and excluding patients to determine whether it impacts on the treatment effect). In the soft tissue sarcoma (Sarcoma Meta-analysis Collaboration, 1997) meta-analysis, exclusion of patients changed the HR for survival from 0.90 ($p = 0.16$) to 0.85 ($p = 0.06$); see Table 14.3. In contrast, sensitivity analyses of prespecified patient exclusions, done uniformly across all trials, made little impact on the estimated effect of chemotherapy (Sarcoma Meta-analysis Collaboration, 1997; Table 14.3), even although these were the same reasons used in different combinations by investigators to exclude patients from their own trial analyses.

Although not all trials may keep records of excluded patients, our experience in the cancer field has been good and we have been able to recover information from the majority of excluded patients. For example, in the soft tissue sarcoma IPD meta-analysis we were able to recover data on 99 % of the 344 patients who had been excluded from the investigators' own analyses. In a more recent meta-analysis in high-grade glioma we were able to recover data from 83 % of the 253 patients that had been excluded from the original analyses.

Follow-up bias: bias from differential follow-up

In circumstances where outcome events take place over a period of time, bias could be introduced if the pattern of repeated measurements differs according to the intervention. For example, if cancer patients receiving one treatment were observed

Table 14.3 Hazard ratios and *p*-values for survival in the soft tissue sarcoma meta-analysis (Sarcoma Meta-analysis Collaboration, 1997), based on investigator and prespecified exclusions.

Patients excluded	Events/Patients	Hazard ratio	*p*-value
None	709/1568	0.90	0.16
Patients excluded *ad hoc* by investigators' from their trial analyses	553/1227	0.85	0.06
Prespecified groups of patients (with locally recurrent disease; less than 15 years old; with metastatic disease; or who received induction chemotherapy) excluded uniformly across trials	597/1366	0.91	0.28

Reproduced from 'Investigating patient exclusion bias in meta-analysis', Jayne F. Tierney and Lesley A. Stewart, in *International Journal of Epidemiology*. Reprinted by permission of Oxford University Press.

more frequently or observed for longer than those allocated to another treatment, then we would be more likely to observe recurrences on the first treatment and analyses would be biased in favour of the other. Similarly, if patients are 'lost to follow-up' from the trial at differential rates across treatment arms, then this could introduce bias, for example, if it was the most ill patients who did not return for hospital visits or observation. In many IPD meta-analyses a standard part of the checking procedures is to look at the censoring patterns over time and check that they are consistent across treatment arms. Where possible, trialists are asked to bring information up to date, and if possible to locate data from missing patients through hospital records or population-based death registries.

Example

In the IPD meta-analysis of soft tissue sarcoma (Sarcoma Meta-analysis Collaboration, 1997), already described, follow-up was quite well reported. The median follow-up of living patients could be obtained from publications for seven of the trials with a range of medians across trials of 16–64 months. However, when investigators supplied IPD they were in some cases able to supply many years of extra follow-up information, thus giving a range of medians of 74–204 months (Figure 14.4). This provided a unique opportunity to look at long-term outcomes in a relatively young group of patients and avoided the potential biases of early trial reporting.

Time-point bias: bias from timing of analyses

If events happen over a prolonged period, then the time point at which trial results are reported could be influential – for example, selection of time points based on maximal or minimal difference between treatments would introduce bias. IPD meta-analyses, at least those interested in time-to-event outcomes, usually request that, as far as possible, data are brought up to date. For such time-dependent outcomes IPD also allows the most appropriate and informative analysis of data derived from the

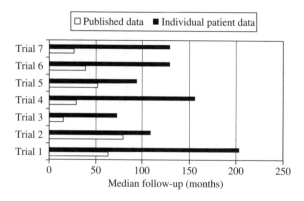

Figure 14.4   Median follow-up of published and individual patient data for trials included in the soft tissue sarcoma meta-analysis (Sarcoma Meta-analysis Collaboration, 1997).

whole period of follow-up. Rather than summarizing the overall number of events at a particular point in time (as in an odds ratio), the time at which each individual event takes place can be used (to calculate a hazard ratio).

The IPD meta-analysis comparing ovarian ablation with control in the treatment of breast cancer (Clarke and Godwin, 1998) highlights the potential effect of maturity of data and timing of analysis in trial publications. A meta-analysis of published data did not suggest any significant effect of ovarian ablation on survival in women with breast cancer, while a meta-analysis of updated IPD showed a highly significant 10.2 % effect on survival at 15 years. The survival curves produced from the IPD revealed that this effect was not apparent in the first few years post-randomization, with an increasing benefit as time passed.

This issue may be less important in situations where we are interested only in dichotomous outcomes. However there are still potential problems that can be associated with the timing and reporting of events. For example, although death is an unequivocal outcome, if used as a binary event, it can be measured at different times and the 'correct' point at which to measure it will depend on the specific research question, the mechanisms and timing of the disease and the study design (Cohen et al., 2001). For example, in trials of the treatment of sepsis (life-threatening inflammatory response produced by the body in response to severe infection), patients are usually ventilator-dependent and the normally 'hard' endpoint of death can becomes less well defined for those who do not recover. There are anecdotal reports suggesting that life support systems can be switched off the day after the end of the trial period. This could obviously be a source of bias if decisions to end the life support were influenced by the trial treatment, and one that might not be immediately apparent to a reviewer. A publication could report survival rates at 14 days, but the rate at 15 days could be quite different. Requesting the actual date of death from the researchers would reveal such problems, even if the reviewer wished simply to analyse event rates (and not time-to-event).

Outcome-reporting bias: bias associated from selective reporting of outcomes

There are a myriad of ways that trial data can be analysed and presented. Clinical trials often explore multiple outcomes and carry out multiple subgroup analyses. They may explore certain data using different analytical tools. Paper journals do not have space to present results of all the analyses that have been done. Selection is inevitable and if they present only the most favourable, exciting or significant outcomes, this may be biased and inconsistent with the trial Protocol. Such reports and systematic reviews that incorporate them may therefore be unreliable (Chan et al., 2004). IPD analyses can avoid this problem by obtaining data for outcomes not reported in publications and can present the same outcomes or subgroup analyses across trials.

Availability bias: bias from differential availability of trial data

The basic premise of the IPD approach is that access to the 'raw' data enables us to get closer to the true answer to an important question because we are able to check, thoroughly explore and reanalyse the data from each trial in a consistent way. We do not have to rely on interpreting information and analyses presented in published reports or be constrained by summary data provided in tabular format. However, the success and validity of the approach require that data from all or nearly all trials will be available. There is an issue about trials for which IPD are not available, for whatever reason (e.g., loss or destruction of data, or unwillingness to collaborate). If unavailability is related to the trial results – for example, if trialists are keen to supply data from trials with promising results but reluctant to provide data from those that were less encouraging – then ignoring the unavailable trial could bias the results of the IPD analysis. It is important, therefore, not to introduce more problems through trial-availability bias than are solved by avoiding the other biases related to the publication process already described. If a large proportion of the data have been obtained, perhaps 90 % or more, we can be relatively confident of the results, but with less information we need to be suitably circumspect in drawing conclusions. Sensitivity analysis, combining the results of any unavailable trials (as extracted from publications) and comparing these with the main IPD results, is a useful aid to interpreting the data. As for other types of systematic review, IPD meta-analyses should clearly state what trials were not included and the reasons why. Clearly, if only a limited number of trials are able to provide IPD for analysis, then the validity of the approach is questionable. Fortunately, our own experience and that of others working in cancer has been good, and in most cases perseverance has led to data being available from a high proportion of eligible trials. Table 14.4 gives details of the proportion of trials and patients obtained for our most recent meta-analyses.

Table 14.4 Proportion of randomized evidence obtained in the UK MRC Clinical Trials Unit's most recently completed IPD meta-analyses.

Cancer	Trials obtained/identified	Total % randomized evidence obtained (% patients)
Advanced ovarian[a]	37/40	95
Bladder[b]	10/11	88
Cervix[c]	21/24	94
Glioma[d]	12/19	81
Lung[e]	52/58	95
Oesophagus[f]	5/5	98
PORT[g]	9/9	99
Soft tissue sarcoma[h]	14/17	98

[a] Advanced Ovarian Cancer Trialists' Group (1998).
[b] Advanced Bladder Cancer (ABC) Meta-analysis Collaboration (2003).
[c] Neoadjuvant Chemotherapy for Cervix Cancer Meta-analysis (NACCCMA) Collaboration (2003).
[d] Glioma Meta-analysis Trialists (GMT) Group (2002).
[e] Non-small Cell Lung Cancer Collaborative Group (1995).
[f] Arnott et al. (1998)
[g] PORT Meta-analysis Trialists Group (1998).
[h] Sarcoma Meta-analysis Collaboration (1997).

DECIDING WHEN OBTAINING IPD IS LIKELY TO BE HELPFUL (IN AVOIDING BIAS)

We have illustrated ways in which using the IPD approach to meta-analysis can help minimize the many potential biases associated with trial publication (and also how the data obtained for IPD meta-analyses can provide a valuable resource for exploring trial methodology). However, it is not always necessary to go to the lengths of collecting IPD. For example, if one is interested in dichotomous outcomes that occur relatively quickly, where all relevant trials are published and data are presented in a comprehensive and compatible way, then the most straightforward type of meta-analysis, based on data presented in trial publications, is probably all that is required. Such an extreme is likely to be rare, and usually unpublished trials will need to be assessed and trialists contacted to provide at least some additional summary data. When initiating a systematic review it is useful to consider carefully which approach and which type of data will be most appropriate. Particular thought should be given to factors that are likely to introduce bias to the systematic review. There may be cases where the benefits turn out to be marginal, and others where it could be vital. The difficulty is that it is often not possible to assess the likely impact of the many possible biases that may be introduced in the transformation from raw data to published report without obtaining the further information. For example, many publications make no reference to whether or not any participants have been excluded from reported analyses. This may reflect the fact that none have

Table 14.5 Factors to consider when deciding whether an IPD meta-analysis is required. Factors relating to publication biases are shaded.

When IPD may be beneficial	When IPD may not be beneficial
Many trials unpublished or published in the grey literature	All or nearly all trials published in peer-reviewed journals (provided sufficient information presented)
Poor reporting of trials. Information inadequate, selective or ambiguous	Detailed and clear and unbiased reporting of trials to CONSORT standards (Begg *et al.*, 1996; Moher *et al.*, 2001)
Long-term outcomes	Short-term outcomes
Time-to-event outcome measures	Binary outcome measures
Multivariate or other complex analyses	Univariate or simple analyses
Differently defined outcome measures	Outcome measures defined uniformly across trials
Subgroup analyses of patient-level characteristics important	Patient subgroups not important
IPD available for high proportion of trials/individuals	IPD available for only a limited number of trials
High proportion of individuals excluded from analyses	Reported analyses include almost all individuals

Modified and reproduced from *Evaluation and the Health Professions*, Vol. 25(1), 76–97, © 2002. Reprinted by Permission of Sage Publications, Inc.

been excluded, or that a considerable number have been removed but that this has not been stated. With the IPD approach we can specifically ask trialists whether any participants have been excluded from the analyses. We can also query any likely exclusions that have been noted during data checking – for example, if there are identification numbers missing from a sequential list. Spotting if individuals have been excluded from trials should in time become easier as more trial reports use the checklist and flow diagram associated with the CONSORT guidelines (Begg *et al.*, 1996; Egger *et al.*, 2001; Moher *et al.*, 2001). However, it cannot help with older trials published in the pre-CONSORT era. Of course, even if there have been substantial numbers of participants excluded, these may not necessarily introduce bias to the trial analyses or subsequent results. The problem is that without access to the data from such participants, we are unable to assess their influence.

Table 14.5 lists a number of factors that should be considered at the outset of a systematic review which may help in deciding whether or not an IPD meta-analysis is required.

PUBLICATION AND TIME-LAG BIAS OF IPD META-ANALYSES

Well-designed and properly conducted IPD meta-analyses may represent a 'gold standard' of secondary clinical research that is less likely than other types of review to be influenced by biases associated with the publication of individual trials. It is interesting to explore whether these IPD reviews may also be subject to the same sort of publication bias as primary clinical research. Are systematic reviews based on IPD with statistically significant results being published more readily, more quickly and in higher-impact journals than similar projects that reveal negative or inconclusive results?

Example

We surveyed those responsible for all known IPD meta-analysis projects in cancer using a questionnaire to determine descriptive characteristics of the meta-analysis, the nature of the results and details of the publication history. Data were available on 38 of 44 potentially eligible IPD meta-analyses in cancer (including updates) and at the time of the survey 30 had been published in full (Tierney *et al.*, 2000). The studies varied considerably in the number of trials, patients, treatment comparisons, outcomes and cancer sites or types.

Unlike previous studies of clinical trials (Dickersin, 1990; Dickersin *et al.*, 1992; Easterbrook *et al.*, 1991) there was no good evidence to suggest that publication of IPD meta-analyses in cancer is biased by the statistical significance or perceived importance of the results. Nonetheless, there was some suggestion of time-lag bias whereby manuscripts containing results that are not statistically significant (Figure 14.5(a) and (b)) or not regarded as clinically significant (Figure 14.5(c)) take longer to publish (Figure 14.5), an aspect of publication bias that has been observed for other clinical studies (Ioannidis, 1998). Such IPD reviews

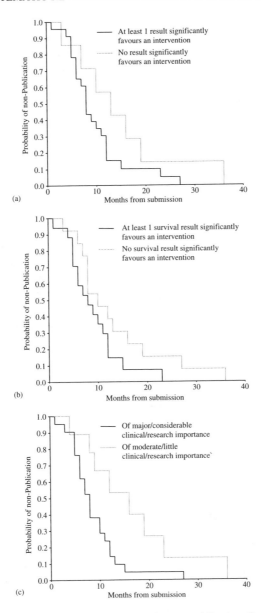

Figure 14.5 Results of IPD meta-analyses and time to publication, from manuscript submission, by (a) the significance of results, (b) the significance of the survival results, and (c) the meta-analysts' impression of results. There is one censored observation, a meta-analysis that was submitted but unpublished at the time of the survey. (Reproduced from Jayne Tierney, Mike Clarke and Lesley A. Stewart, 'Is there bias in the publication of individual patient date meta-analyses?' (2000). *International Technology Assessment in Health Care*, Cambridge University Press.)

(Stern and Simes, 1997) also tended to be published in lower-impact journals. Overall, however, there seemed to be no strong association between the results of IPD meta-analyses in cancer and publication. The general high regard of IPD meta-analyses in cancer, together with the resource-intensive, highly collaborative and international nature of these projects, may ensure that publication arises eventually, irrespective of the results.

CONCLUDING REMARKS

Systematic reviews and meta-analyses that approach researchers directly to collect 'raw' data have been used across a range of health-care fields, and offer a number of advantages that could be very important in testing particular hypotheses.

Collecting IPD enables time-to-event analyses – for example, time to onset of symptoms, time to recovery, time free of seizures, time to conception. Indeed, one of the main reasons why IPD meta-analyses have been so important in the cancer field is that time-to-event analysis of survival is vital in evaluating therapies. For most tumours we can anticipate only a prolongation of survival rather than a cure. Therefore, it is important to measure not only whether a death happens, but also the time at which it takes place. Collecting IPD is also the only practical way to carry out subgroup analyses, to investigate whether any observed effect of an intervention is consistent across well-defined groups of participants – for example, whether women gain a smaller or larger benefit from treatment than men. In conventional analyses using aggregate data from publications, it is usually extremely difficult to extract sufficient compatible data to undertake meaningful subgroup analyses, especially if we want to characterize individuals by two or more factors at a time. Subgroup analyses can be done using tabulated data provided for each category of interest for each trial. However, this can be both cumbersome and time-consuming for those researchers providing the information. In contrast, IPD permits straightforward categorization of individuals for subgroup analysis (stratified by trial) defined by single or multiple factors.

Meta-regression has been proposed as an alternative method of identifying significant relationships between the treatment effect and covariates of interest. However, not all publications report the covariates of interest, limiting power of the analyses. Furthermore (when IPD are not available), the unit of regression is restricted to the trial. Covariates that summarize attributes of the trial population such as median age may be difficult to interpret, as what is true for a trial with a median participant age of 60 may not necessarily be true for a 60-year-old patient. A simulation study comparing meta-regression with IPD subgroup analyses of RCTs found that the statistical power of meta-regression was dramatically and consistently lower than that of the IPD, with little agreement between the results obtained by the two methods on the same data set (Lambert et al., 2002). Collecting IPD is therefore the most practical and reliable means of assessing covariate treatment interactions to determine whether there are particular types of individual who benefit more or less from the intervention in question. Access to the IPD also permits an in-depth exploration of patient characteristics themselves, irrespective of the intervention. For example, the large data sets collected can be used in the construction of prognostic indices that may be able to predict outcome based on patient characteristics (International Germ Cell Cancer Collaborative Group, 1997).

Collecting data at the level of individual participants can also allow translation between different staging, grading, ranking or other scoring systems, and therefore may also enable pooling of data from studies that would not be possible at the trial level. In bypassing dependence on published reports we may also be able to avoid problems related to inadequate or ambiguous reporting of individual trials. The detailed checking that is possible with IPD should also help ensure the quality of the data.

We have shown some empirical evidence that excluding unpublished trials or those included in the grey literature, or using incomplete data sets that have excluded certain participants, can influence the results of a systematic review. The direction of influence is not always predictable. The unavailability of certain information, owing to selective presentation of results in published reports, could also limit the data available for analysis and introduce bias. Other factors, such as the maturity of results or points in time at which results are presented, may also be influential in particular situations. Although individually the effects of some these factors may be small, when brought together in a systematic review the cumulative effect could be quite large as in the ovarian and breast cancer examples discussed previously, and could lead to different interpretations. If we are able to collect all or nearly all of the total randomized evidence in an IPD meta-analysis, then we should be able to circumvent such problems and obtain a reliable and in-depth investigation of our research question. The approach cannot, generally, help avoid biases associated with trial design or conduct. However, if such problems are discovered through data checking procedures or discussions with the researchers, we may be able to find a solution, as in the multiple myeloma example, or alternatively we may decide to exclude the trial from the review. However, we must ensure that by restricting analyses to those trials that can supply IPD, we do not introduce bias through skewed trial availability.

We believe that the IPD approach to systematic review offers an alternative, and in many cases better, approach, to using statistical methods to detect bias and impute the results that would have been obtained if a systematic review had been able to obtain all the research data. Rather, we make considerable efforts to actually obtain all the research data. Of course, there are situations where the approach will simply not be feasible, because data have been destroyed or because, despite every effort, researchers do not wish to collaborate. There may also be others where it may not be necessary – for example, if we are sure that all the required data are readily available in a suitable format within publications. In certain fields it may be easier than in others to ascertain that all or nearly all of the research evidence has been identified, and to obtain data from the relevant studies. To date, cancer and cardiovascular disease are areas where the approach has been used extensively and extremely successfully to obtain answers to important health-care questions. Use is extending into other areas of health care, and there is no reason why the approach or aspects of the approach cannot be usefully adapted to the fields of social sciences and education.

REFERENCES

Advanced Bladder Cancer (ABC) Meta-analysis Collaboration (2003). Neoadjuvant chemotherapy in invasive bladder cancer: A systematic review and meta-analysis. *Lancet*, **361**, 1927–1934.

Advanced Bladder Cancer Overview Collaboration (1995). Does neo-adjuvant cisplatin-based chemotherapy improve the survival of patients with locally advanced bladder cancer? A meta-analysis of individual patient data from randomized clinical trials. *British Journal of Urology*, **75**, 206–213.

Advanced Ovarian Cancer Trialists Group (1991). Chemotherapy in advanced ovarian cancer: An overview of randomized clinical trials. *British Medical Journal*, **303**, 884–893.

Advanced Ovarian Cancer Trialists' Group (1998). Chemotherapy in advanced ovarian cancer: Four systematic meta-analyses of individual patient data from 37 randomized trials. *British Journal of Cancer*, **78**, 1479–1487.

Altman, D.G. (1991). Randomisation. *British Medical Journal*, **302**, 1481–1482.

Arnott, S.J., Duncan, W., Gignoux, M., David, G.J., Hansen, H.S., Launois, B. *et al.* (1998). Preoperative radiotherapy in esophageal carcinoma: A meta-analysis using individual patient data (Oesophageal Cancer Collaborative Group). *International Journal of Radiation Oncology Biology Physics*, **41**, 579–583.

Artemether-Quinine Meta-analysis Study Group (2001). A Meta-analysis using individual patient data of trials comparing artemether with quinine in the treatment of severe falciparum malaria. *Transactions of the Royal Society of Tropical Medicine and Hygiene*, **95**, 637–650.

Barnett, P.G., Rodgers, J.H. and Bloch, D.A. (2001). A meta-analysis comparing buprenorphine to methadone for treatment of opiate dependence. *Addiction*, **96**, 683–690.

Begg, C., Cho, M., Eastwood, S., Horton, R., Moher, D., Olkin, I. *et al.* (1996). Improving the quality of reporting of randomized controlled trials. *Journal of the American Medical Association*, **276**, 637–639.

Burdett, S., Stewart, L.A. and Tierney, J.F. (2003). Publication bias and meta-analysis A practical example. *International Journal of Technology Assessment in Health Care*, **19**, 129–134.

Chalmers, I., Sackett, D. and Silagy, C. (1997). The Cochrane Collaboration. In A. Maynard and I. Chalmers (eds), *Non-random Reflections on Health Services Research* (pp. 231–249). London: BMJ Publishing Group.

Chan, A.W., Hrobjartsson, A., Haahr, M.T., Gøtzsche, P.C. and Attman, D.G. (2004). Empirical evidence for selective reporting of outcomes in randomised trials. *Journal of American Medical Association*, **291**, 2457–2465.

Chang, A.E., Kinsella, T., Glatstein, E., Baker, A.R., Sindelar, W.F., Lotze, M.T. *et al.* (1988). Adjuvant chemotherapy for patients with high-grade soft-tissue sarcomas of the extremity. *Journal of Clinical Oncology*, **6**, 1491–1500.

Clarke, M. and Godwin, J. (1998). Systematic reviews using individual patient data: A map for the minefields? *Annals of Oncology*, **9**, 827–833.

Clarke, M., Stewart, L., Pignon, J.P. and Bijnens, L. (1998). Individual patient data meta-analyses in cancer. *British Journal of Cancer*, **77**, 2036–2044.

Cohen, J., Guyatt, G., Bernard, G.R., Calandra, T., Cook, D., Elbourne, D. *et al.* on behalf of a UK Medical Research Council International Working Party (2001). New strategies for clinical trials in patients with sepsis and septic shock. *Critical Care Medicine*, **29**, 880–886.

D'Amico, R., Pifferi, S., Leonetti, C., Torri, V., Liberati, A., and on behalf of the study investigators (1998). Effectiveness of antibiotic prophylaxis in critically ill adult patients: Systematic review of randomised controlled trials. *British Medical Journal*, **316**, 1275–1285.

Dickersin, K. (1990). The existence of publication bias and risk factors for its occurrence. *Journal of the American Medical Association*, **263**, 1385–1389.

Dickersin, K., Min, Y.I. and Meinert, C.L. (1992). Factors influencing publication of research results. *Journal of the American Medical Association*, **267**, 374–378.

Diehl, L.F. and Perry, D.J. (1986). A comparison of randomized concurrent control groups with matched historical controls groups. Are historical controls valid? *Journal of Clinical Oncology*, **4**, 1114–1120.

Easterbrook, P.J., Berlin, J.A., Gopalan, R. and Matthews, D.R. (1991). Publication bias in clinical research. *Lancet*, **337**, 867–872.

Egger, M., Juni, P. and Bartlett, C., for the CONSORT Group (2001). Value of flow diagrams in reports of randomized controlled trials. *Journal of the American Medical Association*, **285**, 1996–1999.

EU Hernia Trialists Collaboration (2000a). Laparoscopic compared with open methods of groin hernia repair – systematic review of randomized controlled trials. *British Journal of Surgery*, **87**, 860–867.

EU Hernia Trialists Collaboration (2000b). Mesh compared with non-mesh methods of open groin hernia repair – systematic review of randomized controlled trials. *British Journal of Surgery*, **87**, 854–859.

Glenn, J., Kinsella, T., Glatstein, E., Tepper, J., Baker, A., Sugarbaker, P. *et al.* (1985a). A randomized prospective trial of adjuvant chemotherapy in adults with soft tissue sarcomas of the head and neck, breast, and trunk. *Cancer*, **55**, 1206–1214.

Glenn, J., Sindelar, W.F., Kinsella, T., Glatstein, E., Tepper, J., Gosta, J. *et al.* (1985b). Results of multimodality therapy of resectable soft-tissue sarcomas of the retroperitoneum. *Surgery*, **97**, 316–325.

Glioma Meta-analysis Trialists (GMT) Group (2002). Chemotherapy in adult high-grade glioma: A systematic review and meta-analysis of individual patient data from 12 randomised trials. *Lancet*, **359**, 1011–1018.

HIV Trialists Collaborative Group (1999). Zidovudine, didanosine, and zalcitabine in the treatment of HIV infection: Meta-analyses of the randomized evidence. *Lancet*, **353**, 2014–2025.

Holt, S., Suder, A., Weatherall, M., Cheng, S., Shirtcliffe, P. and Beasley, R. (2001). Dose–response relation of inhaled fluticasone propionate in adolescents and adults with asthma: Meta-analysis. *British Medical Journal*, **323**, 1–7.

International Germ Cell Cancer Collaborative Group (1997). International germ cell consensus classification: A prognostic factor-based staging system for metastatic germ cell cancers. *Journal of Clinical Oncology*, **15**, 594–603.

Ioannidis, J.P.A. (1998). Effect of the surgical significance of results on the time to completion and publication of randomized efficacy trials. *The Journal of the American Medical Association*, **279**, 281–286.

Jeng, G.T., Scott, J.R. and Burmeister, L.F. (1995). A comparison of meta-analytic results using literature versus individual patient data: Paternal cell immunization for recurrent miscarriage. *Journal of the American Medical Association*, **274**, 830–836.

Lachin, J.M. (2000). Statistical considerations in the intent-to-treat principle. *Controlled Clinical Trials*, **21**, 167–189.

Lambert, P.C., Sutton, A.J., Abrams, K.R. and Jones, D. R. (2002). A comparison of summary patient-level covariates in meta-regression with individual patient data meta-analysis. *Journal of Clinical Epidemiology*, **55**, 86–94.

Lewis J.A. and Machin, D. (1993). Intention to treat – who should use ITT? *British Journal of Cancer*, **68**, 647–650.

Marson, A.G., Williamson, P.R., Clough, H., Hutton, J.L. and Chadwick, D.W., on behalf of the Epilepsy Monotherapy Trial Group (2002). Carbamazepine versus valproate monotherapy for epilepsy: A meta-analysis. *Epilepsia*, **43**, 505–513.

McAuley, L., Pham, B., Tugwell, P. and Moher, D. (2000). Does the inclusion of grey literature influence estimates of intervention effectiveness reported in a meta-analysis? *Lancet*, **356**, 1228–1231.

Moher, D., Schulz, K.F. and Altman, D., for the CONSORT Group (2001). The CONSORT Statement: Revised recommendations for improving the quality of reports of parallel-group randomized trials. *Journal of the American Medical Association*, **285**, 1987–1991.

Neoadjuvant Chemotherapy for Cervix Cancer Meta-analysis (NACCCMA) Collaboration (2003). Neoadjuvant chemotherapy for locally advanced cervical cancer: A systematic review and meta-analysis of individual patient data from 21 randomized trials. *European Journal of Cancer*, **39**, 2470–2486.

Non-small Cell Lung Cancer Collaborative Group (1995). Chemotherapy in non-small cell lung cancer: A meta-analysis using updated data on individual patients from 52 randomised clinical trials. *British Medical Journal*, **311**, 899–909.

Pignon, J.P., and Arriagada, R. (1993). Meta-analysis. *Lancet*, **341**, 964–965.

PORT Meta-analysis Trialists Group (1998). Postoperative radiotherapy in non-small-cell lung cancer: Systematic review and meta-analysis of individual patient data from nine randomized controlled trials. *Lancet*, **352**, 257–263.

Qizilbash, N., Whitehead, A., Higgins, J., Wilcock, G., Schneider, L. and Farlow, M. (1998). Cholinesterase inhibition of Alzheimer disease: A meta-analysis of the tacrine trials. Dementia Trialists' Collaboration. *Journal of the American Medical Association*, **280**, 1777–1782.

Sacks, H., Chalmers, T.C. and Smith, H.J. (1982). Randomized versus historical controls for clinical trials. *American Journal of Medicine*, **72**, 233–240.

Sarcoma Meta-analysis Collaboration (1997). Adjuvant chemotherapy for localized resectable soft tissue sarcoma in adults: Meta-analysis of individual patient data. *Lancet*, **350**, 1647–1654.

Sarcoma Meta-analysis Collaboration (1999). Adjuvant chemotherapy for localized resectable soft tissue sarcoma in adults. In *The Cochrane Library* Issue 1, 1999. Oxford: Update Software.

Schulz, K.F. and Grimes, D.A. (2002). Sample size slippages in randomized trials: Exclusions and the lost and wayward. *Lancet*, **359**, 781–785.

Schulz, K.F., Chalmers, I., Hayes, R.J. and Altman, D.A. (1995). Empirical evidence of bias. Dimensions of methodological quality associated with estimates of treatment effects in controlled trials. *Journal of the American Medical Association*, **273**, 408–412.

Stern, J.M. & Simes, R.J. (1997). Publication bias: Evidence of delayed publication in a cohort study of clinical research projects. *British Medical Journal*, **315**, 640–645.

Sterne, J.A.C., Bartlett, C., Jüni, P. and Egger, M. (2000). Do we need comprehensive literature searches? A study of publication and language bias in meta-analysis of controlled trials. In *Proceedings of the 3rd Symposium on Systematic Reviews: Beyond the basics* (http://www.ihs.ox.ac.uk/csm/Oralabstract2K.htm#Sterne).

Stewart, L.A. (2003). Systematic Reviews and Meta-analysis. In D.J. Girling, M.K.B. Parmar, S.J. Stenning, R.J. Stephens and L.A. Stewart (eds), *Clinical Trials in Cancer: Principles and Practice*. Oxford: Oxford University Press.

Stewart, L.A., Clarke, M. J., on behalf of the Cochrane Working Party Group on Meta-analysis using Individual Patient Data (1995). Practical methodology of meta-analyses (overviews) using updated individual patient data. *Statistics in Medicine*, **14**, 2057–2079.

Stewart, L.A. and Parmar, M.K.B. (1993). Meta-analysis of the literature or of individual patient data: Is there a difference? *Lancet*, **341**, 418–422.

Stewart, L.A. and Tierney, J.F. (2002). To IPD or not to IPD? *Evaluation and the Health Professions*, **25**, 76–97.

Storosum, J.G., Elferink, A.J., van Zwieten, B.J., van den Brink, W., Gersons, B.P., van Strik, R. and Broekmans, A.W. (2001). Short-term efficacy of tricyclic antidepressants revisited: A meta-analytic study. *European Neuropsychopharmacology*, **11**, 173–80.

Tierney, J.F. and Stewart, L.A. (2005). Investigating patient exclusion bias in meta-analysis. *International Journal of Epidemiology*, **34**, 79–87.

Tierney, J.F., Mosseri, V., Stewart, L.A., Souhami, R.L. and Parmar, M.K.B. (1995). Adjuvant chemotherapy for soft-tissue sarcoma: Review and meta-analysis of the published results of clinical randomical trials. *British Journal of Cancer*, **72**, 469–475.

Tierney, J.F., Clarke, M. and Stewart, L.A. (2000). Is there bias in the publication of individual patient data meta-analyses? *International Journal of Technology Assessment in Health Care*, **16**, 657–667.

Tramèr, M.R., Reynolds, D.J.M., Moore, R.A. and McQuay, H.J. (1997). Impact of covert duplicate publication on meta-analysis: A case study. *British Medical Journal*, **315**, 635–640.

CHAPTER 15

Differentiating Biases from Genuine Heterogeneity: Distinguishing Artifactual from Substantive Effects

John P.A. Ioannidis

Department of Hygiene and Epidemiology, University of Ioannina School of Medicine, Greece, and Tufts-New England Medical Center, USA

KEY POINTS

- Publication bias may often be confused with other factors that cause other types of bias or genuine heterogeneity across studies. Funnel plot asymmetry may not be due to publication bias, but due to these other biases or genuine heterogeneity.
- Key factors that may be misinterpreted as publication bias include time-lag bias, time-dependent effects, control-rate effects, null bias in large studies, study-design effects (including quality aspects), and evolving study populations.
- It is usually difficult to dissect bias (artifactual effects) from true hetero-geneity (substantive effects), but modelling of individual-level data may help in some circumstances.
- More than one factor with or without superimposed publication bias may be operating to create heterogeneity in a particular meta-analysis.

Publication Bias in Meta-Analysis – Prevention, Assessment and Adjustments Edited by H.R. Rothstein, A.J. Sutton
and M. Borenstein © 2005 John Wiley & Sons, Ltd

INTRODUCTION

Publication bias may often be difficult to separate from other factors that cause other types of bias or genuine heterogeneity in the study results. Several sources of heterogeneity can result in funnel plot asymmetry, and may be misinterpreted as publication bias unless the evidence is carefully scrutinized. In this chapter, I shall discuss some of the major issues that one needs to consider, starting with a discussion of time-lag bias and continuing to cover time-dependent effects, control-rate effects, null bias in large studies, study-design effects, and evolving study populations. Finally, I shall briefly discuss the potential for dissecting bias from heterogeneity by modeling individual-level data and whether these issues can be generalized across various scientific fields.

PUBLICATION BIAS VERSUS TIME-LAG BIAS

Definition of time-lag bias

Time-lag bias refers to the phenomenon where the time it takes to complete and publish a study is influenced by its results. In the typical time-lag bias, studies with 'negative' results are published with some delay compared with studies with 'positive' results (Ioannidis, 1998). One may compare time-lag bias with the traditional definition of publication bias. In publication bias, the relative chances of a trial being ever published are influenced by its results: small 'negative' studies have a smaller chance of ever being published then 'positive' studies of similar sample size, while larger studies are likely to be published regardless of their results. Time-lag bias may become identical to publication bias when some small studies with 'negative' results are not only relatively delayed in publication, but eventually are never published.

Empirical evidence

Empirical evidence suggests that this phenomenon is quite widespread in the medical literature and may be prevalent in other scientific domains as well. In an empirical evaluation of 109 randomized efficacy trials (Ioannidis, 1998), the median time from start of enrollment to publication was 4.3 years for 'positive' studies and 6.5 years for 'negative' studies. 'Negative' studies were completed almost as quickly as 'positive' studies, with some exceptions where the conduct of 'negative' studies was protracted. However, typically 'negative' studies were substantially delayed on their way to publication after their completion (median 3.0 years, vs. 1.7 years for 'positive' studies); See Figure 15.1. This delay may reflect mostly investigator delay, rather than editorial preference, at least for major journals (Dickersin *et al.*, 2002). Similar evidence was produced by Stern and Simes (1997) for studies that had been approved by the Royal Prince Alfred Hospital Ethics Committee over a period of 10 years. The median time to publication for clinical trials was 4.7 vs. 8.0 years depending on whether results were 'positive' or 'negative'.

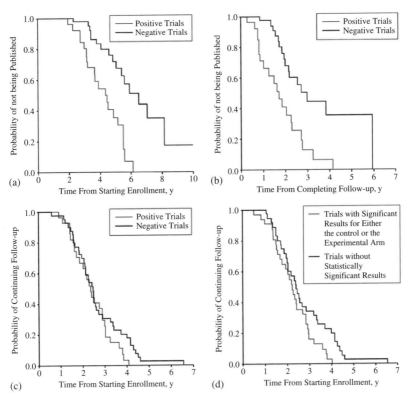

Figure 15.1   Time-lag bias for 'positive' versus 'negative' studies. 'Positive' studies are those with formal statistical significance in favor of the experimental arm. The terminology carries no connotation for the study quality. The Kaplan–Meier plots show: (a) time from start of enrollment to publication; (b) time from study completion to publication; (c) time from start of enrollment to publication; (d) same as (c), but with trials separated according to whether any formal statistical significance was observed (in favor of the experimental or the control arm) or not (Ioannidis, 1998). (Reprinted from *The Journal of the American Medical Association* (1998), Vol. 279, 281–286.)

Consequences of time-lag bias

The consequences of time lag are that more favorable results appear first, but then they are followed by less favorable results that paint a more conservative picture of the efficacy of an intervention. Thus meta-analyses may have to be interpreted cautiously, if there are still ongoing studies with pending results. Some investigators have proposed including in a meta-analysis only these studies that have started before a specific point in time (Clarke and Stewart, 1998). This '*terminus ante quem*' solution may help mitigate time-lag effects, but at the expense of ignoring substantial evidence. Time-lag bias may manifest with continuously decreasing treatment effects in the recursive cumulative meta-analysis framework (Ioannidis & Lau, 2001).

Time lag is probably an ubiquitous phenomenon across biomedical research and beyond. There is no empirical evidence as to whether it is becoming more or less prominent over time. The scale of the time lag may vary in different scientific disciplines with markedly different timing of the writing and publication process (few months or a year in some basic sciences, up to several years or even decades in some theoretical scientific domains). Nevertheless, time-lag bias may entirely replace publication bias if admonitions for the publication of all studies are heard and effective mechanisms (e.g., links to funding or administrative support) ensure that all studies are eventually published, regardless of their results. With greater sensitization to the problem of publication bias, time-lag bias may become more important in the literature.

Relationship of time lag with study parameters

Time lag may not be related only to the nature of the findings of a study. Studies evaluating long-term outcomes may take longer to complete than studies addressing short-term outcomes. Certain interventions may show differential efficacy when short-term vs. long-term outcomes are considered. A typical scenario is interventions that seem to work when surrogate endpoints are considered, but fail to show efficacy in the long term for hard endpoints such as survival (De Gruttola *et al.*, 1997; Fleming and DeMets, 1996).

There is empirical evidence also that early enrollment in randomized trials may largely correlate not only with the time that trials take to complete, but also with the time that they take to get published and with the possibility that they will find statistically significant results (Haidich and Ioannidis, 2001). Early enrollment can be measured either by the absolute number of patients enrolled in the first two months after a trial starts recruitment or as the relative percentage of the target enrollment that these patients represent. The enhanced proportion of statistically significant results for trials with rapid early enrollment was not explained by a better ability to reach the required sample size eventually. This suggests that even though randomized trials are designed theoretically upon the principle of equipoise, physicians and patients are attracted rapidly to trials of experimental interventions that eventually prove to be significantly effective. One potential explanation for this phenomenon may be that the profile of a treatment and its potential eventual efficacy can now be often predetermined or suspected with considerable predictability based on the results of preclinical evaluations, and early clinical studies with surrogate endpoints. Thus enrollment is more rapid in studies where the experimental interventions turn out to be most effective.

Time lag in non-randomized research

Time lag may be even more prominent in the domain of non-randomized studies. Empirical evidence from genetic epidemiology shows that it has been very common for proposed genetic associations to be refuted, when additional evidence from further studies is being published (Ioannidis *et al.*, 2001; Lohmueller *et al.*, 2003). Given the rapidity of publication of positive findings in the basic sciences and the hesitation or delay in publishing 'negative' results, the phenomenon may be quite widespread. The term 'winner's curse' has been used to describe the fact that early published studies in genetics are those that show the most inflated results

(Goring *et al.*, 2001). It is estimated that only 20–30% of proposed molecular genetic associations with various diseases are likely to be true ones (Ioannidis *et al.*, 2001; Lohmueller *et al.*, 2003).

Early stopping

Early stopping may confer a special type of time-lag bias. Some studies are designed with prescheduled interim analyses aiming to ensure that they will not continue if big differences are observed between the compared arms. Special adjustments are used for interpreting *p*-values in the case of such multiple looks (Lan *et al.*, 1993). Nevertheless, even when the interim analyses are performed appropriately, it has been demonstrated that early stopping tends to result in inflated effect sizes (Hughes & Pocock, 1988). The inflation can be dramatic in the case of continuous monitoring of the accumulating data seeking the crossing of a specific *p*-value threshold. The early-appearing inflated effects may contribute to the appearance of time-lag bias.

PUBLICATION BIAS VERSUS TIME-DEPENDENT EFFECTS

Time-limited treatment effects and epidemiologic associations

For certain interventions, efficacy is time-limited. Thus studies with relatively similar design and the same outcomes may reach opposing conclusions, if the follow-up differs. A similar phenomenon may be seen for epidemiologic associations with time-varying strength. For discrete outcomes, this situation will result in heterogeneity and possibly also in funnel plot asymmetry, because the studies with the longer follow-up will likely have more events than studies with more limited follow-up. A study with more events would also have a larger weight, when a traditional metric such as the risk ratio for having the event is used. Clearly in these cases there may be absolutely no publication bias, but funnel plot asymmetry may be misinterpreted as publication bias.

Prerequisites for evaluation of time dependence

It is typically easier to demonstrate an effect that is assumed to be time-constant than one that is time-dependent. Most studies are scarcely sufficiently powered to show time-constant effects, let alone time dependence. False positive findings of time dependence may also have to be considered. However, there are some classic examples that have been well documented.

Examples from randomized trials and epidemiologic research

Antiretroviral monotherapy is one intervention where efficacy is clearly time-limited. Patients with human immunodeficiency type (HIV-1) infection receiving zidovudine monotherapy were found to experience large, statistically significant benefits in terms of delayed disease progression in several trials conducted in the late 1980s and early 1990s with follow-up up to 1–1.5 years. Two subsequent large trials with follow-up exceeding 3 years and many more deaths (thus also larger weight) showed no lasting benefit. A meta-analysis of these data (Ioannidis *et al.*, 1995, 1997) shows clear funnel plot asymmetry with both Kendall's tau and regression publication bias statistics being highly statistically significant ($p < 0.01$); see Figure 15.2. However, the difference is due to the fact that the largest studies are the ones with longest follow-up, and thus with dissipated treatment effect. This was also clearly demonstrated in a subsequent meta-analysis of the individual-level data of these studies (HIV Trialists' Collaborative Group, 1999). It is now well established that drug resistance develops in patients receiving monotherapy and thus the time-limited efficacy is completely anticipated.

Epidemiologic associations may also have a time-limited aspect. For example, we have shown that while some genetic polymorphisms of the *CCR5* and *CCR2* chemokine receptor genes have an overall effect on the rate of HIV-1 disease progression, this effect is probably time-limited (Mulherin *et al.*, 2003), probably because over time the virus may change its use of specific chemokine receptors. Demonstration of the time-dependent effects has required a meta-analysis of individual patient data with detailed information on the outcome and genotype of each patient (Mulherin *et al.*, 2003).

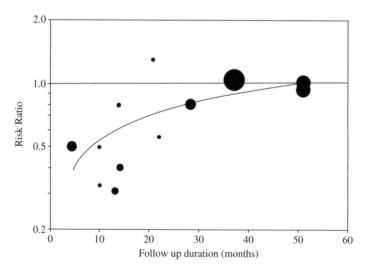

Figure 15.2 Diminishing treatment effect over time that would result in funnel plot asymmetry: a meta-analysis of zidovudine for HIV-infected patients. Studies are plotted with sizes proportional to their weight (inverse variance). Long-term follow-up studies carry the greatest weights and show no effect. The line represents a regression of the logarithm of the risk ratio on the duration of follow-up, weighted by the inverse variance (Ioannidis *et al.*, 1997). (Reprinted from *Controlled Clinical Trials* (1998), Vol. 279, 1089–1093.)

Time-dependence may be modeled either with meta-regressions or dummy variables that capture time intervals, or, if data are available at the individual level, with spline modeling of the hazard function (Mulherin *et al.*, 2003; Rosenberg, 1995). In general, time dependence may be difficult to decipher based on summary group data, especially when the effects are subtle.

PUBLICATION BIAS VERSUS CONTROL-RATE EFFECTS

Definition of control rate and its properties

For studies with binary outcomes, the term 'control rate' refers to the overall baseline risk for the event of interest in the control group of a study. Even when compared at the group level, studies on the same topic may show a large diversity in the baseline risk of their populations. Theoretically, patients at different levels of risk may experience different treatment benefits or harms (Glasziou and Irwig, 1995; Ioannidis and Lau, 1997). Studies with high control rates are likely to have also higher weights than studies with low control rates. Thus a situation of control-rate heterogeneity and differential treatment effects in high- vs. low-risk populations may result in false positive diagnostic tests for publication bias.

Patients at different levels of risk may experience different benefit

Control-rate effects have been demonstrated in meta-analyses of several studies (Schmid *et al.*, 1998) as well as within single large multicenter trials, where clinical centers may be considered as different units of observation with different control rates (Ioannidis *et al.*, 1999). In an empirical sample of 115 meta-analyses (Schmid *et al.*, 1998), control-rate effects were seen in 14 %, when the odds ratio was used as the metric of choice for the treatment effect. Control-rate effects are obviously dependent on the nature of the metric that is used to describe the treatment effect. Theoretically a constant risk ratio or odds ratio translates to a different risk difference at different levels of risk. The converse may also occur (constant risk difference, variable risk ratio or odds ratio). Significant control-rate effects were seen in 13 % and 31 % of the same 115 meta-analyses, when the risk ratio or risk difference were used as metrics for expressing the treatment effect.

Technical issues in control-rate meta-regressions

Description of the models that may be used for evaluation of control-rate effects is beyond the scope of this chapter. The reader is referred to several relevant references (Ioannidis and Lau, 1997; McIntosh, 1996; Sharp and Thompson, 2000; Thompson and Higgins, 2002). Modeling of the control-rate effects may pose some methodological problems. First, the control rate is an ecologic variable, that is, it reflects the average risk of a population, but this average may not be representative of each participating subject and the relationship between risk and treatment effect may be different once individual patient risk is considered. This is a manifestation of the ecological fallacy (Sharp and Thompson, 2000). Second, control rate is measured with error, since it is based on a limited sample of patients. Moreover, the treatment effect metrics are defined using the control rate in their definition. This introduces an in-built correlation, even when there is none present. Simple weighted regressions of the treatment effect metrics against the control rate may find spurious associations and lead to serious misinterpretations. Empirical evidence suggests that simple weighted regressions may claim significant associations between the treatment effect and control rate twice more often than more appropriate models (Schmid *et al.*, 1998).

PUBLICATION BIAS VERSUS NULL BIAS

Why large trials may find null results

Traditionally large studies are considered to be more reliable than smaller studies on the same topic (Yusuf *et al.*, 1984). Moreover, it is often assumed that large studies may be better designed and may be conducted by larger multicenter organizations according to stricter quality standards. However, this may not necessarily be the case. Large studies may require the participation of many centers that are less experienced in clinical research and that have less expertise. Measurement error in the outcomes of interest may be larger. Large measurement error may result in dilution of the observed effect size. Protocol deviations and lack of professional application of the experimental intervention may similarly decrease the observed effect size. These parameters may result in some large trials showing no effect, when an effect does exist, hence null bias. Since large studies carry larger weights than smaller ones, null bias may manifest as false positive diagnostic tests for publication bias.

The debate on large-scale evidence

Null bias has been evoked as a potential explanation of the discrepancies observed between some large trials with null results and smaller trials with 'positive' findings. For example, it has been evoked to explain the discrepancy in trials addressing the effectiveness of intravenous magnesium in the management of acute myocardial infarction, where the ISIS-4 mega-trial showed no benefit while earlier trials had shown spectacular benefits (Woods, 1995). However, these explanations have typically been *post hoc* and may not be reliable. In general, it may be useful to scrutinize parameters relating to the conduct of both small and large trials, to assess whether some features may have accentuated or dissipated the observed treatment effects in either case (Ioannidis *et al.*, 1998). A series of empirical evaluations (Cappelleri *et al.*, 1996; LeLorier *et al.*, 1997; Villar *et al.*, 1995) has demonstrated that large studies usually agree with small studies anyhow, but exceptions do occur (Figure 15.3). Many scientific domains are lacking large-scale evidence, be it single large trials or a plethora of smaller ones. The debate over whether it is better to have many small trials or a single large one may be an artificial one on most occasions, since there is often no real choice as to what can be performed. Large trials require extensive coordination, funding, and infrastructure and may often be infeasible.

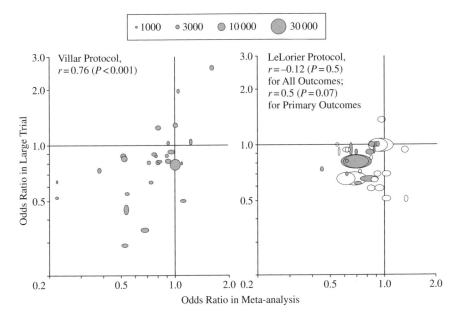

Figure 15.3 Empirical comparisons of the results of large studies vs. meta-analyses of smaller studies. White ellipses represent secondary endpoints in the LeLorier protocol. (Reprinted from *The Journal of the American Medical Association*, Vol. 18(5), Ioannidis, J.P. *et al.*, 'The relationship between study design, results, and reporting of randomized clinical trials of HIV infection', 431–444 (©1997), with permission from Elsevier.)

PUBLICATION BIAS VERSUS STUDY-DESIGN EFFECTS

Quality parameters potentially related to the magnitude of the treatment effect

The study design may affect the observed treatment effect through many different mechanisms that go beyond null bias. Empirical evaluations have targeted various components of study quality as potential correlates of effect size (Balk *et al.*, 2002; Jüni *et al.*, 2001; Kjaergard *et al.*, 2001; Schulz *et al.*, 1995; Sterne *et al.*, 2002). Empirical investigations have suggested that often lack of allocation concealment, lack of double blinding and lack or specification of the randomization mode may be associated with larger treatment effects. These associations are based on the examination of reported study quality.

Empirical evidence suggests that quality associations may be domain-specific

Reported study quality may not accurately reflect the true study quality (Huwiler-Muntener *et al.*, 2002; Ioannidis and Lau, 1998a). Thus a study that does not mention the mode of allocation concealment may well have been adequately concealed. Conversely, a study that describes some kind of adequate allocation concealment may actually have deviated from the anticipated standard during its conduct and the deviation might have been unrecorded. Furthermore, recent empirical evidence suggests that the associations of specific quality parameters with treatment effect are unlikely to be consistently observed across all medical domains and all topics (Balk *et al.*, 2002). For example, in a set of infectious disease meta-analyses, lack of reported allocation concealment was actually related to a smaller treatment effect, the opposite of what has been observed in other domains (Balk *et al.*, 2002). In other fields, lack of allocation concealment had no apparent relationship at all with the treatment effect (Balk *et al.*, 2002).

Quality effects resulting in funnel plot asymmetry

Quality effects may result in funnel plot asymmetry, if the quality characteristics are correlated with the weight that a study carries. The potential quality parameters that one may consider are generally too many (Moher *et al.*, 1995). Moreover, there is little consensus among experts on the exact definition of even the major quality items (Balk *et al.*, 2002; Moher *et al.*, 1995). Correlations between quality and weight may occur for specific topics, although this statement cannot be generalized. In such cases, the results of quality–effect associations should be interpreted cautiously. While it is unquestionably important to conduct studies of high quality and to improve high-quality reporting of research, different quality parameters may be important to consider in each research topic and reported data may not adequately capture the exact reality of the design and conduct of a study. Weighting studies by measures of their quality has been proposed in the conduct of meta-analysis (Moher *et al.*, 1998), but these adjustments are probably very subjective and perilous and they cannot really correct the flaws of each study.

PUBLICATION BIAS VERSUS EVOLVING STUDY POPULATIONS

Evolving heterogeneity of study populations

For some diseases, there may be a shift in the composition and case mix of the study populations over time. Event rates may decrease over time, if patients are generally at lower risk for suffering the outcome of interest. In the case of medical therapies, lower risks for bad outcomes may be associated with improved background management or the advent of other effective adjunct treatments. A shift in the diagnostic criteria for the disease could have a similar effect, if the newer definition also includes lower-risk patients (DECODE Study Group, 1998). Alternatively, there could be a shift for the inclusion criteria of studies in the same direction. For example, early chemotherapy studies on patients with advanced lung cancer were more likely to include terminal patients with brain metastases than current studies (Breathnach *et al.*, 2001). The relative proportion of patients with poor performance status at baseline has declined in these trials over the last three decades (Ioannidis *et al.*, 2003). Lower-risk patients will have lower event rates, thus more recent studies may have smaller weights compared with older ones, when the risk ratio for the event is used as the metric for the analysis. Under improved circumstances and with lower-risk patients, some treatments may sometimes no longer be effective or as effective as they were in the past. Thus studies of lower weight would show smaller treatment effects. If there is a relationship between the case mix of the population and the effect size, this may cause spurious funnel plot asymmetry.

Modeling the expected composition of study populations for epidemic diseases

A special situation arises in the case of epidemic diseases with variable levels of epidemic activity over time (Ioannidis *et al.*, 1996). For example, the HIV pandemic has followed a different dynamic course in various countries and groups of subjects. It has been shown (Ioannidis *et al.*, 1996) that even if background management and treatment were to remain unaltered, and definitions and entry criteria remained the same, the composition of a population of HIV-infected patients at a given disease stage would gradually become a population of lower-risk patients over time, provided that the epidemic of new infections had been curtailed. The reason for this is that longer-surviving individuals are likely to represent lower-risk people, while high-risk patients who transit rapidly through a disease stage are dying early. This situation is analogous to the well-known length bias that affects observational prevalence studies. In this form, it may affect the composition of even randomized trial populations with unknown consequences on the evolution of the observed effect sizes for various interventions.

MODELING TRUE HETEROGENEITY

Individual-level data

Meta-analyses of individual patient data (Stewart and Clarke, 1995; Stewart and Parmar, 1993) – or meta-analyses of individual participant data (MIPD), when subjects are not ill (Ioannidis *et al.*, 2002) – have several advantages for modeling true heterogeneity. However, such analyses are nevertheless still subject to publication bias. The disadvantages of MIPD include the high cost, organizational requirements and the time required to run these projects (Ioannidis *et al.*, 2002). Obviously, MIPD are not recommended for answering questions that are too clearly and unequivocally answered by group data, and in most cases meta-analysts would not have an easy option for conducting such analyses.

Estimating patient-level diversity

The availability of individual-level covariates allows in MIPD the estimation of effect sizes for subgroups of subjects with clearly specified characteristics. Even in the MIPD setting, subgroup analyses should be used and interpreted with caution, since they may suffer the low replication potential that characterizes all subgroup analyses (Oxman and Guyatt, 1992). Besides subgroup analyses, a more appropriate approach may be to model the individual risk in MIPD as a continuous score and to evaluate the relationship between the risk score and the effect size (Trikalinos and Ioannidis, 2001). Modeling approaches have recently been presented in detail and compared (Trikalinos and Ioannidis, 2001). This practice may overcome the limitations of control-rate meta-regressions and may provide insights about the relative effect sizes in subjects at different levels of risk.

Empirical evidence suggests that MIPD agglomerate studies with patients at very different levels of risk. The extreme quartile odds ratio (EQuOR) has been proposed as a metric to describe this heterogeneity (Ioannidis and Lau, 1998b). This is defined as the odds of having an event of interest among subjects in the predicted upper risk quartile divided by the odds of having an event of interest among subjects in the predicted lower risk quartile. Meta-analyses of several studies with individual-level data may have EQuOR values of 250 or more, and the distribution of risks may be far from normal (Trikalinos and Ioannidis, 2001); see Figure 15.4. The wide spectrum of baseline risk involved leaves ample room for the possibility of different effects at different levels of risk.

Provided that the appropriate information is captured in the individual-level data, MIPD are ideal for identifying many of the genuine sources of heterogeneity that may be misinterpreted as publication bias at the group level, such as time dependence, control-rate effects, differences in case mix and evolution of study populations over time.

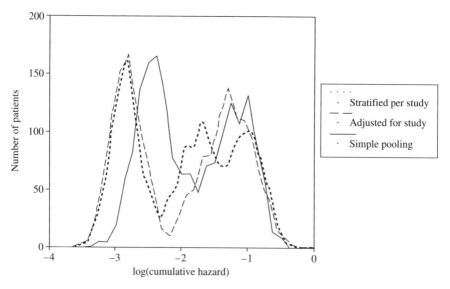

Figure 15.4 Estimated distribution of predicted risks according to different modeling approaches (pooling, adjustment for study, study stratification) in individual subjects in a meta-analysis of individual patient data of acyclovir in HIV infection. The log cumulative hazard represents an absolute measure of the estimated risk of death at 6 months. (Reprinted from *Journal of Clinical Epidemiology*, Vol. 54(3), Trikalinos, T.A. and Ioannidis, J.P., 'Predictive modeling and heterogeneity of baseline risk in meta-analysis of individual patient data', 245–252 (© 2001), with permission from Elsevier.)

SUMMARY AND GENERALIZABILITY

This chapter has discussed several factors that may occasionally result in funnel plot asymmetry without representing publication bias. These factors can be separated into: substantive effects that cause a real heterogeneity in effect sizes; and artifactual effects that only give the appearance of heterogeneity, while the true effect sizes are not heterogeneous. These factors may also be classified into time-dependent, population-dependent, and study design-dependent types (Table 15.1). It is conceivable that more than one factor may be operating to create heterogeneity in a particular meta-analysis.

These considerations are based on evidence derived mostly from the biomedical sciences. As discussed above, the amount of empirical evidence for the relative contribution in the broader picture of heterogeneity may vary for each factor. Evidence from other scientific disciplines (e.g., psychology or education) about the importance of these factors may be even more limited. Publication dynamics and biases may vary between different scientific fields and may present themselves in different forms. For example, in industrial psychology, time-dependent effects may be seen in skill acquisition experiments (Smith and Glass, 1977): the evolution of study populations over time may be related to variables such as changing workforce demographics, organizational contexts and cultures, wealth, and psychological employment contracts; while it may be more difficult to find an equivalent of control-rate effects. Similarly, the extent of time-lag bias may be more limited in scientific fields where there is no marketing pressure to produce and publish results (especially 'positive' results) on a topic within a restricted time-frame. Null bias is difficult to assess in many scientific fields, where very large studies are uncommon or non-existent. Finally, the important study design factors that one needs to consider may differ across sciences. One of the earliest classic papers on meta-analysis in psychology (Kanfer and Ackerman, 1989) evaluated differences in the effect sizes of psychotherapies depending on the source of the study (books, journals, dissertations, and unpublished reports). Quality factors may need to be measured differently in various scientific fields. In all, heterogeneity should be worth exploring in all cases, and publication bias should not be evoked automatically as a prime explanation for the presence of heterogeneity in meta-analyses.

Table 15.1   Substantive and artifactual effects causing heterogeneity

Type of Factor	Type of Effect Caused by Factor	
	Substantive Effects	Artifactual Effects
Time-dependent	Time-limited effects	Time-lag bias
Population-dependent	Control-rate effects	
	Evolving study populations	
Study design-dependent		Null bias
		Quality design effects

REFERENCES

Balk, E.M., Bonis, P.A., Moskowitz, H., Schmid, C.H., Ioannidis, J.P., Wang, C. and Lau, J. (2002). Correlation of quality measures with estimates of treatment effect in meta-analyses of randomized controlled trials. *Journal of the American Medical Association*, **287**, 2973–2982.

Breathnach, O.S., Freidlin, B., Conley, B., Green, M.R., Johnson, D.H., Gandara, D.R., O'Connell, M., Shepherd, F.A. and Johnson, B.E. (2001). Twenty-two years of phase III trials for patients with advanced non-small-cell lung cancer: Sobering results. *Journal of Clinical Oncolology*, **19**, 1734–1742.

Cappelleri, J.C., Ioannidis, J.P., Schmid, C.H., de Ferranti, S.D., Aubert, M., Chalmers, T.C. and Lau, J. (1996). Large trials vs meta-analysis of smaller trials: How do their results compare? *Journal of the American Medical Association*, **276**, 1332–1338.

Clarke, M. and Stewart, L. (1998). Time lag bias in publishing clinical trials. *Journal of the American Medical Association*, **279**, 1952.

DECODE Study Group, on behalf of the European Diabetes Epidemiology Study Group (1998). Will new diagnostic criteria for diabetes mellitus change phenotype of patients with diabetes? Reanalysis of European epidemiological data. *British Medical Journal*, **317**, 371–375.

De Gruttola, V., Fleming, T., Lin, D.Y. and Coombs, R. (1997). Perspective: Validating surrogate markers – are we being naive? *Journal of Infectious Diseases*, **175**, 237–246.

Dickersin, K., Olson, C.M., Rennie, D., Cook, D., Flanagin, A., Zhu, Q., Reiling, J. and Pace, B. (2002). Association between time interval to publication and statistical significance. *Journal of the American Medical Association*, **287**, 2829–2831.

Fleming, T.R. and DeMets, D.L. (1996). Surrogate end points in clinical trials: Are we being misled? *Annals of Internal Medicine*, **125**, 605–613.

Glasziou, P.P. and Irwig, L.M. (1995). An evidence based approach to individualising treatment. *British Medical Journal*, **311**, 1356–1359.

Goring, H.H., Terwilliger, J.D. and Blangero, J. (2001). Large upward bias in estimation of locus-specific effects from genomewide scans. *American Journal of Human Genetics*, **69**, 1357–1369.

Haidich, A.B. and Ioannidis, J.P. (2001). Effect of early patient enrollment on the time to completion and publication of randomized controlled trials. *Amercian Journal of Epidemiology*, **154**, 873–880.

HIV Trialists' Collaborative Group (1999). Meta-analyses of the randomised evidence on zidovudine, didanosine and zalcitabine in the treatment of HIV infection. *Lancet*, **353**, 2014–2025.

Hughes, M.D. and Pocock, S.J. (1988). Stopping rules and estimation problems in clinical trials. *Statistics in Medicine*, **7**, 1231–1242.

Huwiler-Muntener, K., Jüni, P., Junker, C. and Egger, M. (2002). Quality of reporting of randomized trials as a measure of methodologic quality. *Journal of the American Medical Association*, **287**, 2801–2804.

Ioannidis, J.P. (1998). Effect of the statistical significance of results on the time to completion and publication of randomized efficacy trials. *Journal of the American Medical Association*, **279**, 281–286.

Ioannidis, J.P. and Lau, J. (1997). The impact of high-risk patients on the results of clinical trials. *Journal of Clinical Epidemiology*, **50**, 1089–1098.

Ioannidis, J.P. and Lau J. (1998a). Can quality of clinical trials and meta-analyses be quantified? *Lancet*, **352**, 590–591.

Ioannidis, J.P. and Lau, J. (1998b). Heterogeneity of the baseline risk within patient populations of clinical trials: A proposed evaluation algorithm. *American Journal Epidemiology*, **148**, 1117–1126.

Ioannidis, J.P. and Lau, J. (2001). Evolution of treatment effects over time: Empirical insight from recursive cumulative meta-analyses. *Proceedings of the National Academy of Sciences of the United States*, **98**, 831–836.

Ioannidis, J.P., Cappelleri, J.C., Lau, J., Skolnik, P.R., Melville, B., Chalmers, T.C. and Sacks, H.S. (1995). Early or deferred zidovudine therapy in HIV-infected patients without an AIDS-defining illness. *Annals of Internal Medicine*, **122**, 856–866.

Ioannidis, J.P., Cappelleri, J.C., Schmid, C.H. and Lau, J. (1996). Impact of epidemic and individual heterogeneity on the population distribution of disease progression rates. An example from patient populations in trials of human immunodeficiency virus infection. *American Journal of Epidemiology*, **144**, 1074–1085.

Ioannidis, J.P., Cappelleri, J.C., Sacks, H.S. and Lau, J. (1997). The relationship between study design, results, and reporting of randomized clinical trials of HIV infection. *Controlled Clinical Trials*, **18**, 431–444.

Ioannidis, J.P., Cappelleri, J.C. and Lau, J. (1998). Issues in comparisons between meta-analyses and large trials. *Journal of the American Medical Association*, **279**, 1089–1093.

Ioannidis, J.P., Dixon, D.O., McIntosh, M., Albert, J.M., Bozzette, S.A. and Schnittman, S.M. (1999). Relationship between event rates and treatment effects in clinical site differences within multicenter trials: An example from primary *Pneumocystis carinii* prophylaxis. *Controlled Clinical Trials*, **20**, 253–266.

Ioannidis, J.P., Ntzani, E.E., Trikalinos, T.A. and Contopoulos-Ioannidis, D.G. (2001). Replication validity of genetic association studies. *Nature Genetics*, **29**, 306–309.

Ioannidis, J.P., Rosenberg, P.S., Goedert, J.J. and O'Brien, T.R. (2002). Commentary: Meta-analysis of individual participants' data in genetic epidemiology. *American Journal of Epidemiology*, **156**, 204–210.

Ioannidis, J.P., Polycarpou, A., Ntais, C., and Pavlidis, N. (2003). Randomized trials comparing chemotherapy regimens for advanced non-small cell lung cancer: Biases and evolution over time. *European Journal of Cancer*, **39**, 2278–2287.

Jüni, P., Altman, D.G. and Egger, M. (2001). Systematic reviews in health care: Assessing the quality of controlled clinical trials. *British Medical Journal*, **323**, 42–46.

Kanfer, R. and Ackerman, P.L. (1989). Motivation and cognitive abilities – an integrative aptitude treatment interaction approach to skill acquisition. *Journal of Applied Psychology*, **74**, 657–690.

Kjaergard, L.L., Villumsen, J. and Gluud, C. (2001). Reported methodologic quality and discrepancies between large and small randomized trials in meta-analyses. *Annals of Internal Medicine*, **135**, 982–989.

Lan, K.K., Rosenberger, W.F. and Lachin, J.M. (1993). Use of spending functions for occasional or continuous monitoring of data in clinical trials. *Statistics in Medicine*, **12**, 2219–2231.

LeLorier, J., Gregoire, G., Benhaddad, A., Lapierre, J. and Derderian, F. (1997). Discrepancies between meta-analyses and subsequent large randomized, controlled trials. *New England Journal of Medicine*, **337**, 536–542.

Lohmueller, K.E., Pearce, C.L., Pike, M., Lander, E.S. and Hirschhorn, J.N. (2003). Meta-analysis of genetic association studies supports a contribution of common variants to susceptibility to common disease. *Nature Genetics*, **33**, 177–182.

McIntosh, M.W. (1996). The population risk as an explanatory variable in research synthesis of clinical trials. *Statistics in Medicine*, **15**, 1713–1728.

Moher, D., Jadad, A.R., Nichol, G., Penman, M., Tugwell, P. and Walsh, S. (1995). Assessing the quality of randomized controlled trials: An annotated bibliography of scales and checklists. *Controlled Clinical Trials*, **16**, 62–73.

Moher, D., Pham, B., Jones, A., Cook, D.J., Jadad, A.R., Moher, M., Tugwell, P. and Klassen, T.P. (1998). Does quality of reports of randomized trials affect estimates of intervention efficacy reported in meta-analyses? *Lancet*, **352**, 609–613.

Mulherin, S.A., O'Brien, T.R., Ioannidis, J.P., Goedert, J.J., Buchbinder, S.P., Coutinho, R.A. *et al.* (2003). International Meta-Analysis of HIV Host Genetics. Effects of CCR5-Delta32 and CCR2-64I alleles on HIV-1 disease progression: The protection varies with duration of infection. *AIDS*, **17**, 377–387.

Oxman, A.D. and Guyatt, G.H. (1992). A consumer's guide to subgroup analyses. *Annals of Internal Medicine*, **116**, 78–84.

Rosenberg, P.S. (1995). Hazard function estimation using B-splines. *Biometrics*, **51**, 874–887.

Schmid, C.H., Lau, J., McIntosh, M.W. and Cappelleri, J.C. (1998). An empirical study of the effect of the control rate as a predictor of treatment efficacy in meta-analysis of clinical trials. *Statistics in Medicine*, **17**, 1923–1942.

Schulz, K.F., Chalmers, I., Hayes, R.J. and Altman, D.G. (1995). Empirical evidence of bias. Dimensions of methodological quality associated with estimates of treatment effects in controlled trials. *Journal of the American Medical Association*, **273**, 408–412.

Sharp, S.J. and Thompson, S.G. (2000). Analyzing the relationship between treatment effect and underlying risk in meta-analysis: Comparison and development of approaches. *Statistics in Medicine*, **19**, 3251–3274.

Smith, M.L. and Glass, G.V. (1977). Meta-analysis of psychotherapy outcome studies. *American Psychology*, **32**, 752–760.

Stern, J.M. and Simes, R.J. (1997). Publication bias: Evidence of delayed publication in a cohort study of clinical research projects. *British Medical Journal*, **315**, 640–645.

Sterne, J.A., Jüni, P., Schulz, K.F., Altman, D.G., Bartlett, C. and Egger, M. (2002). Statistical methods for assessing the influence of study characteristics on treatment effects in 'meta-epidemiological' research. *Statistics in Medicine*, **21**, 1513–1524.

Stewart, L.A. and Clarke, M.J. (1995). Practical methodology of meta-analyses (overviews) using updated individual patient data. Cochrane Working Group. *Statistics in Medicine*, **14**, 2057–2079.

Stewart, L.A. and Parmar, M.K. (1993). Meta-analysis of the literature or of individual patient data: Is there a difference? *Lancet*, **341**, 418–422.

Thompson, S.G. and Higgins, J.P. (2002). How should meta-regression analyses be undertaken and interpreted? *Statistics in Medicine*, **21**, 1559–1573.

Trikalinos, T.A. and Ioannidis, J.P. (2001). Predictive modeling and heterogeneity of baseline risk in meta-analysis of individual patient data. *Journal of Clinical Epidemiology*, **54**, 245–252.

Villar, J., Carroli, G. and Belizan, J.M. (1995). Predictive ability of meta-analyses of randomized controlled trials. *Lancet*, **345**, 772–776.

Woods, K.L. (1995). Mega-trials and management of acute myocardial infarction. *Lancet*, **346**, 611–614.

Yusuf, S., Collins, R. and Peto, R. (1984). Why do we need some large, simple randomized trials? *Statistics in Medicine*, **3**, 409–422.

CHAPTER 16

Beyond Conventional Publication Bias: Other Determinants of Data Suppression

Scott D. Halpern

Department of Medicine, Center for Clinical Epidemiology and Biostatistics, Center for Education and Research on Therapeutics, and Center for Bioethics, University of Pennsylvania, USA

Jesse A. Berlin

Center for Clinical Epidemiology and Biostatistics, Department of Biostatistics and Epidemiology, University of Pennsylvania School of Medicine, USA

KEY POINTS

- In order for meta-analyses and systematic reviews of studies investigating a common research question to produce accurate results, the full spectrum of evidence must be accessible to those conducting the reviews.
- The selective suppression of certain types of data due to both financial and non-financial competing interests likely represents an underappreciated source of bias in meta-analyses.

Publication Bias in Meta-Analysis – Prevention, Assessment and Adjustments   Edited by H.R. Rothstein, A.J. Sutton and M. Borenstein   © 2005 John Wiley & Sons, Ltd

> - Data have been suppressed to serve financial, political, ideological, professional, or other interests of investigators, research sponsors, and journal editors alike.
> - Though data suppression may never be fully eliminated, both its existence and its effects on the results of meta-analyses may be estimated using established meta-analytic diagnostic tests such as funnel plots and meta-regressions.

INTRODUCTION

Conventional publication bias may manifest when decisions to publish research findings are influenced by the statistical significance or direction of the observed effects (Begg and Berlin, 1988). When significant or 'positive' studies are selectively published, meta-analyses of the available literature may produce biased summary effect estimates (e.g., summary odds ratios). While the significance of results repeatedly has been shown to influence decisions to publish (Begg and Berlin, 1989; Cooper *et al.*, 1997; Dickersin, 1990; Dickersin *et al*, 1992; Easterbrook *et al.*, 1991; Ioannidis, 1998), less commonly considered influences on publication decisions might also produce bias.

This chapter reviews evidence suggesting that other, non-conventional forms of data suppression might affect the summary effect measures produced by meta-analyses. We group these non-conventional forms of data suppression into those arising from financial competing interests (such as a company's attempt to suppress data not favorably reflecting on its product) and those arising from non-financial competing interests (such as political interests, investigators' personal interests, and others). We distinguish financial from non-financial competing interests because the causes of and potential solutions to these problems differ substantially (Levinsky, 2002; Thompson, 1993); we distinguish both from traditional publication bias because, whereas traditional publication bias is widely known to occur and several methods exist to account for it (Begg and Mazumdar, 1994; Duval and Tweedie, 2000a, 2000b; Egger *et al.*, 1997; Sterne *et al.*, 2001), other forms of data suppression are less commonly considered and methods to account for them are lacking. Our goals are to increase awareness of the frequency and magnitude of non-conventional sources of data suppression so as to encourage both efforts to prevent it from occurring, and efforts to understand how different causes of data suppression may alter the results of meta-analyses.

Though we focus our attention on sources of data suppression, we acknowledge that other sources of bias, such as those attributable to multiple publication of positive results, or those arising from the design or analysis stages of research (Schulz *et al.*, 1995), may also contribute to the larger problem of difficulty in interpreting meta-analyses. We also acknowledge that our review may be weighted disproportionately towards examples of data suppression occurring in the testing of new medical therapies. This reflects the greater attention these examples have received in the literature; we do not mean to imply that these problems occur most commonly in medicine. Finally, we wish to highlight that because cases of data suppression, almost by definition, are rarely revealed in the literature, we have no means of documenting the true prevalence of related problems.

SUPPRESSION OF DATA DUE TO FINANCIAL COMPETING INTERESTS

Companies producing interventions (e.g., drugs, weight-loss strategies, or social or educational programs) may have several incentives to prevent the results of research on these interventions from being publicly disseminated via publication. Such financial competing interests may lead to the suppression of data on the safety or efficacy of those interventions. Similarly, companies that manufacture products commonly found in the environment (e.g., asbestos, pesticides, or cellular telephones) may have reason to suppress data deriving from epidemiologic studies of these products' associations with disease. In this section we discuss these reasons for data suppression due to financial competing interests.

Suppression of drug efficacy data

Although US federal regulations require that all available data on drug efficacy be presented to authorities such as the Food and Drug Administration, sponsors may be more selective in the data they promote via publication or marketing to doctors and the public. To exemplify the possibility of suppressing data regarding an intervention's efficacy, suppose a pharmaceutical company conducted several clinical trials of a new therapeutic agent, and that the results of these trials were mixed. That company may have tremendous financial incentives to selectively promote those trials showing promising health benefits from the intervention. By contrast, trials suggesting null or adverse effects may be suppressed.

These phenomena may account for the consistent evidence that clinical trials' results (Davidson, 1986; Wahlbeck and Adams, 1999) and conclusions (Djulbegovic *et al.*, 1999, 2000; Kjaergard and Als-Nielsen, 2002) are significantly more favorable towards experimental interventions when funded by for-profit organizations than when funded by non-profit organizations. Some may suggest an alternative explanation for these findings – industry tends naturally to fund studies of interventions likely to succeed. However, it is difficult to predict which interventions will ultimately succeed. Furthermore, this argument must be considered in light of recent evidence that industry may tip the odds of success in their favor by more commonly comparing interventions to inferior controls than do non-profit organizations (Bekelman *et al.*, 2003; Djulbegovic *et al.*, 2000; Johansen and Gotzsche, 1999; Rochon *et al.*, 1994).

How might financial competing interests influence the accessibility of efficacy data? We suggest that there are at least five related mechanisms. First, when faced with accumulating evidence that an intervention is unlikely to be effective, sponsoring companies may choose to terminate an ongoing study early (Davidson, 1986) or fail to release unfavorable data to the investigators for analysis (Davidoff *et al.*,

2001; Easterbrook *et al.*, 1991; Kahn *et al.*, 2000; Lauritsen *et al.*, 1987). A particularly concerning example occurred in 1999 after a data safety and monitoring board terminated a trial of HIV-1 Immunogen due to its failure to produce a clinical benefit. When a disagreement developed between the investigators and the sponsor, the Immune Response Corp., regarding the analysis of the data, the sponsor decided to withhold the complete data set from the lead investigators (DeAngelis, 2000; Kahn *et al.*, 2000). Occasionally, companies have justified this practice by stating that the research was intended only for product license applications (Easterbrook *et al.*, 1991). However, if meta-analyses are to meet their goal of synthesizing the universe of evidence regarding a particular research question, then motivations for research should be unrelated to the subsequent accessibility of the results.

A second way in which financial competing interests can lead to data suppression is that sponsoring companies may choose to terminate an ongoing study if the prospects for financial gain from the study's outcome appear grim. For example, in 1996, Hoechst Marion Roussel (HMR) withdrew financial support from a European trial evaluating the efficacy of its drug, Pimagedine (aminoguanidine), in delaying the progression of early renal disease in diabetic patients (Anonymous, 1997; Viberti *et al.*, 1997). HMR's decision to withdraw support was based on corporate priorities relating to the projected commercial interest of all agents being evaluated in sponsored studies at the time (Viberti *et al.*, 1997). Because HMR's partner company in the United States, Alteon, had insufficient funds to maintain this trial independently, the study was terminated, and the data that had accumulated were never published.

Another well-documented example occurred in 1999, when Novartis discontinued a trial of its antihyperlipidemic drug Lescol (fluvastatin) in the elderly (Boyd, 2001; Evans and Pocock, 2001; Lièvre *et al.*, 2001). Novartis based their decision to terminate the study on fears that an ongoing study of a competitor's drug, pravastatin, would document pravastatin's efficacy among elderly patients first. Because this would deprive Novartis of a substantial market share, the company stated that it was 'necessary to reallocate resources from Lescol to newer growth assets' (Lièvre *et al.*, 2001). This decision was made despite the fact that 1208 participants had already been enrolled and randomized in the trial; these accumulated data remain unavailable.

At least two other examples of trial discontinuation for financial reasons have been reported. In 1997, HMR closed a trial of its antihypertensive drug Cardizem (diltiazem), because Cardizem's patent was soon to expire and generic diltiazem would soon be released (Langer, 1997). Also in 1997, Pfizer and The Liposome Company discontinued a trial of their chemotherapeutic preparation, liposomal doxorubicin, for what appeared to be purely financial reasons (Hopf, 1997). These examples of data suppression may result in bias – specifically, an overestimation of the drugs' efficacy – in subsequent meta-analyses or systematic reviews if the decisions to stop trials are based in part on accumulated evidence from the research. However, if stopping decisions were based purely on competing priorities, without knowledge of, or at least without regard to the nature of, any data that had already accumulated, then such suppression may be random and result instead in reduced statistical precision.

A third mechanism by which competing financial interests may lead to data suppression is that investigators receiving sponsorship from for-profit organizations may be less likely to submit studies for publication when results are not favorable to that organization's product (Davidson, 1986; Easterbrook *et al.*, 1991). It is clear that the conclusions of articles published by investigators with competing financial interests differ from those of investigators without such interests (Azimi and Welch, 1998; Bekelman *et al.*, 2003; Cho and Bero, 1996; Djulbegovic *et al.*, 2000; Stelfox *et al.*, 1998; Yaphe *et al.*, 2001); research sponsorship may have similar influence on investigators' decisions to submit their work for publication (Bekelman *et al.*, 2003; Blumenthal *et al.*, 1997). Because the likelihood of such data suppression may be related to how favorable the results are to the experimental intervention, such suppression may directly bias the results of subsequent meta-analysis of the literature.

Fourth, for-profit organizations may attempt to block investigators from publishing results of sponsored studies when the results do not support their products' efficacy. For example, in 1995, Boots Pharmaceuticals, the producer of Synthroid, the dominant synthetic preparation of levothyroxine (used to treat hypothyroidism), blocked publication of a study conducted by Dr. Betty Dong at the University of California, San Francisco, showing that Synthroid was bioequivalent with other levothyroxine preparations (Rennie, 1997; Vogel, 1997). Although the study was eventually published (Dong *et al.*, 1997), legal maneuvering between Boots Pharmaceuticals and the University of California, San Francisco, caused the data to be suppressed for 7 years (Rennie, 1997).

Finally, even when interventions prove effective, financial competing interests may delay or prevent dissemination of research results to allow time for the filing of patent applications or to maintain a competitive edge (Blumenthal *et al.*, 1996a). Both delays in the publication of research results (Blumenthal *et al.*, 1996b, 1997) and full suppression of results – in the form of non-publication or refusal to share information with colleagues (Blumenthal *et al.*, 1996a; Campbell *et al.*, 2002) – occur more commonly among studies funded by for-profit organizations, as opposed to non-profit organizations. Both delays in publication and outright suppression of results may lead to biased estimates of these interventions' efficacies if the delayed and suppressed results differ systematically from those that become accessible more rapidly, or at all.

Suppression of drug safety data

The presence of financial incentives may also govern decisions to publish post-marketing, pharmacoepidemiologic studies of drug safety. Companies may push harder to publish studies showing minimal or no increased risk of adverse events following administration of their agent than they would for studies suggesting the agent confers serious or common risks. Additionally, companies might conduct studies that are underpowered to document side effects that occur uncommonly, thereby increasing their chances of failing to detect true risks of their products. As a result of these two phenomena, a funnel plot (see Chapter 5) of privately sponsored studies of adverse drug effects might show an atypical pattern in which studies with null effects and large variances (attributable to small sample sizes) are selectively published, and studies with 'positive' findings of adverse effects are absent.

We are aware of no data to indicate how commonly these phenomena occur, but a notorious case highlights the potential problem. In 1998, a Canadian pharmaceutical company, Apotex Inc., threatened legal action against Dr. Nancy Olivieri of Toronto's Hospital for Sick Children if she went public with results from an Apotex-sponsored trial suggesting that the company's drug deferiprone caused progression of hepatic fibrosis (Baird *et al.*, 2002; Spurgeon, 1998). The deferiprone study was eventually published (Olivieri *et al.*, 1998). However, like the Synthroid case, the example highlights the importance of investigators ensuring, prior to commencing sponsored research, that they will have complete autonomy over the data and publication decisions. Universities can help prevent such data suppression by prohibiting contracts for clinical research from containing clauses that give the sponsor the ability to suppress publication of the results (Baird *et al.*, 2002). However, this approach may succeed only when work is contracted through an investigator's home university. When investigators function as consultants, their work may be covered under confidentiality agreements, putting the decision to publish or not more squarely in the hands of the company.

Suppression of environmental epidemiology data

A final arena in which financial competing interests may lead to the suppression of data is concerned with environmental exposures. A variety of industries may be motivated to suppress data regarding adverse health effects associated with their products. The tobacco industry, for example, may selectively promote studies concluding that environmental tobacco smoke is not associated with increased risks of lung cancer, cardiovascular disease, or other illnesses (Barnes and Bero, 1998; Bero *et al.*, 1994). Manufacturers of asbestos may be similarly motivated to suppress information suggesting that exposure to asbestos in occupational or domestic settings is associated with mesothelioma or gastrointestinal malignancy (Kotelchuck, 1974). Such selective promotion of studies showing null associations between environmental exposures and adverse outcomes, and suppression of studies concluding that such relations exist, may lead subsequent meta-analyses to underestimate the magnitude of the associated risk, or to provide a false negative conclusion that such associations do not exist.

SUPPRESSION OF DATA DUE TO NON-FINANCIAL COMPETING INTERESTS

Although financial competing interests have received the bulk of attention in discussions of bias and objectivity in research, other competing interests, not directly arising from financial stakes in the study results, also may lead to data suppression. These non-financial competing interests may be grouped broadly into those that arise from investigators' personal agendas and interests, those that arise from editors' personal or organizational interests, and those that arise from third-party organizations who oppose particular research projects on political or ideological grounds.

Data suppression due to investigators' personal agendas

It has been suggested that investigators are little more than bundles of biases held loosely together by a sense of method (Horton, 2000). While this depiction may be overly cynical, the personal agendas and interests of individual investigators may certainly contribute to the selective promotion or suppression of data in ways that may bias subsequent meta-analyses of accessible data. Pressures of promotion and tenure, competition for limited research funds, desires for personal recognition, and deeply held world views may all contribute to the chosen design of studies and to investigators' decisions to pursue publication of their results (Blumenthal *et al.*, 1996a, 1997; Levinsky, 2002; Marshall, 1992; Rosenberg, 1996; Rothman, 1993). Indeed, Rothman (1993) has suggested that non-financial competing interests, such as the rewards of career advancement and future research funding, may influence investigators considerably more than their potential for direct financial gain.

Internal and external pressures on investigators may influence the timing and eventuality of their decisions to publish or otherwise make available the results of their work. For example, the strength of a priori viewpoints (Marshall, 1992) may prevent many investigators from submitting for publication studies whose results are inconsistent with their underlying hypotheses or prior publications (Blumenthal *et al.*, 1997; Smith, 1980). The deleterious results of this internal pressure are perhaps best illustrated in a meta-analysis of published and unpublished studies examining whether counseling and other forms of psychotherapy exert a sex bias against women (Smith, 1980). In this meta-analysis, Smith found that published studies supported the widely held notion that practitioners of counseling and psychotherapy hold biased views against women, but that unpublished studies revealed an equal bias in the opposite direction (i.e., in favor of women). Furthermore, among published studies, those with lower quality showed stronger biases against women, whereas among unpublished studies, quality was generally as good or better, and was unrelated to bias.

Smith (1980) suggests that these data, combined with evidence that studies showing stronger biases against women are more commonly cited than are studies showing smaller or null effects, support the view that research on bias in counseling and psychotherapy has been ideologically driven, with investigators seemingly 'intent on establishing counselor sexism'. As a result, 'studies have been published more often when a sex-bias effect was shown, regardless of the quality of the study itself'. Summaries of the published literature on ideologically driven topics may therefore reveal far greater homogeneity of effects than would summaries of the true universe of evidence, and so meta-analyses would show summary effect measures that are biased to support the investigators' underlying beliefs.

External pressures, such as competition for future research funding, promotion, and even respect from colleagues, also may subvert the free trade of ideas that ought to drive scientific pursuits (Campbell *et al.*, 2002). Such competitive pressures may make basic scientists unwilling to share reagents or assays with those working in similar fields (Blumenthal *et al.*, 1997; Campbell *et al.*, 2002; Rosenberg, 1996), and may make clinical researchers unwilling to release unpublished data to others for inclusion in a meta-analysis of a related topic (M. McDaniel, personal communication, 2002). Many investigators withhold information out of fear that sharing of methods or materials, or release of unpublished data, may pre-empt their own ability to publish similar data in the future (Campbell *et al.*, 2002).

Finally, a review of studies conducted in a US psychology department found that a lack of persistent interest among study investigators, and the frequent conduct of studies for hypothesis generation or other aims not directly leading to publication, may contribute to the low proportion of initiated studies that ultimately get published (Cooper *et al.*, 1997). The probability and magnitude of bias in meta-analyses are directly related to the extent that such withheld or unpublished studies differ systematically from those that are available to meta-analysts.

Data suppression due to editors' agendas

In addition to the agendas of investigators, editors' views regarding what types of studies provide the most useful information, what innovations are most likely to prove important, and what sizes of effects are most deserving of publication may also contribute to systematic data suppression. For example, there has been some concern that because editors' salaries are paid, in part, through journals' advertising revenues, their editorial decisions may be consciously or subconsciously altered (Koshland, 1992).

Although we are not aware of evidence that such financially motivated editorial bias has occurred, we have recently learned of a case in which an editorial decision was based on the editors' views of what effect sizes are sufficiently important to warrant publication. The editors in this case rejected a manuscript documenting the utility of situational judgement tests in aiding personnel selection because the effect shown was small, albeit statistically significant (M. McDaniel, personal communication, 2002).

Investigators must determine for themselves what effect sizes are important in order to plan appropriately for an adequate sample size. Editors' opinions about useful effect sizes may be viewed as providing an additional perspective on investigators' judgements. However, because editors typically make these judge-ments *post hoc*, their judgements may result in a biased sample of effect sizes being published in the literature. Shortly after rejection of the aforementioned manuscript, a competing journal accepted a similarly designed study that happened to show a larger effect (M. McDaniel, personal communication, 2002). Thus, any summary estimate of the effect of this intervention from the published literature will include the large effect and exclude the small effect, thereby overestimating the true effect size.

Data suppression due to organizations' political or ideological views

Whereas investigators, editors, and public and private sponsors have long been key players in determining the conduct of research and dissemination of results, outside organizations with political agendas increasingly have attempted to influence the face of scientific investigation. As Morton Hunt noted in a 1999 book detailing numerous examples of political suppression of research, 'In the course of the past decade and a half, especially the past half dozen years, a variety of social science research projects have been impeded, harmed, and in some cases wholly closed down by individuals and groups opposed to them for ideological or political reasons'. These groups have tried both to prevent the conduct of studies addressing questions they believe ought not to be asked, and to later block dissemination of these studies' results. For example, in 1976, a coalition of newspaper reporters, state officials, and religious lobbyists persuaded the federal government to withdraw funding that had been granted a year earlier to Professor Harris Rubin of the Southern Illinois University School of Medicine to study the effects of alcohol and marijuana on male sexual arousal (Hunt, 1999). Although Rubin eventually published the data he collected on alcohol's effects, the lack of funding prevented him from completing data collection on marijuana's effects; the data he had collected to that point were never published. Other areas of investigation engendering particularly vehement responses from groups on either the extreme left or right of the political spectrum have included studies of racial differences, addiction, and genetic influences on a variety of behaviors. Hunt (1999) provides a more comprehensive discussion of how groups have attempted to suppress inquiry into these areas.

BIAS DUE TO OMITTED RESEARCH

Most of the discourse regarding bias in meta-analysis has focused on factors causing the selective publication of certain results. However, it is also possible that bias may stem from the selective conduct of evaluations likely to yield specified results (Garattini and Liberati, 2000). Financial interests encourage investigators to conduct trials likely to generate the most lucrative results (Blumenthal *et al.*, 1996; Garattini and Liberati, 2000). Furthermore, financial interests may govern the design of trials so as to increase the probability of finding a favorable result. For example, industry-sponsored drug trials are more likely to use placebo (rather than active) controls than are trials sponsored by non-profit organizations (Djulbegovic *et al.*, 2000; Halpern and Karlawish, 2000), probably because comparison to placebo is more apt to shed favorable light on a company's product than is comparison with a competitor's product.

We note that non-financial motives may also contribute to the selective initiation of studies likely to yield statistically significant results. Though less well established in the literature, it seems likely that investigators submitting grant applications to federal funding agencies only choose projects which are likely, at least in principle, to yield significant associations that will help those investigators' careers.

Another arena in which biased meta-analyses may result from omitted research is in the realm of economic evaluations of drugs. Competing interests may dictate that such evaluations be based selectively on clinical trials that are likely to make the drugs appear cost-effective. For example, among economic evaluations of several blockbuster cancer drugs, 89 % were designed retrospectively, using already completed randomized trials as the basis for their data inputs (Friedberg *et al.*, 1999). This gives pharmaceutical companies an 'early look' at the range of conducted trials, allowing them to choose, as the basis of their economic analyses, those trials most likely to make their products look favorable (Friedberg *et al.*, 1999). If economic drug evaluations are based selectively on trials in which the drug either showed above-average efficacy, or required below-average resource utilization, then even if all such analyses were accessible, meta-analyses of these studies still would yield overly favorable estimates of the drug's cost-effectiveness.

Alternatively, it is possible that companies only choose to study the cost-effectiveness of interventions believed to be economically favorable. If so, then the conducted analyses may adequately represent a particular intervention's cost-effectiveness, and subsequent meta-analyses may be free from related biases. However, economic evaluations of interventions predicted to be less cost-effective might never be undertaken, skewing the total fund of knowledge about related interventions' cost-effectiveness.

ETHICAL IMPLICATIONS

Data suppression subverts the free exchange of ideas and information that ought to drive scientific endeavors towards their goal of improving the human condition. In addition, the suppression of data derived from the study of human volunteers breaks an implicit (and often explicit) contract upon which volunteers agreed to participate – namely, that the results of the research might contribute to some greater good (Emmanuel *et al.*, 2000; Freedman, 1987; Halpern *et al.*, 2002a, 2002b) Not only does the suppression of data from a given study prevent the fulfillment of altruistic motives among participants in that study, but it potentially undermines the evidence-based treatment of others in the future by contributing to bias in meta-analyses and systematic reviews.

Whereas these fundamental ethical problems link all forms of data suppression, data suppression attributable to financial competing interests may involve additional impropriety. For example, closure of drug trials for financial reasons breaches bonds of good faith between doctors and patients, undermines the risk–benefit calculus used by ethics committees in deciding to approve studies, and, perhaps most importantly, exposes early participants to the risks and burdens of research without any chance for personal or social benefit (Anonymous, 1997; Viberti *et al.*, 1997). Some may argue that by closing studies of interventions doomed to fail, industry conserves resources that may instead be used to generate evidence of a more socially beneficial intervention. However, even if net societal benefit is what motivates these decisions, the premature closure of trials violates established duties to enrolled participants, and thus undermines central tenets of ethical research (Department of Health and Human Services, 1991; Emmanuel *et al.*, 2000; National Commission for the Protection of Human Subjects of Biomedical and Behavioral Research, 1979).

Perhaps most concerning of all, data suppression could undermine the entire research enterprise by decreasing people's trust in science. As growing media attention makes the scientific process increasingly transparent, publicity of any form of data suppression, whether motivated by financial gain, political ideology, or otherwise, could substantially decrease people's willingness to participate in future studies.

There are no easy solutions to this dilemma. We must ensure that profit motives are kept in check when they conflict with the best interests of study participants or society more generally. However, doing so must not discourage for-profit organizations from continuing to invest in critical research that might not be conducted otherwise. Numerous authors and editors have called for full disclosure of all competing interests in journal articles (Davidoff *et al.*, 2001; DeAngelis *et al.*, 2001; Drazen and Curfman, 2002; James and Horton, 2003; Stelfox *et al.*, 1998), but such improvements will not ensure that all research that is conducted is also accessible. Another improvement is the increasingly common registration of randomized clinical trials, but few requirements exist for the centralized listing of studies with different designs or in different fields. The topic of prospective registration of trials is explored in greater detail in Chapter 3 of this book.

EVALUATING DATA SUPPRESSION DUE TO COMPETING INTERESTS

An ideal solution would be to prevent the suppression of data for any reason. This goal, of course, cannot be achieved. Still, we might substantially improve efforts to synthesize primary research by developing methods to detect the magnitudes of the major forms of data suppression. If such methods could account for the associated biases in meta-analyses, they would improve our ability to summarize and interpret the totality of evidence regarding a particular question.

For example, a preliminary step in evaluating the potential for data suppression due to financial competing interests might be to produce funnel plots (see Chapter 5) of publication bias (Egger *et al.*, 1997; Sterne *et al.*, 2001) that are stratified by funding source. If data suppression due to competing interests were influencing the evidence available for inclusion in a meta-analysis, then simple inspection should reveal that the funnel plot among studies funded by for-profit organizations has greater asymmetry than the plots of either unfunded studies or those funded by non-profit organizations. Alternatively, one might include funding source as an independent variable in 'meta-regression' models (see Chapter 6) that attempt to detect bias by modeling study characteristics that predict publication or the magnitude of treatment effects (Sterne *et al.*, 2001). If such models revealed that effect measures were strongly related to source of funding, this might suggest the need for caution in interpreting the summary effect measures yielded by a meta-analysis. The models might also be used to adjust summary estimates for various sources of bias.

CONCLUSION

We have discussed how both financial and non-financial competing interests contribute to data suppression, and have described many pathways through which suppression might arise. We have focused largely on for-profit organizations, and their roles in the conduct of biomedical research, because much evidence about these relationships has emerged in recent years. We do not intend to imply that for-profit organizations are the only ones to blame for the suppression of data, nor that research in other fields is comparatively spared of the phenomena we describe.

Non-profit organizations like the National Institutes of Health also might have proprietary or other reasons to influence how and whether results are disseminated. Furthermore, investigators, institutional review committees, and journal editors may all contribute to the processes by which access to data is limited. We hope that growing public attention to these problems, and the development of methods to prevent or overcome them, will improve the evidence bases that drive all fields of study.

REFERENCES

Anonymous (1997). A curious stopping rule from Hoechst Marion Roussel. *Lancet*, **350**, 155.

Azimi, N.A. and Welch, H.G. (1998). The effectiveness of cost-effectiveness analysis in containing costs. *Journal of General Internal Medicine*, **13**, 664–669.

Baird, P., Downie, J. and Thompson, J. (2002). Clinical trials and industry. *Science*, **287**, 2211.

Barnes, D.E. and Bero, L.A. (1998). Why review articles on the health effects of passive smoking reach different conclusions. *Journal of the American Medical Association*, **279**, 1566–1570.

Begg, C.B. and Berlin, J.A. (1988). Publication bias: A problem in interpreting medical data. *Journal of the Royal Statistical Society, Series A*, **151**, 419–463.

Begg, C.B. and Berlin, J.A. (1989). Publication bias and the dissemination of clinical research. *Journal of the National Cancer Institute*, **81**, 107–115.

Begg, C.B. and Mazumdar, M. (1994). Operating characteristics of a rank correlation test for publication bias. *Biometrics*, **50**, 1088–1101.

Bekelman, A.B., Li, Y. and Gross, C.P. (2003). Scope and impact of financial conflicts of interest in biomedical research. *Journal of the American Medical Association*, **289**, 454–465.

Bero, L.A., Glantz, S.A. and Rennie, D. (1994). Publication bias and public health policy on environmental tobacco smoke. *Journal of the American Medical Association*, **272**, 133–136.

Blumenthal, D., Campbell, E.G., Causino, N. and Louis, K.S. (1996a). Participation of life-science faculty in research relationships with industry. *New England Journal of Medicine*, **335**, 1734–1739.

Blumenthal, D., Causino, N., Campbell, E.G. and Louis, K.S. (1996b). Relationships between academic institutions and industry in the life sciences – an industry survey. *New England Journal of Medicine*, **334**, 368–373.

Blumenthal, D., Campbell, E.G., Anderson, M.S., Causino, N. and Louis, K.S. (1997). Withholding research results in academic life science: Evidence from a national survey of faculty. *Journal of the American Medical Association*, **277**, 1224–1228.

Boyd, K. (2001). Early discontinuation violates Helsinki principles. *British Medical Journal*, **322**, 605–606.

Campbell, E.G., Clarridge, B.R., Gokhale, M., Birenbaum, L., Hilgartner, S., Holtzman, N.A. and Blumenthal, D. (2002). Data withholding in academic genetics. Evidence from a national survey. *Journal of the American Medical Association*, **287**, 473–480.

Cho, M.K. and Bero, L.A. (1996). The quality of drug studies published in symposium proceedings. *Annals of Internal Medicine*, **124**, 485–489.

Cooper, H., DeNeve, K. and Charlton, K. (1997). Finding the missing science: The fate of studies submitted for review by a human subjects committee. *Psychological Methods*, **2**, 447–452.

Davidoff, F., DeAngelis, C.D., Drazen, J.M., Nicholls, M.G., Hoey, J., Højgaard, L. *et al.* (2001). Sponsorship, authorship, accountability. *Annals of Internal Medicine*, **135**, 463.

Davidson, R.A. (1986). Source of funding and outcome of clinical trials. *Journal of General Internal Medicine*, **1**, 155–158.

DeAngelis, C.D. (2000). Conflict of interest and the public trust. *Journal of the American Medical Association*, **284**, 2237–2238.

DeAngelis, C.D., Fontanarosa, P.B. and Flanagin, A. (2001). Reporting financial conflicts of interest and relationships between investigators and research sponsors. *Journal of the American Medical Association*, **286**, 89–91.

Department of Health and Human Services (1991). Common Rule, 45 CFR 46. Federal policy for the protection of human subjects. http://www.hhs.gov/ohrp/humansubjects/guidance/45cfr46.htm (accessed 16 May 2005).

Dickersin, K. (1990). The existence of publication bias and risk factors for its occurrence. *Journal of the American Medical Association*, **263**, 1385–1389.

Dickersin, K., Min, Y.-I. and Meinert, C.L. (1992). Factors influencing publication of research results: Follow-up of applications submitted to two institutional review boards. *Journal of the American Medical Association*, **267**, 374–378.

Djulbegovic, B., Bennett, C.L. and Lyman, G.H. (1999). Violation of the uncertainty principle in conduct of randomized controlled trials (RCTs) of erythropoieten. *Blood*, **94**, 399a.

Djulbegovic, B., Lacevic, M., Cantor, A., Fields, K.K., Bennett, C.L., Adams, J.R., Kuderer, N.M. and Lyman, G.H. (2000). The uncertainty principle and industry-sponsored research. *Lancet*, **356**, 635–638.

Dong, B.J., Hauck, W.W., Gambertoglio, J.G., Gee, L., White, J.R., Bubp, J.L. and Greenspan, F.S. (1997). Bioequivalence of generic and brand-name levothyroxine products in the treatment of hypothyroidism. *Journal of the American Medical Association*, **277**, 1205–1213.

Drazen, J.M. and Curfman, G.D. (2002). Financial associations of authors. *New England Journal of Medicine*, **346**, 1901–1902.

Duval, S. and Tweedie, R. (2000a). A nonparametric 'trim and fill' method of accounting for publication bias in meta-analysis. *Journal of the American Statistical Association*, **95**, 89–98.

Duval, S. and Tweedie, R. (2000b). Trim and fill: A simple funnel-plot-based method of testing adjusting for publication bias in meta-analysis. *Biometrics*, **56**, 455–463.

Easterbrook, P.J., Berlin, J.A., Gopalan, R. and Matthews, D.R. (1991). Publication bias in clinical research. *Lancet*, **337**, 867–872.

Egger, M., Smith, G.D., Schneider, M. and Minder, C. (1997). Bias in meta-analysis detected by a simple graphical test. *British Medical Journal*, **315**, 629–634.

Emmanuel, E.J., Wendler, D. and Grady, C. (2000). What makes clinical research ethical? *Journal of the American Medical Association*, **283**, 2701–2711.

Evans, S. and Pocock, S. (2001). Societal responsibilities of clinical trial sponsors. *British Medical Journal*, **322**, 569–570.

Freedman, B. (1987). Scientific value and validity as ethical requirements for research: A proposed explication. *IRB: Review of Human Subjects Research*, **9**(6), 7–10.

Friedberg, M., Saffran, B., Stinson, T.J., Nelson, W. and Bennett, C.L. (1999). Evaluation of conflict of interest in economic analyses of new drugs used in oncology. *Journal of the American Medical Association*, **282**, 1453–1457.

Garattini, S. and Liberati, A. (2000). The risk of bias from omitted research. *British Medical Journal*, **321**, 845–846.

Halpern, S.D. & Karlawish, J.H.T. (2000). Industry-sponsored research [letter]. *Lancet*, **356**, 2193.

Halpern, S.D., Karlawish, J.H.T. and Berlin, J.A. (2002a). The continuing unethical conduct of underpowered clinical trials. *Journal of the American Medical Association*, **288**, 358–362.

Halpern, S.D., Karlawish, J.H.T. and Berlin, J.A. (2002b). The ethics of underpowered clinical trials [letter]. *Journal of the American Medical Association*, **288**, 2118–2119.

Hopf, G. (1997). Early stopping of trials. *Lancet*, **350**, 891.

Horton, R. (2000). The less acceptable face of bias. *Lancet*, **356**, 959–960.

Hunt, M. (1999). *The New Know-Nothings. The Political Foes of the Scientific Study of Human Nature*. New Brunswick, NJ: Transaction Publishers.

Ioannidis, J.P.A. (1998). Effect of the statistical significance of results on the time to completion and publication of randomized efficacy trials. *Journal of the American Medical Association*, **279**, 281–286.

James, A. and Horton, R. (2003). The Lancet's policy on conflicts of interest. *Lancet*, **361**, 8–9.

Johansen, H.K. and Gotzsche, P.C. (1999). Problems in the design and reporting of trials of antifungal agents encountered during meta-analysis. *Journal of the American Medical Association*, **282**, 1752–1759.

Kahn, J.O., Cherng, D.W., Mayer, K., Murray, H. and Lagakos, S. (2000). Evaluation of HIV-1 Immunogen, an immunologic modifier, administered to patients infected with HIV having 300 to 549 $\times 10^6$/L CD4 cell counts. *Journal of the American Medical Association*, **284**, 2193–2202.

Kjaergard, L.L. and Als-Nielsen, B. (2002). Association between competing interests and authors' conclusions: Epidemiological study of randomized clinical trials published in the BMJ. *British Medical Journal*, **325**, 249–252.

Koshland, D.E. (1992). Conflict of interest policy. *Science*, **257**, 595.

Kotelchuck, D. (1974). Asbestos research: Winning the battle but losing the war. *Health/PAC Bulletin*, **61** (Nov.–Dec.), 1–27.

Langer, A. (1997). Early stopping of trials. *Lancet*, **350**, 890–891.

Lauritsen, K., Havelund, T., Larsen, L.S. and Rask-Madsen, J. (1987). Withholding unfavourable results in drug company sponsored clinical trials. *Lancet*, **i**, 1091.

Levinsky, N.G. (2002). Nonfinancial conflicts of interest in research. *New England Journal of Medicine*, **347**, 759–761.

Lièvre, M., Ménard, J., Bruckert, E., Cogneau, J., Delahaye, F., Giral, P. *et al.* (2001). Premature discontinuation of clinical trial for reasons not related to efficacy, safety, or feasibility. *British Medical Journal*, **322**, 603–605.

Marshall, E. (1992). The perils of a deeply held point of view. *Science*, **257**, 621–622.

National Commission for the Protection of Human Subjects of Biomedical and Behavioral Research. (1979). *The Belmont Report. Ethical Principles and Guidelines for the Protection of Human Subjects of Research*. Washington, DC: US Government Printing Office.

Olivieri, N.F., Brittenham, G.M., McLaren, C.E., Templeton D.M., Cameron, R.G. and McClelland, R.A. (1998). Long-term safety and effectiveness of iron-chelation therapy with deferiprone for thalassemia major. *New England Journal of Medicine*, **339**, 417–423.

Rennie, D. (1997). Thyroid storm. *Journal of the American Medical Association*, **277**, 1238–1243.

Rochon, P.A., Gurwitz, J.H., Simms, R.W., Fortin, P.R., Felson, D.T., Minaker, K.L. and Chalmers, T.C. (1994). A study of manufacturer-supported trials of nonsteroidal anti-inflammatory drugs in the treatment of arthritis. *Archives of Internal Medicine*, **154**, 157–163.

Rosenberg, S.A. (1996). Secrecy in medical research. *New England Journal of Medicine*, **334**, 393–394.

Rothman, K.J. (1993). Conflict of interest: The new McCarthyism in science. *Journal of the American Medical Association*, **269**, 2782–2784.

Schulz, K.F., Chalmers, I., Hayes, R.J. and Altman, D.G. (1995). Empirical evidence of bias. Dimensions of methodological quality associated with estimates of treatment effects in controlled trials. *Journal of the American Medical Association*, **273**, 408–412.

Smith, M.L. (1980). Sex bias in counseling and psychotherapy. *Psychological Bulletin*, **87**, 392–407.

Spurgeon, D. (1998). Trials sponsored by drug companies: Review ordered. *British Medical Journal*, **317**, 618.

Stelfox, H.T., Chua, G., O'Rourke, K. and Detsky, A.S. (1998). Conflict of interest in the debate over calcium-channel antagonists. *New England Journal of Medicine*, **338**, 101–106.

Sterne, J.A.C., Egger, M. and Smith, G.D. (2001). Investigating and dealing with publication and other biases in meta-analysis. *British Medical Journal*, **323**, 101–105.

Thompson, D.F. (1993). Understanding financial conflicts of interest. *New England Journal of Medicine*, **329**, 573–576.

Viberti, G., Slama, G., Pozza, G., Czyzyk, A., Bilous, R.W., Gries, A. *et al.* (1997). Early closure of European Pimagedine trial. *Lancet*, **350**, 214–215.

Vogel, G. (1997). Long-suppressed study finally sees light of day. *Science*, **276**, 523–525.

Wahlbeck, K. and Adams, C. (1999). Beyond conflict of interest. Sponsored drug trials show more favorable outcomes. *British Medical Journal*, **318**, 465.

Yaphe, J., Edman, R., Knishkowy, B. and Herman, J. (2001). The association between funding by commercial interests and study outcome in randomized controlled drug trials. *Family Practice*, **18**, 565–568.

APPENDICES

APPENDIX A

Data Sets

Three data sets have been used throughout this book to illustrate the application of different analytic approaches to addressing publication bias. In this appendix, we provide a brief description of each data set, and of the meta-analysis publication in which it was originally analysed and described. Meta-analysis results are given for each data set. Note these may differ slightly from those presented in some of the chapters due to differences in the meta-analysis models used (for example, fixed vs. random effects), and differences in methods of parameter estimation.

DATA SET I: STUDIES OF THE EFFECTS OF TEACHER EXPECTANCY ON STUDENT INTELLIGENCE

Raudenbush (1984) and Raudenbush and Bryk (1985) analysed the results of randomized experiments on the effects of teacher expectancy on student intelligence. The theory behind these experiments is that manipulating teachers' expectations of students' ability influences important student outcomes such as intelligence and achievement. In the typical experiment, all students are given an intelligence test, after which a randomly selected sample of students are identified to their teachers early in the school year as 'likely to experience substantial intellectual growth'. Later in the school year, both treated and control group students are administered an intelligence test. The effect size in these studies is the standardized difference between the treatment group mean and the control group mean at post-test. Raudenbush hypothesized that the more contact teachers had with their students before the expectancy manipulation, the less effective it would be. He suggested that teachers who had more contact with students would already have developed expectations of their students, and that these made them less amenable to influence by the experimental manipulation. To test this hypothesis, Raudenbush divided the effect size data into subgroups of studies, a high-contact subgroup where the effects were from studies in which teachers had more than one week of contact with students prior to the expectancy manipulation (11 studies, 3109 subjects, $d = -0.02$, 95 % confidence

Publication Bias in Meta-Analysis – Prevention, Assessment and Adjustments   Edited by H.R. Rothstein, A.J. Sutton and M. Borenstein   © 2005 John Wiley & Sons, Ltd

Table A.1 Studies of the effects of teacher expectancy on student intelligence.

Study	Estimated weeks of teacher student contact prior to expectancy induction	Number in Experimental Group	Number in control group	Standardized mean difference (d)	Standard error of standardized mean difference SE(d)
Rosenthal et al., 1974	2	77	339	0.03	0.126
Conn et al., 1968	21	60	198	0.12	0.147
Jose and Cody, 1971	19	72	72	−0.14	0.167
Pellengrini and Hicks, 1972 [1]	0	11	22	1.18	0.397
Pellegrini and Hicks, 1972 [2]	0	11	22	0.26	0.371
Evans and Rosenthal, 1969	3	129	348	−0.06	0.103
Fielder et al., 1971	17	110	636	−0.02	0.103
Claiborn, 1969	24	26	99	−0.32	0.221
Kester, 1969	0	75	74	0.27	0.165
Maxwell, 1970	1	32	32	0.80	0.260
Carter, 1970	0	22	22	0.54	0.307
Flowers, 1966	0	43	38	0.18	0.223
Keshock, 1970	1	24	24	−0.02	0.289
Henrikson, 1970	2	19	32	0.23	0.291
Fine, 1972	17	80	79	−0.18	0.159
Grieger, 1970	5	72	72	−0.06	0.167
Rosenthal and Jacobson, 1968	1	65	255	0.30	0.139
Fleming and Anttonen, 1971	2	233	224	0.07	0.094
Ginsburg, 1970	7	65	67	−0.07	0.174

Source: Data from Raudenbush and Bryk (1985), but see also Raudenbush (1984).

internal −0.10 to 0.06, fixed effects), and a low-contact subgroup where the effects were from studies in which teachers had a week or less of contact with students prior to the expectancy induction. (8 studies, 772 subjects, $d = 0.35$, 95 % CI 0.19 to 0.50, fixed effects). Hence, Raudenbush's results strongly supported his hypothesis. A summary of the primary studies and their effect sizes is given in Table A.1; see also Figure A.1. Note that a positive standardized mean difference denotes a higher score for the experimental (high-expectancy) group than for the control.

DATA SET II: STUDIES OF THE RELATIONSHIP BETWEEN 'SECOND-HAND' (PASSIVE) TOBACCO SMOKE AND LUNG CANCER

There have been long-running debates about whether inhaling other people's tobacco smoke has harmful effects on health, especially with respect to increased risks of lung cancer. Hackshaw et al. (1997) analysed the results of 37 cohort and case–control studies that compared the risk of developing lung cancer of lifelong non-smoking wives whose husbands currently smoked, with those whose husbands had never smoked. The effect size measure used was the unadjusted odds ratio (although it is reported in the original study as a relative risk since it approximates relative risk. Due to this, authors in this volume have used both odds ratio and

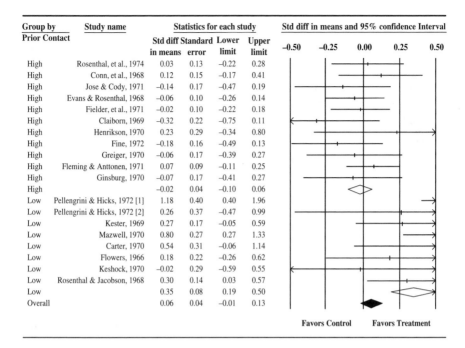

Group by Prior Contact	Study name	Statistics for each study				Std diff in means and 95% confidence Interval
		Std diff in means	Standard error	Lower limit	Upper limit	
High	Rosenthal, et al., 1974	0.03	0.13	−0.22	0.28	
High	Conn, et al., 1968	0.12	0.15	−0.17	0.41	
High	Jose & Cody, 1971	−0.14	0.17	−0.47	0.19	
High	Evans & Rosenthal, 1968	−0.06	0.10	−0.26	0.14	
High	Fielder, et al., 1971	−0.02	0.10	−0.22	0.18	
High	Claiborn, 1969	−0.32	0.22	−0.75	0.11	
High	Henrikson, 1970	0.23	0.29	−0.34	0.80	
High	Fine, 1972	−0.18	0.16	−0.49	0.13	
High	Greiger, 1970	−0.06	0.17	−0.39	0.27	
High	Fleming & Anttonen, 1971	0.07	0.09	−0.11	0.25	
High	Ginsburg, 1970	−0.07	0.17	−0.41	0.27	
High		−0.02	0.04	−0.10	0.06	
Low	Pellengrini & Hicks, 1972 [1]	1.18	0.40	0.40	1.96	
Low	Pellengrini & Hicks, 1972 [2]	0.26	0.37	−0.47	0.99	
Low	Kester, 1969	0.27	0.17	−0.05	0.59	
Low	Mazwell, 1970	0.80	0.27	0.27	1.33	
Low	Carter, 1970	0.54	0.31	−0.06	1.14	
Low	Flowers, 1966	0.18	0.22	−0.26	0.62	
Low	Keshock, 1970	−0.02	0.29	−0.59	0.55	
Low	Rosenthal & Jacobson, 1968	0.30	0.14	0.03	0.57	
Low		0.35	0.08	0.19	0.50	
Overall		0.06	0.04	−0.01	0.13	

Favors Control   Favors Treatment

Figure A.1   Data Set I: Studies of the effects of teacher expectancy on student intelligence (Raudenbush and Bryk, 1984).

relative risk/risk ratio to refer to this outcome). Analyses were conducted using a random-effects model. As reported by the authors, results showed that the pooled odds ratio from the 37 studies was 1.24 (95 % confidence interval 1.13 to 1.36) ($p<0.001$), which can be interpreted as meaning that lifelong non-smoking women with husbands who smoked were 24 % more likely to contract lung cancer than those who lived with husbands who did not smoke.

Hackshaw *et al.* conducted a (non-standard) form of failsafe *N* analysis (see Chapter 7), which indicated that there would need to be over 200 unpublished negative studies to nullify their results. On this basis, they concluded that no publication bias was operating in their data set. A summary of the primary studies and their effect sizes is given in Figure A.2. Note that an effect size greater than 1 indicates greater risk to those living with smokers.

DATA SET III: STUDIES OF THE RELATIONSHIP BETWEEN EMPLOYMENT INTERVIEW PERFORMANCE AND JOB PERFORMANCE

McDaniel *et al.* (1994) analysed the relationship between scores on employment interviews (which were conducted as part of a process to select applicants for a job) and several outcome variables, including job performance, performance in training, and tenure. For our purposes, only those studies with an outcome of job

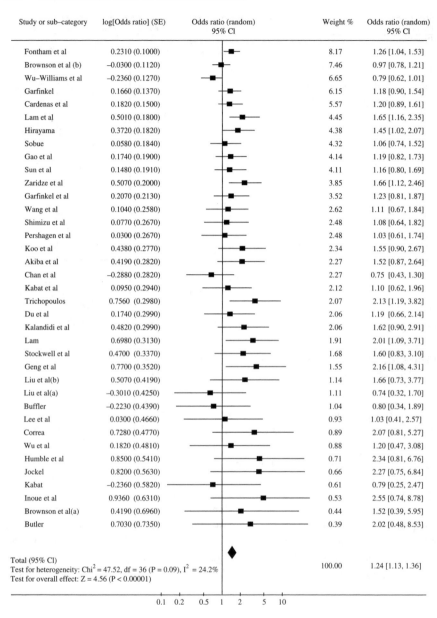

Study or sub–category	log[Odds ratio] (SE)	Odds ratio (random) 95% CI	Weight %	Odds ratio (random) 95% CI
Fontham et al	0.2310 (0.1000)		8.17	1.26 [1.04, 1.53]
Brownson et al (b)	–0.0300 (0.1120)		7.46	0.97 [0.78, 1.21]
Wu–Williams et al	–0.2360 (0.1270)		6.65	0.79 [0.62, 1.01]
Garfinkel	0.1660 (0.1370)		6.15	1.18 [0.90, 1.54]
Cardenas et al	0.1820 (0.1500)		5.57	1.20 [0.89, 1.61]
Lam et al	0.5010 (0.1800)		4.45	1.65 [1.16, 2.35]
Hirayama	0.3720 (0.1820)		4.38	1.45 [1.02, 2.07]
Sobue	0.0580 (0.1840)		4.32	1.06 [0.74, 1.52]
Gao et al	0.1740 (0.1900)		4.14	1.19 [0.82, 1.73]
Sun et al	0.1480 (0.1910)		4.11	1.16 [0.80, 1.69]
Zaridze et al	0.5070 (0.2000)		3.85	1.66 [1.12, 2.46]
Garfinkel et al	0.2070 (0.2130)		3.52	1.23 [0.81, 1.87]
Wang et al	0.1040 (0.2580)		2.62	1.11 [0.67, 1.84]
Shimizu et al	0.0770 (0.2670)		2.48	1.08 [0.64, 1.82]
Pershagen et al	0.0300 (0.2670)		2.48	1.03 [0.61, 1.74]
Koo et al	0.4380 (0.2770)		2.34	1.55 [0.90, 2.67]
Akiba et al	0.4190 (0.2820)		2.27	1.52 [0.87, 2.64]
Chan et al	–0.2880 (0.2820)		2.27	0.75 [0.43, 1.30]
Kabat et al	0.0950 (0.2940)		2.12	1.10 [0.62, 1.96]
Trichopoulos	0.7560 (0.2980)		2.07	2.13 [1.19, 3.82]
Du et al	0.1740 (0.2990)		2.06	1.19 [0.66, 2.14]
Kalandidi et al	0.4820 (0.2990)		2.06	1.62 [0.90, 2.91]
Lam	0.6980 (0.3130)		1.91	2.01 [1.09, 3.71]
Stockwell et al	0.4700 (0.3370)		1.68	1.60 [0.83, 3.10]
Geng et al	0.7700 (0.3520)		1.55	2.16 [1.08, 4.31]
Liu et al(b)	0.5070 (0.4190)		1.14	1.66 [0.73, 3.77]
Liu et al(a)	–0.3010 (0.4250)		1.11	0.74 [0.32, 1.70]
Buffler	–0.2230 (0.4390)		1.04	0.80 [0.34, 1.89]
Lee et al	0.0300 (0.4660)		0.93	1.03 [0.41, 2.57]
Correa	0.7280 (0.4770)		0.89	2.07 [0.81, 5.27]
Wu et al	0.1820 (0.4810)		0.88	1.20 [0.47, 3.08]
Humble et al	0.8500 (0.5410)		0.71	2.34 [0.81, 6.76]
Jockel	0.8200 (0.5630)		0.66	2.27 [0.75, 6.84]
Kabat	–0.2360 (0.5820)		0.61	0.79 [0.25, 2.47]
Inoue et al	0.9360 (0.6310)		0.53	2.55 [0.74, 8.78]
Brownson et al(a)	0.4190 (0.6960)		0.44	1.52 [0.39, 5.95]
Butler	0.7030 (0.7350)		0.39	2.02 [0.48, 8.53]
Total (95% CI)			100.00	1.24 [1.13, 1.36]

Test for heterogeneity: Chi^2 = 47.52, df = 36 (P = 0.09), I^2 = 24.2%
Test for overall effect: Z = 4.56 (P < 0.00001)

0.1 0.2 0.5 1 2 5 10

Figure A.2 Forest plot of studies of 'second-hand' (passive) tobacco smoke and lung cancer.

performance were considered (160 studies, 25 244 individuals). The effect size here is the correlation coefficient, r. For those who wish to compare our results with those of the original meta-analysis, it is important to note that the original meta-analysis by McDaniel *et al.* was conducted using the Hunter-Schmidt method, while our reanalysis uses the formulas and procedures contained in Hedges and Olkin (1985).

The key differences are that in this volume (i) the inverse variance weighting rather than weighting directly by sample size is used; (ii) Fisher's z transform of r is used, rather than r directly in our calculations (here, results are retransformed into r); and (iii) no corrections for statistical artefacts other than sampling error (this is what Hunter-Schmidt would term a 'bare-bones' analysis) were made. This approach was necessary to allow application of some of the publication bias analyses, and to allow standardization of the meta-analytic approach across our three data sets. Under the random-effects model, the point estimate of the validity (measured by the correlation) across 160 studies is 0.23, with a 95 % CI of 0.20 to 0.26.

In addition to estimating the overall validity of employment interviews, McDaniel *et al.* hypothesized that the correlation between interview scores and job performance would be moderated by several variables, most notably interview structure and interview content. It was expected that interviews that were structured to consist of a standard set of questions and a standard scoring system would have a closer relationship with job performance than those that were not structured (traditional interviews). Furthermore, it was expected that interviews whose content was directly related to the job to be performed (either traditional job-related interviews, or situational interviews, based on applicant responses to hypothetical job situations) would be more highly correlated with job performance than those that measured psychological characteristics. Under the random-effects model, the point estimate of validity of the structured interviews was 0.27, with a 95 % CI of 0.22 to 0.32, while the point estimate for unstructured interviews was 0.19, with a 95 % CI of 0.14 to 0.24, providing support for the authors' hypothesis. Their second hypothesis also received support: under the random-effects model the combined point estimate for job-related interviews (job-related and situational) was 0.25, with a 95 % CI of 0.22 to 0.28, while the point estimate for psychological interviews was 0.14, with a 95 % CI of 0.05 to 0.23. A summary of the primary studies and their effect sizes is given in Table A.2 and Figure A.3.

Table A.2 Studies of the relationship between employment interview scores and job performance.

Study	N	r	Content of interview	Interview structure
1	123	0.00	Job-related	Structured
2	95	0.06	Psychological characteristics	Unstructured
3	69	0.36	Job-related	Structured
4	1832	0.15	Job-related	Structured
5	78	0.14	Job-related	Structured
6	329	0.06	Job-related	Structured
7	153	0.09	Job-related	Structured
8	29	0.40	Job-related	Structured
9	29	0.39	Situational	Structured
10	157	0.14	Situational	Structured
11	149	0.36	Situational	Structured
12	92	0.28	Job-related	Unstructured

Table A.2 (Continued).

Study	N	r	Content of interview	Interview structure
13	15	0.62	Job-related	Structured
14	15	0.07	Job-related	Unstructured
15	170	0.18	Job-related	Unstructured
16	19	0.42	Job-related	Structured
17	19	0.08	Job-related	Unstructured
18	68	0.18	Psychological characteristics	Unstructured
19	93	0.43	Job-related	Unstructured
20	57	0.04	Job-related	Unstructured
21	80	−0.04	Psychological characteristics	
22	53	0.05	Psychological characteristics	
23	24	−0.14	Psychological characteristics	
24	57	0.05	Job-related	Structured
25	275	0.35	Job-related	Structured
26	45	−0.08	Psychological characteristics	
27	79	0.24	Psychological characteristics	
28	107	0.16	Psychological characteristics	
29	31	0.25	Job-related	Unstructured
30	407	0.68	Job-related	Structured
31	84	0.61	Job-related	Structured
32	8	0.81	Job-related	Structured
33	6	0.99	Job-related	Structured
34	7	0.66	Job-related	Structured
35	12	0.45	Job-related	Structured
36	14	0.71	Job-related	Structured
37	40	0.27	Job-related	Structured
38	40	−0.02	Job-related	Structured
39	99	0.29	Job-related	Unstructured
40	164	0.13	Job-related	Unstructured
41	67	0.03	Job-related	Unstructured
42	57	0.00	Job-related	Unstructured
43	50	0.09	Job-related	Structured
44	129	−0.03	Job-related	Unstructured
45	49	0.46	Situational	Structured
46	63	0.3	Situational	Structured
47	56	0.33	Situational	Structured
48	238	0.24	Psychological characteristics	
49	20	0.64	Job-related	Structured
50	122	0.12	Job-related	Unstructured
51	51	0.15	Job-related	Unstructured
52	40	0.44	Job-related	Unstructured
53	210	0.00	Job-related	Structured
54	334	0.16	Job-related	Structured
55	310	0.21	Psychological characteristics	
56	180	0.29	Job-related	Structured
57	93	0.19	Job-related	Unstructured
58	472	0.04	Job-related	Unstructured

59	44	0.56	Job-related	Unstructured
60	75	0.14	Job-related	Unstructured
61	68	0.44	Job-related	Unstructured
62	38	0.36	Job-related	Unstructured
63	42	0.34	Job-related	
64	39	0.11	Job-related	
65	49	0.40	Job-related	
66	41	0.23	Job-related	
67	200	0.22	Job-related	Structured
68	850	0.44	Job-related	Structured
69	41	0.27	Job-related	Structured
70	32	0.11	Job-related	Structured
71	65	0.27	Job-related	Structured
72	125	−0.07	Job-related	Structured
73	134	0.32	Job-related	Structured
74	21	0.05	Job-related	Unstructured
75	44	0.20	Job-related	Unstructured
76	170	0.18	Job-related	Unstructured
77	149	0.34	Job-related	Structured
78	296	0.03	Job-related	Structured
79	24	0.45	Situational	Structured
80	312	0.34	Job-related	Structured
81	205	0.51	Job-related	Structured
82	30	0.41	Situational	Structured
83	11	0.37	Situational	Structured
84	22	0.25	Situational	Structured
85	37	−0.17	Job-related	Structured
86	43	0.47	Job-related	Structured
87	72	0.32	Job-related	Structured
88	72	−0.09	Situational	Structured
89	108	0.33	Job-related	Structured
90	73	0.22	Job-related	Structured
91	73	0.27	Situational	Structured
92	117	0.00	Job-related	Structured
93	80	0.41	Job-related	Structured
94	95	0.16	Job-related	Structured
95	182	0.00	Job-related	Structured
96	93	0.03	Job-related	Structured
97	64	0.01	Job-related	Structured
98	370	0.03	Job-related	Structured
99	131	0.14	Job-related	Structured
100	87	0.11	Job-related	Structured
101	80	0.08	Job-related	Structured
102	41	−0.13	Job-related	Structured
103	35	0.13	Job-related	Unstructured
104	106	0.36	Job-related	Structured
105	86	0.06	Job-related	Structured
106	54	0.19	Job-related	Structured
107	393	0.27	Job-related	Structured

Table A.3 (Continued).

Study	N	r	Content of interview	Interview structure
108	102	0.17	Job-related	Structured
109	115	0.34	Job-related	Structured
110	63	0.28	Situational	Structured
111	22	0.11	Job-related	Structured
112	37	0.07	Job-related	Unstructured
113	116	−0.13		
114	416	0.12	Job-related	Unstructured
115	101	0.12	Job-related	Unstructured
116	1359	0.37	Job-related	Unstructured
117	82	0.26	Psychological characteristics	Unstructured
118	32	0.42	Job-related	Structured
119	42	0.37	Job-related	Structured
120	196	0.17	Job-related	Structured
121	44	0.19	Job-related	Structured
122	47	0.32	Situational	Structured
123	37	0.33		
124	12	0.24	Job-related	Structured
125	1807	0.09		
126	73	0.36	Job-related	Structured
127	73	0.26	Situational	Structured
128	70	0.42	Job-related	Structured
129	30	0.62	Job-related	Structured
130	60	0.87	Job-related	Structured
131	38	−0.07	Job-related	Structured
132	12	0.65	Job-related	Structured
133	33	0.17	Job-related	Unstructured
134	33	0.30	Job-related	Unstructured
135	28	0.45	Situational	Structured
136	51	0.24	Psychological characteristics	Unstructured
137	49	0.02	Psychological characteristics	Unstructured
138	164	0.23	Job-related	Structured
139	195	0.17	Job-related	Structured
140	165	0.32	Job-related	Structured
141	40	0.36	Job-related	Structured
142	100	0.09	Psychological characteristics	Structured
143	4195	0.13	Job-related	Unstructured
144	179	0.29	Job-related	Structured
145	74	0.49	Job-related	Structured
146	110	0.40	Job-related	Structured
147	31	0.23	Job-related	Structured
148	70	0.31	Job-related	Structured
149	21	0.46	Job-related	Structured
150	29	−0.12	Job-related	Structured
151	51	0.22	Job-related	Unstructured
152	51	0.59	Job-related	Structured
153	40	0.21	Job-related	Structured

154	40	0.02	Job-related	Structured
155	129	−0.03	Job-related	Structured
156	196	0.28	Job-related	Structured
157	31	−0.04	Job-related	Structured
158	494	0.19	Job-related	Unstructured
159	101	0.23	Job-related	Structured
160	175	0.30	Job-related	Structured

Source: McDaniel *et al.* (1994).
Note: A blank space indicates missing data. Both job-related and situational interviews represent job-related validities; psychological represents an interview that assessed psychological characteristics.

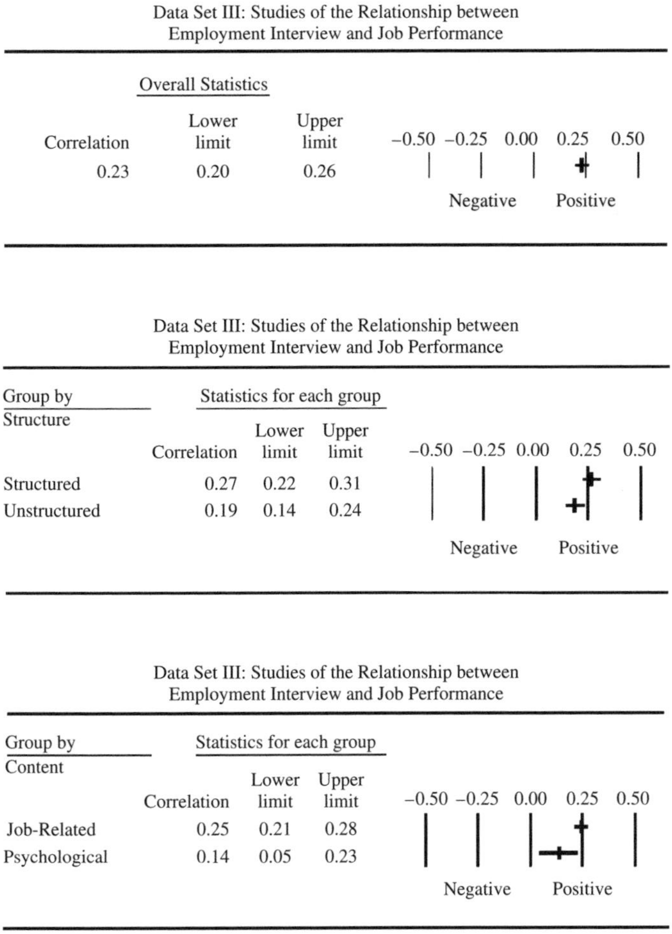

Data Set III: Studies of the Relationship between Employment Interview and Job Performance

Overall Statistics			
Correlation	Lower limit	Upper limit	−0.50 −0.25 0.00 0.25 0.50
0.23	0.20	0.26	
			Negative Positive

Data Set III: Studies of the Relationship between Employment Interview and Job Performance

Group by Structure	Statistics for each group			
	Correlation	Lower limit	Upper limit	−0.50 −0.25 0.00 0.25 0.50
Structured	0.27	0.22	0.31	
Unstructured	0.19	0.14	0.24	
				Negative Positive

Data Set III: Studies of the Relationship between Employment Interview and Job Performance

Group by Content	Statistics for each group			
	Correlation	Lower limit	Upper limit	−0.50 −0.25 0.00 0.25 0.50
Job-Related	0.25	0.21	0.28	
Psychological	0.14	0.05	0.23	
				Negative Positive

Figure A.3   Summary meta-analysis results for of the relationship between employment interview scores and job performance, overall and by structure and content subgroups (data from McDaniel *et al.*, 1994).

REFERENCES

Hackshaw, A.K., Law, M.R. and Wald, N.J. (1997). The accumulated evidence on lung cancer and environmental tobacco smoke. *British Medical Journal*, **315**, 980–988.

Hedges, L.V. and Olkin, I. (1985) *Statistical Methods for Meta-analysis*. London: Academic Press.

McDaniel, M.A., Whetzel, D., Schmidt, F.L. and Maurer, S. (1994). The validity of the employment interview: A comprehensive review and meta-analysis. *Journal of Applied Psychology*, **79**, 599–616.

Raudenbush, S.W. (1984). Magnitude of teacher expectancy effects on pupil IQ as a function of the credibility of expectancy induction. *Journal of Educational Psychology*, **76**, 85–97.

Raudenbush, S.W. and Bryk, A.S. (1985). Empirical Bayes meta-analysis. *Journal of Educational Statistics*, **10**, 75–98.

APPENDIX B

Annotated Bibliography

Hannah R. Rothstein

Department of Management, Zicklin School of Business, Baruch College, New York, USA

Ashley Busing

Department of Psychology, Weissman School of Arts and Sciences, Baruch College, New York, USA

In this bibliography we list and describe a selection of materials that provide a more detailed look at various aspects of publication bias. Although some were selected independently by the authors of the appendix, many were chosen based on conversations with the chapter authors, who had been asked to nominate articles that were either 'classics' or 'cutting-edge'. By 'classics' we meant articles that described the research that established the legitimacy of publication bias as a topic of scientific inquiry, or that provided the methods by which this inquiry could be advanced; by 'cutting-edge' we meant articles that present the newest techniques or findings in the area. The references are presented in chronological order so that the reader will be able to get a sense of the way in which the scientific study of publication bias has developed over time.

Sterling, T.D. (1959). Publication decisions and their possible effects on inferences drawn from tests of significance – or vice versa. *Journal of the American Statistical Association*, **54**, 30–34.

> This is possibly the earliest paper to provide (indirect) evidence of publication bias.

Rosenthal, R. (1979). The 'file drawer problem' and tolerance for null results. *Psychological Bulletin*, **86**, 638–461.

In addition to thanking the chapter authors, we would like to thank Doug Altman and Will Shadish for their suggestions about what to include in this bibliography.

Publication Bias in Meta-Analysis – Prevention, Assessment and Adjustments   Edited by H.R. Rothstein, A.J. Sutton and M. Borenstein   © 2005 John Wiley & Sons, Ltd

This is the first study to suggest a method for assessing the potential impact of publication bias on meta-analytic results. It uses a 'failsafe' approach to test the robustness of a statistically significant effect by providing a formula to calculate the number of 'missing' studies averaging zero effect that would be needed to reduce a significant observed overall effect to non-significance.

Hemminki, E. (1980). Study of information submitted by drug companies to licensing authorities. *British Medical Journal*, **280**, 833–836.

This paper provides early evidence that industry-based clinical trials of new drugs may remain unpublished. It reports on clinical trials that were submitted by drug companies to regulatory authorities in Finland and Sweden during 1965–1975, and were never submitted for publication. Although trial design and quality varied among the studies, the author suggests that valuable information, including information about adverse effects, was lost because these studies were not published.

Smith, M.L. (1980). Publication bias and meta-analysis. *Evaluation in Education*, **4**, 22–24.

This was the first paper to use the term 'publication bias'.

Smith, M.L. (1980). Sex bias in counseling and psychotherapy. *Psychological Bulletin*, **87**, 392–407.

Smith separately meta-analysed published and unpublished studies in order to see whether they provided different answers to the question of whether counsellors and therapists were biased against women. (This was a hot political issue at the time of Smith's research.) Her results provide one of the earliest examples of publication bias. She found that published studies showed a small effect of bias against women, while unpublished studies showed the same magnitude of bias towards women. Her analysis also indicated that the degree of rigour in research design was the same in both published and unpublished studies.

Orwin, R.G. (1983). A failsafe *N* for effect size in meta-analysis. *Journal of Educational Statistics*, **8**, 157–159.

In this paper, Orwin proposes an early method for evaluating the effects of putatively missing studies on meta-analytic results. His method is similar to Rosenthal's file-drawer method, except that it is focused on assessing the influence of missing studies on the magnitude of the combined mean effect rather than on its statistical significance.

Hedges, L.V. (1984). Estimation of effect size under nonrandom sampling: The effects of censoring studies yielding statistically insignificant mean differences. *Journal of Educational Statistics*, **9**, 61–85.

This paper investigates the effects of extreme publication bias (where only statistically significant results are observed) on estimates of the standardized mean difference. It derives the sampling distribution of the effect size estimates under extreme publication bias, shows that the bias depends on sample size and the true effect size, and shows that estimates can be biased by over 200 %. The maximum likelihood estimates of effect size assuming the extreme publication bias are also derived and shown to correspond to shrinking the observed (and biased) effect sizes towards zero. A table is provided to compute effect size estimates 'corrected' for extreme publication bias. Finally, an estimate of effect size derived from counts of positive and negative significant results is obtained which is valid under extreme publication bias. Sampling distributions and standard errors for all of the estimates are presented. Highly statistical.

Simes, R.J. (1986). Publication bias: The case for an international registry of clinical trials. *Journal of Clinical Oncology*, **4**, 529–541.

This paper provides one of the first illustrations of how a trials registry could reduce the problem of publication bias. The paper compares the results of published trials with the result of all trials appearing in a registry, whether published or not. For treatment of ovarian cancer with combination chemotherapy, pooled analysis of published clinical trials demonstrated a significant survival advantage for combination chemotherapy but no significant difference in survival when in the pooled analysis of all registered trials. For multiple myeloma, a pooled analysis of published trials demonstrated a significant survival advantage for combination chemotherapy, as did a pooled analysis of all registered trials; however, for all registered trials the estimated magnitude of the benefit was reduced.

Begg, C.B. and Berlin, J.A. (1988). Publication bias: A problem in interpreting medical data (with discussion). *Journal of the Royal Statistical Society, Series A*, **151**, 419–463.

This is the first comprehensive review of the publication bias literature presented to a medical statistics audience. The authors of this paper may have been the first to propose the idea of looking at sample size versus effect size in a formal, quantitative manner. They suggested that publication bias might be a bigger problem for new than for established areas of inquiry.

Iyengar, S. and Greenhouse, J.B. (1988). Selection models and the file drawer problem. *Statistical Science*, **3**, 109–135.

This is one of the first papers to advance the use of selection modelling to deal with publication bias. The authors use an example to demonstrate the differences between the failsafe N approach to dealing with publication bias and a maximum likelihood selection-modelling approach, and conclude that the maximum likelihood method offers several advantages over the failsafe N approach.

Meinert, C.L. (1988). Toward prospective registration of clinical trials. *Controlled Clinical Trials*, **9**, 1–5.

Along with the papers by Dickersin and Min (1993) and Simes (1986), this paper is among the earliest calls for a registry of clinical trials.

Berlin, J.A., Begg, C.B. and Louis, T.A. (1989) An assessment of publication bias using a sample of published clinical trials. *Journal of the American Statistical Association*, **84**, 381–392.

> This is one of the earliest papers to demonstrate a relationship between sample size and effect size in the published clinical trials literature. The authors found a strong relationship between sample size and treatment effect in published cancer trials for three outcomes: overall patient survival, disease-free survival and tumour response rate. An examination of several study features showed that some were associated with bias, but that none could account for the impact of sample size on treatment effect. The authors present a carefully detailed description of the data set they used, the methodological problems they dealt with, the decisions they made, and the statistical analyses they conducted. Some of this information is statistically demanding, but other parts of the article should be easily understandable by non-statisticians.

Hetherington, J., Dickersin, K., Chalmers, I. and Meinert, C.L. (1989). Retrospective and prospective identification of unpublished controlled trials: lessons from a survey of obstetricians and pediatricians. *Pediatrics*, **84**, 374–380.

> The authors attempted to retrieve information about unpublished randomized trials from obstetricians and paediatricians, and were largely unsuccessful except for trials that had been completed within the most recent two years. They conclude that the problem of publication bias will not be resolved through retrospective identification of unpublished trials, and encourage the use of prospective registries.

Chalmers, I. (1990). Underreporting research is scientific misconduct. *Journal of the American Medical Association*, **263**, 1405–1408.

> In this paper, the author contends that the failure to publish results of rigorously designed clinical trials is a form of scientific misconduct, noting that it can lead to inappropriate treatment decisions. He suggests that researchers, research ethics committees, funders and editors all share the responsibility for this problem, and that all must participate in reducing it. He suggests that prospective registration of clinical trials would ameliorate the problem substantially.

Easterbrook, P.J., Berlin, J.A., Gopalan, R. and Matthews, D.R. (1991). Publication bias in clinical research. *Lancet*, **337**, 867–872.

> This is the first paper to look at a cohort of studies approved by an institutional review board (IRB) and follow them forward to see whether the studies that were eventually published had a greater likelihood of being statistically significant than those that were not published. The results showed that statistically significant studies were about twice as likely to be published as those that did not reach statistical significance. Higher likelihood of publication was also associated with the importance of the study results (as rated by the investigator), and with increasing sample size. The tendency towards publication bias was greater

with observational and laboratory-based experimental studies than with random-ized clinical trials. The authors caution against overinterpretation of conclusions based on a review of only published data, and suggest the need for improved strategies to identify the results of unpublished studies.

Chalmers, I., Dickersin, K. and Chalmers, T.C. (1992). Getting to grips with Archie Cochrane's agenda. *British Medical Journal*, **305**, 786–788.

This short paper describes the work of Archie Cochrane and his focus on using the results of randomized clinical trials to develop health care practices and policies in Great Britain.

Dickersin, K., Min, Y.I. and Meinert, C.L. (1992). Factors influencing publication of research results. Follow-up of applications submitted to two institutional review boards. *Journal of the American Medical Association*, **267**, 374–378.

This is one of the first studies to demonstrate that publication bias originates for the most part with investigators rather than with journal editors. In a study of 737 studies that had received IRB approval, there were significant differences in the probability of publication for studies with significant or non-significant findings, but less than 5 % of the studies that had not been published were reported to have been rejected for publication.

Hedges, L.V. (1992). Modeling publication selection effects in meta-analysis. *Statistical Science*, **7**, 246–255.

This paper investigates the problem of modelling, and correcting for, publication selection in meta-analysis. It distinguishes between the concepts of selection models and effect size models. Selection models describe the probability that an effect size estimate is observed as a function of its level of statistical significance. Effect size models describe what the sampling distribution of effect size estimates would be if there were no selection. If the selection model were known, it would be possible to estimate the effect size model – that is, to estimate what the value of the mean effect size would be if there were no selection. This paper introduces a flexible, non-parametric, selection model that can be estimated (by maximum likelihood) from effect size data simultaneously with the effect size model. This provides information about the selection process that is operating (via the estimated selection model) and what the mean and between-studies variance component of the effect sizes would have been in the absence of selection (via the parameters of the effect size model). Statistical tests for the presence of selection and for mean and between-studies variance of effect sizes are given. Highly statistical.

Dickersin, K. and Min, Y.I. (1993). NIH clinical trials and publication bias. *Online Journal of Current Clinical Trials*, Doc. No. 50.

This is one of the earliest papers to systematically demonstrate that papers with statistically significant findings were more likely to be published than those with non-significant findings. A meta-analysis of data from this and three similar studies yielded results that showed that significant papers were nearly three times as likely to be published as papers that did not reach statistical significance. Also of note is that this is one of the earliest papers to call for a registry of initiated trials as a means of preventing publication bias, as well as for financial support for this effort.

Stewart, L.A. and Parmar, M.K.B. (1993). Meta-analysis of the literature or of individual patient data: is there a difference? *Lancet*, **341**, 418–422.

This is the first paper describing a formal comparison of the results and implications of two different approaches to meta-analysis. Using an example in ovarian cancer, an IPD meta-analysis was compared with a meta-analysis using data extracted from trial publications. Substantially more data were available with the IPD approach. In addition, longer-term follow-up was available and more appropriate time-to-event analysis of survival data could be done with the IPD data. The results of the meta-analysis of data from published reports were more positive, with an absolute estimate of treatment effect that was three times as large as the estimate from IPD. The results based on data extracted from trial reports were also statistically significant, whereas those from the IPD were not. The authors concluded that the results of each approach were likely to have a different clinical interpretation.

Begg, C.B. and Mazumdar, M. (1994). Operating characteristics of a rank correlation test for publication bias. *Biometrics*, **50**, 1088–1101.

This paper introduces the first statistical test for the assessment of publication bias, which is based on the rank correlation between the treatment effect and the standard error for each study. The test is shown to be fairly powerful for large meta-analyses (75 or more studies), but has only moderate power for medium-sized meta-analyses (25 studies).

Bushman, B.J. and Wang, M.C. (1995). A procedure for combining sample correlation coefficients and vote counts to obtain an estimate and a confidence interval for the population correlation coefficient. *Psychological Bulletin*, **117**, 530–546.

In this article, the authors propose supplementing sample size procedures with vote-counting procedures as a means of salvaging studies that are relevant to the meta-analysis, but which are missing effect size estimates. They suggest that this will lead to less biased estimates of the population effect, and a narrower confidence interval around it than ignoring the studies without effect sizes. The article describes three vote-counting procedures that can be used to estimate the population effect for studies with missing sample correlations.

Stewart, L.A. and Clarke, M.J. on behalf of the Cochrane Working Party Group on Meta-analysis using Individual Patient Data (1995). Practical methodology of meta-analyses (overviews) using updated individual patient data. *Statistics in Medicine*, **14**, 2057–2079.

This paper, based on experiences shared at a Cochrane Collaboration workshop, describes the rationale and practical methodology of IPD meta-analyses.

Vevea, J.L. and Hedges, L.V. (1995). A general linear model for estimating effect size in the presence of publication bias. *Psychometrika*, **60**, 419–435.

This paper presents a maximum likelihood based model of estimation of effect sizes in the presence of selection based on one-tailed p-values (publication bias, based on significance of results). It presents a test for the presence of publication bias, and a means of estimated corrected effect values. The authors provide an example based on the psychotherapy effectiveness literature. Highly statistical.

Bushman, B.J. and Wang, M.C. (1996). A procedure for combining sample standardized mean differences and vote counts to estimate the population standardized mean difference in fixed effect models. *Psychological Methods*, **1**, 66–80.

In this paper, the authors outline a procedure for handling missing estimates which combines effect size estimates and vote counts. They recommend this as the method of choice for a meta-analysis when some studies do not provide sufficient information to compute effect size estimates but do present the direction or statistical significance of results.

Hedges, L.V. and Vevea, J.L. (1996). Estimating effect size under publication bias: Small sample properties and robustness of a random effects selection model. *Journal of Educational and Behavioral Statistics*, **21**, 299–332.

In this paper, the authors offer a procedure for dealing with publication bias, based on modelling its effects on random-effects meta-analytic results. The model, which is based on one-tailed p-values, can be used to assess the plausibility of the existence of publication bias, as well as to estimate effects, corrected for the operation of bias. The authors also provide the results of a simulation study used to test their model. These results indicate that the model is reasonably accurate under plausible conditions. Highly statistical.

Egger, M., Davey Smith, G., Schneider, M. and Minder, C. (1997). Bias in meta-analysis detected by a simple, graphical test. *British Medical Journal*, **315**, 629–634.

This paper introduces a very widely used statistical test for the assessment of funnel plot asymmetry. It presents a comparison of cases in which a single large trial and a meta-analysis were conducted on the same question. In no case where the large trial and the meta-analysis agreed was there evidence of funnel plot asymmetry. However, when the large trial and the meta-analysis disagreed, half the time there was funnel plot asymmetry. The paper also briefly reviews potential causes of asymmetry in addition to publication bias and cautions against assessing funnel plot asymmetry in meta-analysis with few studies.

Stern, J.M. and Simes, R.J. (1997). Publication bias: evidence of delayed publication in a cohort study of clinical research projects. *British Medical Journal*, **315**, 640–645.

This is one of the first studies to demonstrate the impact of publication bias on time to publication as well as on the likelihood of publication. Using a cohort of 743 studies submitted to a hospital ethics committee over 10 years, this study provides clear evidence that in addition to having a lower probability of publication, papers that fail to reach statistical significance experience serious publication delays compared with similar studies with statistically significant findings.

Whitehead, A. (1997). A prospectively planned cumulative meta-analysis applied to a series of concurrent clinical trials. *Statistics in Medicine*, **16**, 2901–2913.

This is one of the first studies to propose that prospective meta-analysis be applied to sets of broadly similar clinical trials that are being conducted at approximately the same time. The author suggests that the advantages of this approach include rapid answers to questions of safety or efficacy, and the facilitation of assessment of subgroup differences in treatment effects, as well as the early identification of adverse effects. The author indicates that a random-effects combined analysis, within a sequential framework, is the appropriate method for analysing these data.

Auger, C.P. (1998). *Information Sources in Grey Literature*, 4th edition. London: Bowker Saur.

This book is considered the classic introductory guide to grey literature. It describes what grey literature is and how it can be identified. The book discusses the types of publications included in the term 'grey literature' and covers collection/acquisition, bibliographic control, cataloguing/indexing, and distribution methods. It also discusses the grey literature in six specific content areas. This book may be of limited interest to those who have substantial experience with grey literature.

Ioannidis, J.P. (1998). Effect of the statistical significance of results on the time to completion and publication of randomized efficacy trials. *Journal of the American Medical Association*, **279**, 281–286.

This paper examines the relationship between statistical significance and time to publication in a series of HIV trials. The author found that statistically significant findings were submitted for publication significantly more rapidly after the trial was completed than was the case for trials that did not achieve statistically significant results, and that such findings, were published more rapidly after they were submitted.

Berlin, J.A. and Colditz, G.A. (1999). The role of meta-analysis in the regulatory process for foods, drugs, and devices. *Journal of the American Medical Association*, **281**, 830–834.

In this paper, the authors discuss the use of meta-analysis in evaluating the efficacy of drugs. The paper provides one of the earliest allusions to prospective meta-analysis. The authors claim that 'Preplanned meta-analysis of individual trials with deliberately introduced heterogeneity may maximize the generalizability of results from randomized trials'. They suggest that meta-analysis may be particularly helpful in identifying adverse effects, and in identifying particular persons, settings and conditions in which a drug may be particularly effective or ineffective.

Duval, S. and Tweedie R. (2000). A nonparametric 'trim and fill' method of accounting for publication bias in meta-analysis. *Journal of the American Statistical Association*, **95**, 89–99.

This paper introduces the trim and fill method as a means of estimating the number of 'missing studies' in a meta-analysis, and for adjusting the mean effect size and confidence intervals accordingly. Examples are given based on several data sets from medicine and epidemiology. Although the method itself is quite easy to grasp conceptually, the article contains a substantial amount of statistics.

Hahn, S., Williamson, P.R., Hutton, J.L., Garner, P. and Flynn, E.V. (2000). Assessing the potential for bias in meta-analysis due to selective reporting of subgroup analyses within studies. *Statistics in Medicine*, **19**, 3325–3336.

In this paper, the authors discuss the potential for bias of meta-analytic results due to selective reporting of subgroup data. The authors present a method of sensitivity analysis that imputes data for missing subgroups that can be used to assess the robustness of the results and conclusions of systematic reviews that analyse subgroup data. They illustrate their method with reference to a published systematic review. The review in question addressed malaria chemoprophylaxis in pregnancy, and had concluded that benefits were limited to women who were pregnant for the first time. This conclusion was based on subgroup analysis using the three trials out of five which reported on subgroups. The authors' reanalysis suggested that the effect size reported in the original review probably overestimated the actual effect, and called into question the conclusion that the treatment benefited only first-time pregnant women.

Hutton, J.L. and Williamson, P.R. (2000). Bias in meta-analysis due to outcome variable selection within studies. *Applied Statistics*, **49**, 359–370.

This article is one of the earliest to focus specifically on publication bias within studies, rather than on entirely missing studies. The authors describe the potential effects of bias due to multiple testing of outcomes, and selective reporting based on significance levels or effect size. The authors demonstrate the operation of selective reporting by reanalysing two meta-analyses, where it was clear that more outcomes had been measured than were reported, and show that in one of the two cases selective reporting bias threatened the conclusions of the original meta-analysis.

McAuley, L., Pham, B., Tugwell, P. and Moher, D. (2000). Does the inclusion of grey literature influence estimates of intervention effectiveness reported in meta-analyses? *Lancet*, **356**, 1228–1231.

This paper examines whether the inclusion or exclusion of grey literature in meta-analyses affects the estimates of effects of interventions assessed in randomized trials. The authors found that one-third of the meta-analyses they examined included some grey literature, but that there was a large amount of variation in the proportion of 'grey' studies across meta-analyses. On average, published studies yielded significantly larger estimates of effects than did the 'grey' studies. The authors conclude that 'the exclusion of grey literature from meta-analyses can lead to exaggerated estimates of intervention effectiveness', and suggest that systematic reviewers should attempt to search for and include grey literature in their meta-analyses.

Sterne, J.A.C., Gavaghan, D. and Egger, M. (2000). Publication and related bias in meta-analysis: Power of statistical tests and prevalence in the literature. *Journal of Clinical Epidemiology*, **53**, 1119–1129.

This paper describes the results of simulations performed to assess the power of a weighted regression method and a rank correlation test in the presence of no bias, moderate bias and severe bias. The power to detect bias increased with increasing numbers of trials for both methods. The rank correlation test was less powerful than the regression method for both moderate and severe bias. On the other hand, the regression method produced higher than desirable false positive rates under some conditions. The authors suggest that when evidence of small-study effects is found, publication bias should not be the only possible explanation considered.

Sutton, A.J., Duval, S.J., Tweedie, R.L., Abrams, K.R. and Jones, D.R. (2000). Empirical assessment of effect of publication bias on meta-analyses. *British Medical Journal*, **320**, 1574–1577.

This study analysed 48 reviews from the Cochrane Database of Systematic Reviews that contained 10 or more individual studies and used a binary endpoint. The authors found that, using the trim and fill method, approximately half the reviews were missing studies and, in nearly 20 % of the reviews, the number missing was significant. In four cases, statistical inferences about the intervention's effect were altered after publication bias was adjusted for. Although in most cases publication biases did not affect the review's conclusions, the authors recommended that researchers should routinely check whether the conclusions of systematic reviews are robust to the operation of these biases.

Copas, J.B. and Shi, J.Q. (2001). A sensitivity analysis for publication bias in systematic reviews. *Statistical Methods in Medical Research*, **10**, 251–265.

This paper proposes a sensitivity analysis in which different patterns of selection bias can be tested against the fit to the funnel plot. The authors illustrate with two medical examples: passive smoking and coronary heart disease; and the effectiveness of prophylactic antibiotics in critically ill adults. An appendix lists the S-Plus code needed for carrying out the analysis. Highly statistical.

Lefebvre, C. and Clarke, M. (2001). Identifying randomised trials. In M. Egger, G.D. Smith and D.G. Altman (eds). *Systematic Reviews in Healthcare: Meta-analysis in Context*, pp. 69–86. London: BMJ Books.

This book chapter provides a useful overview of the sources that have contributed to the Cochrane Collaboration's main database, the Central Register of Controlled Trials. In addition, it describes how these sources have been searched. The chapter serves as a practical account of the issues to keep in mind when searching for randomized trials for systematic reviews.

Macaskill, P., Walter, S.D. and Irwig, L. (2001). A comparison of methods to detect publication bias in meta-analysis. *Statistics In Medicine*, **20**, 641–654.

This study compares the performance of three methods of testing for bias: a rank correlation method; a simple linear regression of the standardized estimate of treatment effect on the precision of the estimate; and a regression of the treatment effect on sample size. These methods were tested using simulated meta-analyses of studies with binary endpoints. The results indicated that there was no 'winning' method. Performance varied depending upon the magnitude of the true treatment effect, distribution of study sizes and whether one- or two-tailed tests of significance test were used. In general, all methods suffered from low power in meta-analyses where the number of studies was typical of those found in the medical literature. Higher power was related to higher type I error rates. According to the authors, regressing the treatment effect on sample size, weighted by the inverse variance of the logit of the pooled proportion, is the preferred means of testing for bias.

Pham, B., Platt, R., McAuley, L., Klassen, T.P. and Moher, D. (2001). Is there a 'best' way to detect and minimize publication bias? An empirical evaluation. *Evaluation and the Health Professions*, **24**, 109–125.

This paper compares the performance of file-drawer analysis, the Begg test, the Egger test, trim and fill, and weighted estimation methods to assess the robustness of meta-analytic findings, and to detect and minimize publication bias effects in meta-analyses. The authors found that different approaches to dealing with publication bias reached different conclusions, when applied to the same set of meta-analytic data. This paper does not require advanced knowledge of statistics.

Pigott, T.D. (2001). Missing predictors in models of effect size. *Evaluation and the Health Professions*, **24**, 277–307.

The author of this paper reviews commonly used methods for dealing with missing data and their application to meta-analysis, such as complete case analysis and mean substitution, and suggests that they often yield biased estimates. The article briefly reviews the effects of missing predictors on the results of meta-analyses, discusses the strengths and weaknesses of commonly used missing-data methods, and suggests more desirable ways of handling missing predictors.

Specifically, the author recommends the use of maximum likelihood methods for multivariate normal data and multiple imputation methods.

Jennions, M.D. and Møller, A.P. (2002). Publication bias in ecology and evolution: an empirical assessment using the 'trim and fill' method. *Biological Review*, **77**, 211–222.

This article demonstrates the existence of publication bias in systematic reviews in ecology and evolution. The authors used the trim and fill method to examine the results of 40 published meta-analyses in this field. They found that for random-effects meta-analyses, 38 % had a significant number of 'missing' studies, and that after correcting for potential publication bias, approximately 20 % of weighted mean effects were no longer statistically significant. The authors conclude that in ecology and evolution, publication bias may affect the main conclusions of at least 15–20 % of published meta-analyses, and suggest that researchers routinely examine their results for the possible effects of publication bias.

Little, R.J.A. and Rubin, D.B. (2002). *Statistical Analysis with Missing Data*, 2nd edition. New York: John Wiley.

This is the classic reference for dealing with missing data. The authors focus on imputation, and therefore present a variety of imputation methods, in addition to the bootstrap, jackknife, and similar techniques. The theory behind each technique, as well as algorithms for implementation, is clearly laid out. It is written at the level of a graduate-school textbook. Although no consideration of how such techniques could be applied to meta-analysis is given, the book describes a lot of the statistical ideas underlying missing-data problems.

Manheimer, E. and Anderson, D. (2002). Survey of public information about ongoing clinical trials funded by industry: evaluation of completeness and accessibility. *British Medical Journal*, **325**, 528–531.

This paper reports a study of the completeness of on-line US-based trials registries. The authors examined whether ongoing trials of experimental drugs for colon and prostate cancer were reported in publicly accessible on-line trials registries. They found that 'a substantial proportion' of these trials were not contained in any of the publicly available registries. The authors concluded that there is a need for a comprehensive on-line registry of trials that includes Phase 3 industry-sponsored research.

Olson, C.M., Rennie, D., Cook, D., Dickersin, K., Flanagin, A., Hogan, J., Zhu, Q., Reiling, J. and Pace, B. 2002. Publication bias in editorial decision making. *Journal of the American Medical Association*, **287**, 825–828.

This paper provides evidence that a major reason for publication bias in medical research is that researchers are less likely to submit manuscripts reporting non-significant results to journals. Based on a sample of papers submitted to *JAMA*, the authors found no evidence that publication bias occurs after manuscripts have been submitted for publication.

Scherer R.W. and Langenberg, P. (2002). Full publication of results initially presented in abstracts (Cochrane Methodology Review). In *The Cochrane Library*, Issue 3, 2002. Oxford: Update Software.

The authors synthesized the results of 46 reports that examined the publication of articles based on abstracts that had been presented at scientific conferences. Overall, fewer than half of the abstracts were eventually published as full articles. Full publication of studies was more likely among abstracts with statistically significant results than among those with non-significant results. Abstracts of results of clinical research were less likely to be published than abstracts of the results of basic research, while abstracts of studies using randomized designs were published at higher rates than were other types of studies.

Song, F., Kahn, K.S., Dinnes, J. and Sutton, A.J. (2002). Asymmetric funnel plots and publication bias in meta-analyses of diagnostic accuracy. *International Journal of Epidemiology*, **31**, 88–95.

This is one of the first studies to assess the potential effects of publication bias in studies of diagnostic test research – empirical studies of publication having mainly focused on studies of treatment effect. The authors examined a sample of 28 meta-analyses of diagnostic accuracy from the Database of Abstracts of Reviews of Effectiveness (DARE). They found that, in general, the authors of the meta-analyses had not sufficiently considered either literature search strategies that would minimize publication bias or the impact of possible publication bias. Results showed that in a substantial proportion of the meta-analyses evaluated, smaller sample sizes were associated with greater diagnostic accuracy and greater funnel plot asymmetry. In addition, the fewer the literature databases searched, the greater the funnel plot asymmetry in meta-analyses. The authors suggest that authors of systematic reviews of diagnostic tests should perform more comprehensive literature reviews, and assess the likelihood and severity of publication bias.

Stewart, L.A. and Tierney, J.F. (2002). To IPD or not to IPD? *Evaluation and the Health Professions*, **25**, 76–97.

This paper describes the history, rationale and the pros and cons of the individual patient data approach to meta-analysis. These are discussed in terms of issues relating to data quality, analysis, and the organizational and collaborative approach. The authors conclude that reviewers should, at the outset, consider the methodological factors likely to influence or bias the outcome of their review, together with time and resource considerations, in order to take an active decision about the most appropriate approach.

Egger, M., Jüni, P., Bartlett, C., Holenstein, F. and Sterne, J. (2003). How important are comprehensive literature searches and the assessment of trial quality in systematic reviews? Empirical study. *Health Technology Assessment*, **7**, 1–76.

This paper describes the importance of comprehensive literature searches which cover the grey literature and all relevant databases and languages. It also makes an argument for the importance of assessing trial quality in systematic reviews.

Hopewell, S., McDonald, S., Clarke, M. and Egger, M. (2003). Grey literature in meta-analyses of randomized trials of health care interventions (Cochrane Methodology Review). In *The Cochrane Library*, Issue 4, 2003. Chichester: John Wiley and Sons, Ltd.

This paper describes a systematic review of studies comparing the impact of grey versus published literature on the overall results of meta-analyses. It shows that the exclusion of studies from the grey literature may lead to an exaggeration of the effects of treatment.

MacLean, C.H., Morton, S.C., Ofman, J.J., Roth, E.A. and Shekelle, P.G. (2003). How useful are unpublished data from the Food and Drug Administration in meta-analysis? *Journal of Clinical Epidemiology*, 56, 44–51.

This paper uses the example of research on non-steroidal anti-inflammatory drugs (NSAIDs) to examine whether studies summarized in Food and Drug Administration (FDA) reviews are eventually published, and to compare key characteristics of FDA-reviewed studies with those reported in published, peer-reviewed literature. The authors found that of 37 studies described in the FDA reviews, only one was published. They also found that there were no meaningful sample or methodological differences between FDA and published studies, and that the effect sizes found in both FDA and published studies were neither 'significantly [n]or practically' different. They concluded that FDA reviews could be a source of data for systematic reviews.

Melander, H., Ahlqvist-Rastad, J., Meijer, G. and Beermann, B. (2003). Evidence b(i)ased medicine – selective reporting from studies sponsored by pharmaceutical industry: review of studies in new drug applications. *British Medical Journal*, **326**, 1171–1173.

These authors examined publication bias related to multiple publication, selective publication and selective reporting in 42 studies of five selective serotonin reuptake inhibitors that were conducted by pharmaceutical companies in Sweden between 1983 and 1999. They compared reports of 42 studies as they were submitted to the Swedish drug regulatory authority with the published versions of these studies. Their results showed that 21 of the studies contributed to at least two publications each; that studies with significant effects were published as stand-alone publications more frequently than those with non-significant results; that many publications ignored the results of intention to treat analyses and presented only the more positive per-protocol analyses. However, the degree of multiple publication, selective publication and selective reporting differed across drugs. The authors concluded that publicly available data are likely to be biased.

Terrin, N., Schmid, C.H., Lau, J. and Olkin, I. (2003). Adjusting for publication bias in the presence of heterogeneity. *Statistics in Medicine*, 22, 2113–2126.

This study uses simulation to evaluate the accuracy of the trim and fill method when studies are heterogeneous. The results indicate that when studies are heterogeneous, trim and fill may wrongly adjust for publication bias when there is none. The authors suggest that funnel plots may not be appropriate for heterogeneous meta-analyses, and suggest that in cases of heterogeneity, selection modelling may be a better approach. The authors report that a selection model was superior to trim and fill in their simulations, although the results converged at times, under some conditions.

Bennett, D.A., Latham, N.K., Stretton, C. and Anderson, C.S. (2004). Capture–recapture is a potentially useful method for assessing publication bias. *Journal of Clinical Epidemiology*, **57**, 349–57.

This paper compares several approaches to the assessment of publication bias, including capture–recapture, visual examination of funnel plots, the Egger test, a funnel plot regression, trim and fill, and a selection model approach. In the illustrative example, all methods employed yielded broadly consistent results. Capture–recapture estimated that three relevant studies were missed, while trim and fill estimated that there were 16 missing studies. Both the Egger test and a funnel plot regression approach indicated the presence of publication bias, while selection modelling suggested that the observed funnel plot asymmetry observed was not entirely the result of publication bias. The authors suggest that capture–recapture is a potentially useful means of assessing publication bias, but that further simulation studies on all the methods should be conducted.

Chan, A.-W., Hrobjartsson, A., Haahr, M.T., Gøtzsche, P.C. and Altman, D.G. (2004). Empirical evidence for selective reporting of outcomes in randomized trials: Comparison of protocols to published articles. *Journal of the American Medical Association*, **291**, 2457–2465.

In this paper, the authors demonstrate the prevalence of incompletely reported outcomes in Danish randomized trials from 1994–95, by comparing protocols with journal articles and by surveying researchers. They identified 102 trials with 122 published journal articles and 3736 outcomes. Overall, half of efficacy outcomes and nearly two-thirds of harm outcomes per trial were incompletely reported. Statistically significant outcomes were much more likely to be fully reported than non-significant outcomes for both efficacy and harm. When published articles were compared with their protocols, 62 % of trials altered, added or omitted at least one primary outcome. Eighty-six per cent of survey respondents (42/49) denied the existence of unreported outcomes, despite the evidence. The authors conclude that reporting of trial outcomes is often incomplete, biased and inconsistent with protocols. As a result, they caution, published articles, as well as reviews that include them, may overestimate the benefits of interventions. The authors issue a call for the registration of planned trials and public availability of protocols before trial completion.

Trikalinos, T.A., Churchill, R., Ferri, M., Leucht, S., Tuunainen, A., Wahlbeck, K. and Ioannidis, J.P.; EU-PSI project (2004). Effect sizes in cumulative meta-analyses of mental health randomized trials evolved over time. *Journal of Clinical Epidemiology*, **57**, 1124–1130.

The authors investigated whether the certainty and estimates of effect size of mental health interventions change over time, as additional trials appear on the same topic. This sort of evolution of effect sizes over time had previously been found in trials in genetics, but had not been examined in mental health intervention trials. Using cumulative meta-analysis and recursive cumulative meta-analysis, they examined 100 meta-analyses containing five or more trials each, published in at least three different years. Outcomes included death, relapse, failure or dropout. The authors found that eight meta-analyses reached statistical significance at some point, but became non-significant as more trials were published. In general, large effect sizes in early trials were reduced as further evidence accrued. The authors concluded that, in mental health as in other areas, evidence based on a small number of randomized subjects should be interpreted cautiously, and that early estimates of treatment effectiveness may be overly optimistic.

Chan, A.-W. and Altman, D.G. (2005). Identifying outcome reporting bias in randomised trials on PubMed: Review of publications and survey of authors. *British Medical Journal*, **330**, 753.

In this paper, the authors examine the extent of selective reporting of outcomes in published randomized trials in health care by reviewing publications, and then surveying their authors. They conclude that the published medical literature represents a biased subset of study outcomes. The sample for this study consisted of all journal articles reporting on randomized trials from journals that are indexed in PubMed, and which were published in December 2000. The authors identified 519 trials with 553 publications and 10 557 outcomes. They found that authors who responded to their survey were often unreliable about the presence of unreported outcomes. On average, over 20 % of the outcomes per trial were incompletely reported. The reasons most commonly given for omitting outcomes included space limitations, and lack of clinical or of statistical significance.

Ioannidis, J.P.A. and Trikalinos, T.A. (2005). Early extreme contradictory estimates may appear in published research: The Protens phenomenon in molecular genetics research and randomized trials. *Journal of Clinical Epidemiology*, **58**, 543–549.

In this paper, the authors make the argument that early hypothesis-generating research in areas where a lot of data can be generated quickly is prone to produce early studies with extreme, opposite results, due to the preferences of authors and editors.

GLOSSARY

Allocation concealment: In randomized controlled trials, the situation where the participant's assignment to the treatment or control group is not known to the investigator. When allocation is not concealed, the randomization process is potentially undermined, since knowledge of which group the next participant will be randomized to may influence who is recruited. Examples of methods of assignment where allocation is adequately concealed are use of sealed opaque envelopes or when randomization is done by someone other than the investigator.

Begg and Mazumdar's test: A test for funnel plot asymmetry based on the rank correlation between the effect estimates and their sampling variances.

Cohen's *d*: A specific standardized mean difference estimate where the difference between two group means is usually divided by the standard deviation pooled across both groups (although either group's standard deviation is sometimes used when the data are homogeneous).

Conventional publication bias: A systematic bias in meta-analyses or systematic reviews of studies regarding a specific research question that manifests when decisions to publish research findings are influenced by the statistical significance, magnitude or direction of the observed effects.

Correlation coefficient: A measure of the degree of linear relationship between two continuous variables. See also Pearson's *r*.

Cumulative meta-analysis: The repeated performance of meta-analysis sequentially including an extra study. Studies are often ordered chronologically, allowing an examination of how the pooled effect size changes over time; however, other factors can be used to order the studies. For example, study precision can be used to assess whether the pooled effect size changes as larger/smaller studies are added and hence give an indication of whether publication bias may be present.

Domain: A generic term describing wide areas of a specialty or a discipline.

Domain-wide plot: A scatter plot where all the relative changes pertaining to meta-analyses of a whole domain are plotted against a function of the corresponding cumulative sample size N, such as $\log(N)$.

Publication Bias in Meta-Analysis – Prevention, Assessment and Adjustments   Edited by H.R. Rothstein, A.J. Sutton and M. Borenstein   © 2005 John Wiley & Sons, Ltd

Effect size: A measure of the magnitude of a relationship or the difference in effect between groups. Common measures of effect size include the standardized mean difference statistic, the correlation coefficient, the odds ratio, the risk difference, and the relative risk (sometimes called risk ratio or rate ratio).

Egger's test: A test for funnel plot asymmetry based on a linear regression of a standard normal deviate (defined as the effect size divided by its standard error) on its precision (defined as the reciprocal of its standard error). The test is equivalent to a weighted regression of the effect on its standard error.

Failsafe N: The number of additional null results it would take to reduce the Stouffer sum of Zs test (an overall significance test) to non-significance. The original failsafe N was proposed by Rosenthal in 1979.

File-drawer problem: A situation in which researchers are presumed not to have published results that did not reach statistical significance, and those unpublished articles remain in the researchers' file drawers. In such cases, conclusions drawn based on reviews of published studies are likely to be biased. (Often used to indicate conventional publication bias.)

Fixed-effects meta-analysis model: A statistical model used to combine effect size estimates from multiple studies. A key feature of the model is that it assumes all studies are estimating the same underlying, true effect size and therefore observed study results are assumed to differ only due to random variation.

Funnel plot: Funnel plots are simple scatter plots of the effects estimated from individual studies against a measure of study size. The name 'funnel plot' is based on the fact that the precision in the estimation of the underlying effect will increase as the sample size of component studies increases, and hence, when no publication bias is present, appear funnel-shaped. Departures from this expected shape may indicate that publication bias is present in a data set.

Hazard function: A function that describes the instantaneous risk any given individual has of experiencing the outcome of interest.

Hedges' g: The difference between the means of two groups divided by the pooled within-group standard deviation, and corrected for small-sample bias.

Heterogeneity: Statistical heterogeneity between effect sizes from studies included in a meta-analysis exists if there is more variability between results than would be expected by random variation, or chance, alone. A statistical test for heterogeneity exists, but has low power when only a small number of studies are included in a meta-analysis.

Individual patient data (IPD) meta-analyses: Systematic reviews/meta-analyses that centrally locate, obtain, verify and combine data obtained directly from those responsible for the primary studies. Analysis is carried out at the individual level, rather than at the aggregated study level usually used in conventional meta-analysis.

Kendall's tau (τ): A non-parametric, ranked correlation coefficient, used for paired data.

Maximum likelihood: A method of parameter estimation in a statistical model based on the assumption that the best estimate of a parameter is the one that gives the highest probability of yielding the set of measurements that was actually obtained.

Meta-analysis: The statistical analysis of quantitative results from multiple individual studies for the purpose of integrating their findings. Often carried out as part of a systematic review.

Meta-regression: An extension of either the fixed- or random-effects meta-analysis models in which (study-level) covariates are added to the meta-analysis model in an attempt to explain differences (heterogeneity) in estimated effects from the included studies.

Non-conventional publication bias: A systematic bias in meta-analyses or systematic reviews of studies regarding a specific research question that manifests when decisions to publish (or otherwise make available) research findings are influenced by financial, political, ideological, professional, or other interests that may or may not be directly related to the direction or significance of the results.

Non-parametric method: Statistical procedure for testing hypotheses or estimating parameters, where no assumptions are made about the nature or shape of the distribution from which the data were obtained. (Often used for ranked (ordinal) data.)

Null bias: Especially for large, multicentre trials, it is possible that some centres (with limited experience in clinical research) measure outcomes so poorly that the (non-systematic) errors that are introduced become greater that the magnitude of the intervention effects. In this case, the (usually small) true effect is lost in the noise of poor measurement, and a misleading conclusion of null effect may result.

Odds ratio: Defined as the ratio of the odds of an event occurring in one group to the odds of it occurring in another group. It is used as an effect size measure when comparing groups and the outcome is dichotomous (e.g. dead/alive).

Pearson's *r*: A correlation coefficient which measures the strength of a linear relationship.

Power (of a study): In the context of study design, the power refers to the *statistical* power of the study. It is the ability of the study to detect statistically an effect, if an effect exists. In general, power is expressed as the probability, prior to the conduct of the study, that the sample size will be large enough to detect a particular, pre-specified magnitude of effect. Power depends directly on the size of the study – other things being equal, larger studies have more power than smaller studies to detect the same effect. A power of 80 % is the usual minimally acceptable value among researchers.

Principle of equipoise: To conduct a randomized controlled trial, one must be truly uncertain about which of the trial interventions is superior; otherwise it is unethical to subject people to less effective or more hazardous interventions. The principle of equipoise states that, at the inception of the trial, the compared interventions are equally likely to be the more effective one.

Random-effects meta-analysis model: A statistical model used to combine effect size estimates from multiple studies. It relaxes the assumption of the fixed-effect model that all studies are estimating exactly the same underlying effect. Instead, it is assumed that underlying effects from studies can differ and that such effects are drawn from a distribution of effects, commonly assumed to be a normal distribution. Random-effects models estimate the average population effect, but also estimate the between-study variability of effects.

Randomized controlled trial: A true experiment, in which participants are randomly assigned to two or more groups that are then given different interventions. Used extensively to ascertain the effectiveness of (new) health-care interventions, and less extensively to ascertain the effectiveness of (new) interventions in education, social welfare, criminal justice and psychology.

Rate ratio: See relative risk.

Register/Registry: An organized database of ongoing, planned or completed studies, generally sharing some common characteristic. A *prospective* registry enrolls trials prior to knowledge of their results.

Regression: A general technique used to discover a mathematical relationship between two variables using a set of data points. A commonly used specific model is simple linear regression where the relationship between two variables, both measured on a continuous scale, is estimated. The fitted regression line depicts the way in which a change in one of the variables is related to a change in the second variable. The *slope* describes the rate of change; the *intercept* represents what the value of the second variable would be if the first variable's value was zero.

Relative risk: Defined as the ratio of the probability of an event occurring in one group to the probability of it occurring in another group. It is used as an effect size measure when comparing groups and the outcome is dichotomous (e.g. dead/alive).

Risk ratio: See relative risk.

Small-study effects: The tendency for the smaller studies in a meta-analysis to show larger treatment effects. Small-study effects may result from publication bias, from other types of bias, or may be due to true treatment effects being different in smaller studies than in larger ones.

Spline modelling: Spline modelling allows the estimation of a relationship between a dependent variable and a predictor as a piecewise function. This means that one fits different functions (splines) for successive predefined intervals of the independent variable range. Adjacent splines are connected in a so-called 'knot'. In this way one can capture changes in the relationship between the variables for different levels of the independent variable.

Standardized mean difference: A class of effect size estimators used to measure the effect between groups when the outcome is measured on a continuous scale. It is standardized by dividing by some measure of the variability in the data, which allows meta-analysis to be conducted when individual studies did not use the same outcome measure. Cohen's *d* and Hedges' *g* are both examples of standardized mean difference estimates.

Systematic review: A review of a clearly formulated question that uses systematic and explicit methods to identify, select and critically appraise relevant research in an attempt to minimize bias. Statistical methods (meta-analysis) may or may not be used to analyse and summarize the results of the studies included.

Time-lag bias: The accelerated or delayed publication of studies, depending on the nature and the direction of the results.

Vote-counting methods: Procedures for combining results of studies in which the researcher evaluates a research area by tallying the positive, negative and null results of studies conducted. It is considered a basic form of meta-analysis.

Wilcoxon matched pairs signed rank sum test: Non-parametric statistical test that is used to examine the degree of correspondence of paired observations.

'Winner's curse' phenomenon: In scientific fields where the production of data is rapid and copious, such as genetic epidemiology, the first published reports are very often optimistically biased, and their results are not replicated by subsequent research. According to an often-quoted metaphor, the 'winners' (the first) are 'cursed' (not to be validated in the future).

z-value: The number of standard deviations away from the mean of a normal distribution. Often used as a test statistic to test how extreme data are given a null hypothesis by computing a *p*-value based on the z-value. Also called the 'normal score'.

Index